Landscape Ecology in Theory and Practice

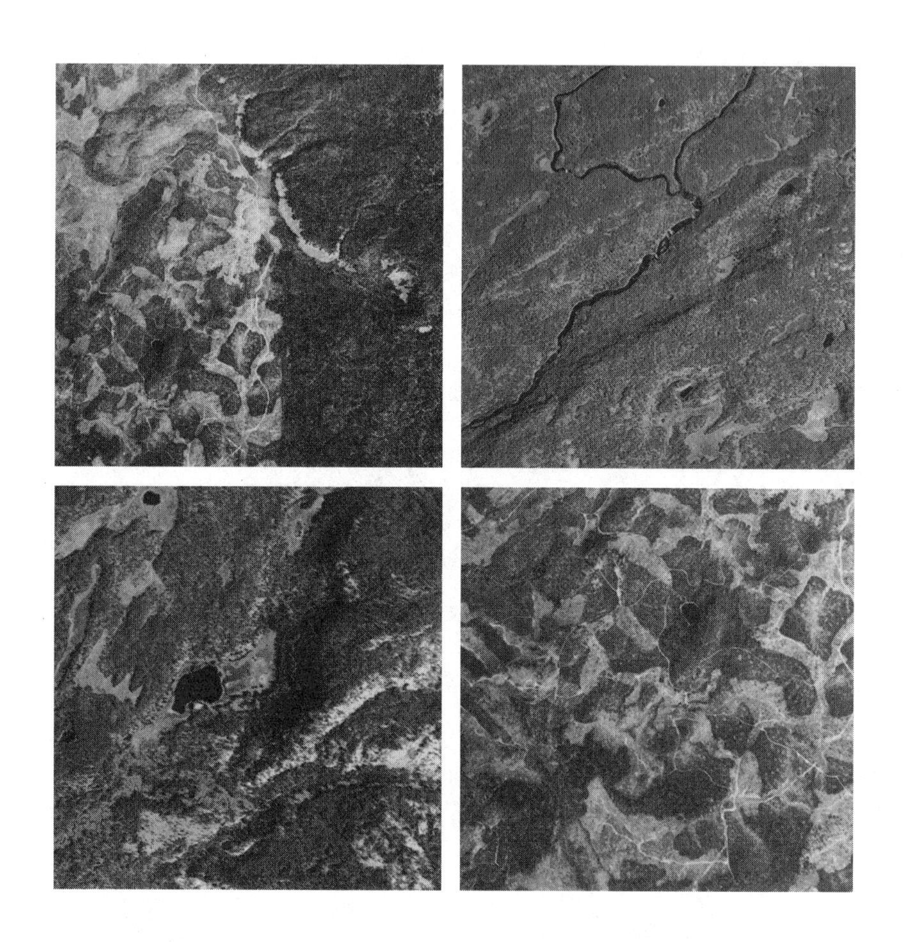

Monica G. Turner

Robert H. Gardner

Robert V. O'Neill

Landscape Ecology in Theory and Practice

Pattern and Process

Springer

MONICA G. TURNER
Department of Zoology
Birge Hall
University of Wisconsin
Madison, WI 53706
USA
mgt@mhub.zoology.wisc.edu

ROBERT H. GARDNER
Appalachian Laboratory
University of Maryland
301 Braddock Road
Frostburg, MD 21532
USA
gardner@al.umces.edu

ROBERT V. O'NEILL
Environmental Sciences Division
Oak Ridge National Laboratory
Oak Ridge, TN 37831-6036
USA
eoneill@attglobal.net

Printed on acid-free paper.

Production coordinated by WordCrafters Editorial Services, Inc., Sterling, VA, and managed by Steven Pisano.
Manufacturing supervised by Jacqui Ashri.
Typeset by Matrix Publishing Services, York, PA.
Printed and bound by Edwards Brothers, Inc., Ann Arbor, MI.

Library of Congress Cataloging-in-Publication Data
Landscape ecology in theory and practice : pattern and process / Monica G. Turner, Robert H. Gardner, Robert V. O'Neill.
p. cm.
Includes bibliographical references (p.).
ISBN 0-387-95122-9 (alk. paper)
ISBN 0-387-95123-7 (softcover : alk. paper)
1. Landscape ecology. I. Gardner, R.H.
II. O'Neill, R.V. (Robert V.), 1940– III. Title.
QH541.15.L35T87 2001
577—dc21 00-047094

Printed in the United States of America.

9 8 7 6 5 4 3 2 1

ISBN 0-387-95122-9 SPIN 10778231 (hardcover)
ISBN 0-387-95123-7 SPIN 10778249 (softcover)

Springer-Verlag New York Berlin Heidelberg
A member of BertelsmannSpringer Science+Business Media GmbH

Preface

Landscape ecology is not a distinct discipline or simply a branch of ecology, but rather is the synthetic intersection of many related disciplines that focus on the spatial-temporal pattern of the landscape.

Risser et al., 1984

The emergence of landscape ecology as a discipline has catalyzed a shift in paradigms among ecologists, . . . resource managers and land-use planners. Having now seen the faces of spatial pattern and scale . . . we can never go back to the old ways of viewing things.

Wiens, 1999

This book presents the perspective of three ecologists on the concepts and applications of landscape ecology, a discipline that has shown expansive growth during the past two decades. Although landscape ecology is a multidisciplinary subject involving components as diverse as economics and sociology, the earth sciences and geography, remote sensing and computer applications, we focus here on what ecologists need to know about landscapes.

Landscape ecology served as the integrating theme of our collaborative research for nearly 15 years, including a 7-year period during which we worked together at Oak Ridge National Laboratory. We became acquainted in January 1986 at the first annual United States Landscape Ecology symposium held at the University of Georgia and organized by Monica Turner and Frank Golley. Landscape ecology was, at that time, a new subject in the United States. The first U.S. workshop on landscape ecology, organized by Paul Risser, Richard Forman, and Jim Karr, had occurred less than 3 years prior (Risser et al., 1984). One of us (O'Neill) was a participant in that workshop, and two of us (O'Neill and Gardner) were research scientists in the Environmental Sciences Division of Oak Ridge National

Laboratory (ORNL) who had collaborated for several years on many aspects of ecosystem ecology and ecological modeling. Turner had a newly minted Ph.D. and was continuing as a postdoctoral research associate at the University of Georgia, where Frank Golley (also a participant at the 1993 workshop and Turner's Ph.D. advisor) was actively engaging his colleagues in the developing ideas in landscape ecology. The mutual interests shared by Turner, Gardner, and O'Neill, coupled with the excitement and challenge of working in a newly emerging branch of ecology, led to Turner's move to ORNL in July 1987. The subsequent seven years in which we collaborated so closely were among the most exciting times that any of us have had in our careers. There are times and places at which creativity seems to be fostered more than others, and that time and place had it. The writing of this book was precipitated, in part, by the fact that although we are now located at different institutions, we shared the desire to provide a synthesis of the field in which we have worked so closely together.

As ecologists embraced the challenges of understanding spatial complexity, landscape ecology moved from being a tangential subdiscipline in the early 1980s to one that is now mainstream. Indeed, a landscape approach, or the landscape level, is now considered routinely in all types of ecological studies. It is our hope that this text will provide a synthetic overview of landscape ecology, including its development, the methods and techniques that are employed, the major questions addressed, and the insights that have been gained. The enclosed CD contains all figures for this book, including the color images. We hope that this will enhance the utility of the book, especially for teaching. The companion volume (Gergel and Turner, 2001) provides opportunities for hands-on learning of many of the methods and concepts employed by landscape ecologists. It is our hope that our books might serve to inspire others to embark on landscape ecological studies, for there is much yet to be learned. As we begin this new century, we look forward to the many contributions that landscape ecologists will make in the future and to the continued growth of this exciting discipline.

Acknowledgments

Research that we have conducted over the past 15 years and that led to the development of this book has been funded by a variety of agencies, and we gratefully acknowledge research support from the National Science Foundation (Long-term Ecological Research, Ecosystem Studies, and Ecology programs), Department of Energy, USDA Competitive Grants Program, National Geographic Society,

Environmental Protection Agency (EMAP and STAR programs), and the University of Wisconsin-Madison Graduate School.

Our ideas have evolved over the years and been shaped by fruitful and often spirited discussions with many colleagues. Among the most memorable of these were discussions with Don DeAngelis, Jeff Klopatek, John Krummel, and George Sugihara in the "prelandscape" years at ORNL that crystallized much of the philosophy and approach adopted in our research. Virginia Dale, Kim With, Scott Pearson, and Bill Romme have been regular collaborators as well as supportive friends. Although it is impossible to mention everyone at ORNL who contributed ideas and assisted us with their expertise, we would be remiss not to acknowledge the valuable contributions of Steve Bartell, Antoinette Brenkert, Carolyn Hunsaker, Tony King, and Robin Graham. While at Oak Ridge we hosted a number of visitors from other institutions, including Bill Romme, Linda Wallace, Bruce Milne, Tim Kratz, Sandra Lavorel, Tim Allen, Eric Gustafson, and Roy Plotnick. These colleagues made substantial contributions to and lasting impacts on our ideas, and we thank them all for engaging interactions and fruitful collaborations.

Special thanks are due to Richard Forman, Frank Golley, and John Wiens for longstanding collegial relationships, the sharing of their ideas (and students!) that often challenged our thinking, and their invaluable reviews and critiques over the years. Turner also sincerely thanks Hazel Delcourt (University of Tennessee) and David Mladenoff (University of Wisconsin), with whom she has jointly taught landscape ecology courses over the past decade; co-teaching has been inspiring and fun and has certainly helped shape her thinking.

This book benefited tremendously from valuable critical comments provided by numerous colleagues. We especially thank David Mladenoff and Sarah Gergel, who both read nearly the entire manuscript and provided constructive criticism that has been enormously helpful. David Mladenoff actually read the whole manuscript twice, and Turner especially thanks him for being such a good colleague. In addition, we are grateful to the following friends and colleagues for reviewing one or more chapters: Jeff Cardille, Steve Carpenter, F. S. (Terry) Chapin, Mark Dixon, Tony Ives, Dan Kashian, Jim Miller, Bill Romme, Tania Schoennagel, Steve Seagle, Emily Stanley, Dan Tinker, Phil Townshend, and Kim With. Comments on draft chapters from the students in Principles of Landscape Ecology, taught at the University of Wisconsin-Madison during the spring 1999 semester, were also very helpful.

The graphics and illustrations for this book were prepared by Michael Turner, and we are indebted to him for greatly improving the visual communication of the concepts and examples in this book. We are delighted with the clarity and consistency of the figures throughout the text. We thank Kandis Elliot (University of Wisconsin) and Michael Mac (Biological Resources Division, U.S. Geological Survey) for sharing visual resources. Sandi Gardner and Sally Tinker provided valuable editorial assistance in the final stages of manuscript preparation. Finally, we thank the two editors with whom we worked at Springer-Verlag, initially Rob Garber and then Robin Smith, for their patience and support of this effort, especially given the time it has taken us to complete it.

MONICA G. TURNER
Madison, Wisconsin

ROBERT H. GARDNER
Frostburg, Maryland

ROBERT V. O'NEILL
Oak Ridge, Tennessee

REFERENCES

GERGEL, S. E., AND M. G. TURNER, editors. 2001. *Learning Landscape Ecology: A Practical Guide to Concepts and Techniques.* Springer-Verlag, New York.

RISSER, P. G., J. R. KARR, AND R. T. T. FORMAN. 1984. *Landscape Ecology: Directions and Approaches.* Special Publication Number. Illinois Natural History Survey, Champaign, Illinois.

WIENS, J. A. 1999. The science and practice of landscape ecology, *in* J. M. Klopatek and R. H. Gardner, eds. *Landscape Ecological Analysis: Issues and Applications*, pp. 371–383. Springer-Verlag, New York.

Contents

Preface v

Chapter 1 ~ Introduction to Landscape Ecology 1

What Is Landscape Ecology? 2
Why Landscape Ecology Has Emerged as a Distinct Area of Study 7
The Intellectual Roots of Landscape Ecology 10
Objectives of This Book 20
Summary 21
Discussion Questions 22
Recommended Readings 23

Chapter 2 ~ The Critical Concept of Scale 25

Scale Terminology and Its Practical Application 27
Scale Problems 32

Scale Concepts and Hierarchy Theory 34
Identifying the "Right" Scale(s) 38
Reasoning About Scale 40
Scaling Up 40
Summary 43
Discussion Questions 44
Recommended Readings 45

CHAPTER 3 ~ INTRODUCTION TO MODELS 47

What are Models and Why Do We Use Them? 47
Steps in Building a Model 56
Landscape Models 64
Caveats in the Use of Models 66
Summary 67
Discussion Questions 68
Recommended Readings 69

CHAPTER 4 ~ CAUSES OF LANDSCAPE PATTERN 71

Abiotic Causes of Landscape Pattern 73
Biotic Interactions 83
Human Land Use 86
Disturbance and Succession 90
Summary 90
Discussion Questions 92
Recommended Readings 92

CHAPTER 5 ~ QUANTIFYING LANDSCAPE PATTERN 93

Why Quantify Pattern? 93
Data Used in Landscape Analyses 95
Caveats for Landscape Pattern Analysis, or "Read This First" 99
Metrics for Quantifying Landscape Pattern 108
Geostatistics or Spatial Statistics 125

Summary 132
Discussion Questions 133
Recommended Readings 134

CHAPTER 6 ~ NEUTRAL LANDSCAPE MODELS 135

Random Maps: The Simplest Neutral Model 138
Maps with Hierarchical Structure 147
Fractal Landscapes 149
Neutral Models Relating Pattern to Process 153
General Insights from the Use of NLMs 153
Summary 155
Discussion Questions 156
Recommended Readings 156

CHAPTER 7 ~ LANDSCAPE DISTURBANCE DYNAMICS 157

Disturbance and Disturbance Regimes 159
Influence of the Landscape on Disturbance Pattern 162
Influence of Disturbance on Landscape Pattern 174
Concepts of Landscape Equilibrium 188
Summary 196
Discussion Questions 198
Recommended Readings 199

CHAPTER 8 ~ ORGANISMS AND LANDSCAPE PATTERN 201

Conceptual Development of Organism-Space Interactions 204
Scale-Dependent Nature of Organism Responses 221
Effects of Spatial Pattern on Organisms 229
Spatially Explicit Population Models 240
Summary 243
Discussion Questions 246
Recommended Readings 247

Chapter 9 ∾ Ecosystem Processes in the Landscape 249

Spatial Heterogeneity in Ecosystem Processes 251
Effects of Landscape Position on Lake Ecosystems 261
Land–Water Interactions 265
Linking Species and Ecosystems 280
Searching for General Principles 284
Summary 285
Discussion Questions 287
Recommended Readings 288

Chapter 10 ∾ Applied Landscape Ecology 289

Land Use 290
Forest Management 307
Regional Risk Assessment 314
Continental-Scale Monitoring 319
Summary 321
Discussion Questions 324
Recommended Readings 325

Chapter 11 ∾ Conclusions and Future Directions 327

What Have We Learned? 328
Research Directions 329
Conclusion 331
Discussion Questions 332
Recommended Readings 332

References 333

Index 389

CHAPTER 1

Introduction to Landscape Ecology

Landscape ecology offers new concepts, theory, and methods that are revealing the importance of spatial patterning on the dynamics of interacting ecosystems. Landscape ecology has come to the forefront of ecology and land management and is still expanding very rapidly. The last decade has seen a dramatic growth in the number of studies and variety of topics that fall under the broad banner of landscape ecology. Interest in landscape studies has been fueled by many factors, the most important being the critical need to assess the impact of rapid, broad-scale changes in our environment.

Most of us have an intuitive sense of the term *landscape*; we think of the expanse of land and water that we observe from a prominent point and distinguish between agricultural and urban landscapes, lowland and mountainous landscapes, natural and developed landscapes. Any of us could list components of these landscapes, for example, farms, fields, forests, wetlands, and the like. If we consider how organisms other than humans may see their landscape, our own sense of landscape may be broadened to encompass components relevant to a honey bee, beetle, vole, or bison. In all cases, our intuitive sense includes a variety of different elements that comprise the landscape, change through time, and influence ecological dynamics. In his 1983 editorial in *BioScience*, Richard T. T. Forman used tangible examples to bring these ideas to the attention of ecologists:

> What do the following have in common? Dust-bowl sediments from the western plains bury eastern prairies, introduced species run rampant through native ecosystems, habitat destruction upriver causes widespread flooding down river, and acid rain originating from distant emissions wipes out Canadian fish. Or closer to home: a forest showers an adjacent pasture with seed, fire from a fire-prone ecosystem sweeps through a residential area, wetland drainage decimates nearby wildlife populations, and heat from a surrounding desert desiccates an oasis. In each case, two or more ecosystems are linked and interacting. (Forman, 1983)

In this chapter, we define landscape ecology, discuss the importance of landscape studies within ecology, briefly review the intellectual roots of landscape, and present an overview of the remainder of the book. In addition, some commonly used terms in landscape ecology are defined in Table 1.1.

WHAT IS LANDSCAPE ECOLOGY?

Landscape ecology emphasizes the interaction between spatial pattern and ecological process, that is, the causes and consequences of spatial heterogeneity across a range of scales. The term *landscape ecology* was introduced by the German biogeographer Carl Troll (1939), arising from the European traditions of regional geography and vegetation science and motivated particularly by the novel perspective offered by aerial photography. Landscape ecology essentially combined the spatial approach of the geographer with the functional approach of the ecologist (Naveh and Lieberman, 1984; Forman and Godron, 1986). During the past two decades, the focus of landscape ecology has been defined in various ways:

> Landscape ecology . . . focuses on (1) the spatial relationships among landscape elements, or ecosystems, (2) the flows of energy, mineral nutrients, and species among the elements, and (3) the ecological dynamics of the landscape mosaic through time. (Forman, 1983)

> Landscape ecology focuses explicitly upon spatial patterns. Specifically, landscape ecology considers the development and dynamics of spatial hetero-

Table 1.1.
Definition of commonly used terms in landscape ecology

Configuration: Specific arrangement of spatial elements; often used synonymously with spatial structure or patch structure.

Connectivity: Spatial continuity of a habitat or cover type across a landscape.

Corridor: Relatively narrow strip of a particular type that differs from the areas adjacent on both sides.

Cover type: Category within a classification scheme defined by the user that distinguishes among the different habitats, ecosystems, or vegetation types on a landscape.

Edge: Portion of an ecosystem or cover type near its perimeter and within which environmental conditions may differ from interior locations in the ecosystem; also used as a measure of the length of adjacency between cover types on a landscape.

Fragmentation: Breaking up of a habitat or cover type into smaller, disconnected parcels.

Heterogeneity: Quality or state of consisting of dissimilar elements, as with mixed habitats or cover types occurring on a landscape; opposite of homogeneity, in which elements are the same.

Landscape: Area that is spatially heterogeneous in at least one factor of interest.

Matrix: Background cover type in a landscape, characterized by extensive cover and high connectivity; not all landscapes have a definable matrix.

Patch: Surface area that differs from its surroundings in nature or appearance.

Scale: Spatial or temporal dimension of an object or process, characterized by both grain and extent.

Adapted from Forman, 1995.

geneity, spatial and temporal interactions and exchanges across heterogeneous landscape, influences of spatial heterogeneity on biotic and abiotic processes, and management of spatial heterogeneity. (Risser et al., 1984)

Landscape ecology is motivated by a need to understand the development and dynamics of pattern in ecological phenomena, the role of disturbance in ecosystems, and characteristic spatial and temporal scales of ecological events. (Urban et al., 1987)

Landscape ecology emphasizes broad spatial scales and the ecological effects of the spatial patterning of ecosystems. (Turner, 1989)

Landscape ecology deals with the effects of the spatial configuration of mosaics on a wide variety of ecological phenomena. (Wiens et al., 1993)

Landscape ecology is the study of the reciprocal effects of spatial pattern on ecological processes; it promotes the development of models and theories of spatial relationships, the collection of new types of data on spatial pattern and dynamics, and the examination of spatial scales rarely addressed in ecology. (Pickett and Cadenasso, 1995)

Collectively, this set of definitions clearly emphasizes two important aspects of landscape ecology that distinguish it from other subdisciplines within ecology. First, *landscape ecology explicitly addresses the importance of spatial configuration for ecological processes.* Landscape ecology is not only concerned with how much there is of a particular component, but also with how it is arranged. The underlying premise of landscape ecology is that the explicit composition and spatial form of a landscape mosaic affect ecological systems in ways that would be different if the mosaic composition or arrangement were different (Wiens, 1995). Most ecological understanding previously had implicitly assumed an ability to average or extrapolate over spatially homogeneous areas. Ecological studies often attempted to achieve a predictive knowledge about a particular type of system, such as a salt marsh or forest stand, without consideration of its size or position in a broader mosaic. Considered in this way, with its emphasis on spatial heterogeneity, landscape ecology is applied across a wide range of scales (Figure 1.1). Studies might address the response of a beetle to the patch structure of its environment within square meters (e.g., Johnson et al., 1992a), the influence of topography and vegetation patterns on ungulate foraging patterns (e.g., Pearson et al., 1995), or the effects of land-use arrangements on nitrogen dynamics in a watershed (e.g., Kesner and Meentemeyer, 1989).

Second, *landscape ecology often focuses on spatial extents that are much larger than those traditionally studied in ecology,* often, the landscape as seen by a human observer (Figure 1.2). In this sense, landscape ecology addresses many kinds of ecological dynamics across large areas such as the Southern Appalachian Mountains, Yellowstone National Park, the Mediterranean, or the rain forests of Rondonia, Brazil. However, it is important to note that, although these areas are typically larger than those used in most community- or ecosystem-level studies, the spatial scales are not absolutes. We deal with issues of scale in the next chapter and throughout this book, but suffice it to say here that landscape ecology does not define, a priori, specific spatial scales that may be universally applied; rather, the emphasis is to identify scales that best characterize relationships between spatial heterogeneity and the processes of interest. These two aspects, explicit treat-

FIGURE 1.1.

The concept of landscape as a spatial mosaic at various spatial scales: (a) An example of a microlandscape, or landscape complexity from the perspective of a grasshopper. Grass cover is *Bouteloua gracilis* and *Buchloe dactyloides,* and vegetation cover in the ~4 m^2 microlandscape is occasionally disrupted by bare ground. (Photo by Kimberly A. With.) (b) Set of experimental microlandscapes used to explore relative effects of habitat abundance and fragmentation on arthropod communities in an agroecosystem. System consists of a replicated series of 12 plots (each 16 m^2) that vary in habitat abundance and spatial contagion based on fractal neutral landscape models (With et al., 1999). (Photo by Kimberly A. With.) (c) Clones of Gambel oak (*Quercus gambelii*) in Colorado illustrating heterogeneity within approximately 1 km^2. (Photo by Sally A. Tinker.) (d) Aerial view of a muskeg and string bog landscape, Alaska. (Photo by John A. Wiens.) (Refer to the CD-ROM for a four-color reproduction of this figure.)

FIGURE 1.2.
Different types of landscapes across relatively large areas in the western United States: (a) Undeveloped mountainous landscape in the Front Range of Colorado, USA. (Photo by Monica G. Turner.) (b) Landscape mosaic of forest and agricultural land south of Santiago, Chile. (Photo by John A. Wiens.) (c) Urbanizing landscape outside Denver, Colorado. (Photo by John A. Wiens.) (d) Aerial view of clear-cuts in a coniferous (lodgepole pine, *Pinus contorta*) landscape, Targhee National Forest, Idaho. Postharvest slash piles scheduled for burning can be seen in the clear-cuts. (Photo by Dennis H. Knight.) (Refer to the CD-ROM for a four-color reproduction of this figure.)

ment of spatial heterogeneity and a focus on broad spatial scales, are not mutually exclusive and encompass much of the breadth of landscape ecology.

The role of humans, obviously a dominant influence on landscape patterns worldwide, is sometimes considered an important component of a definition of landscape ecology. Indeed, in the landscape ecology approaches characteristic of China, Europe, and the Mediterranean region, human activity is perhaps the central factor in landscape ecological studies. Landscape ecology is sometimes considered to be an interdisciplinary science dealing with the interrelation between human society and its living space—its open and built up landscapes (Naveh and Lieberman, 1984). Landscape ecology draws from a variety of disciplines, many of which emphasize social sciences, including geography, landscape architecture, regional planning, economics, forestry, and wildlife ecology. Throughout this book, the role of humans in shaping and responding to landscapes will be considered in many ways. The scientific contributions of landscape ecology are essential for land-management and land-use planning. However, we do not think it necessary to include a human component explicitly in the definition of landscape ecology, because humans are but one of the factors creating and responding to spatial heterogeneity.

What, then, is a landscape? We suggest a general definition that does not require an absolute scale: *a landscape is an area that is spatially heterogeneous in at least one factor of interest.* Although at the human scale we may observe "a kilometers-wide mosaic over which local ecosystems recur" (Forman, 1995), it is important to recognize that landscape ecology may deal with landscapes that extend over tens of meters rather than kilometers, and a landscape may even be defined in an aquatic system. In addition, we might observe a landscape represented by a gradient across which ecosystems do not necessarily repeat or recur. Thus our definition is general enough to permit consideration of both aspects of landscape ecology described above.

WHY LANDSCAPE ECOLOGY HAS EMERGED AS A DISTINCT AREA OF STUDY

The recent emergence of landscapes as appropriate subjects for ecological study resulted from three main factors: (1) broad-scale environmental issues and land-management problems, (2) the development of new scale-related concepts in ecology, and (3) technological advances, including the widespread availability of spatial data, the computers and software to manipulate these data, and the rapid rise in computational power.

Broad-Scale Environmental Issues

Demand for the scientific underpinnings of managing large areas and incorporating the consequences of spatial heterogeneity into land-management decisions has been growing since the 1970s and is now enormous. The paradigm of *ecosystem management*, for example, carries with it an implicit focus on the landscape (Agee and Johnson, 1988; Slocombe, 1993; Christensen et al., 1996). Applied problems and resource-management needs have clearly helped to catalyze the development and emergence of landscape ecology. For example, questions of how to manage populations of native plants and animals over large areas as land use or climate changes, how to mediate the effects of habitat fragmentation or loss, how to plan for human settlement in areas that experience a particular natural disturbance regime, and how to reduce the deleterious effects of nonpoint source pollution in aquatic ecosystems all demand basic understanding and management solutions at landscape scales. Federal agencies concerned with conservation in the United States are faced with many of these challenges. The cumulative loss of wetlands and riparian forests from many landscapes poses challenges for the management of animal populations and of water flow and quality. The U.S. Forest Service continues to wrestle with resource-management questions regarding fragmentation of contiguous old-growth forests in the northwestern United States. The patchwork quilt of overgrazed lands in the western United States poses management difficulties for the Bureau of Land Management that extend over multiple states. The National Park Service must attempt to determine whether existing parklands are of sufficient size to sustain biotic populations and natural processes over the long term. These problems require a spatially explicit, broad-scale approach, yet much of ecology had focused on mechanistic studies in relatively small homogeneous areas over relatively short time periods. Landscape ecology provides concepts and methods that complement those that have been traditionally employed in ecology.

Concepts of Scale

The importance of scale (see Chapter 2) became widely recognized in ecology only in the 1980s, despite a long history of attention to the effect of quadrat size on measurements and recognition of species–area relationships. The development of conceptual frameworks focused on scale (Allen and Starr, 1982; Delcourt et al., 1983; O'Neill et al., 1986; Allen and Hoekstra, 1992) prompted ecologists to think hard about the patterns and processes that were important at different scales of space and time. It became clear that no single scale was appropriate for the study of all ecological problems. Some problems required focus on an individual organism and its physiological response to environmental changes. Other problems required study of how numbers of individuals or species change with competition for a limited resource. Still other problems required study of communities and the potential for stable configurations of interacting populations. And still other problems required focus on the arrangement of communities in space and how they interact with heterogeneous patterns of resources on the landscape.

The theory of scale and hierarchy that emerged in the 1980s emphasized that attention should be focused directly on the scale at which a phenomenon of interest occurs. It demonstrated that the insights gained at one scale could not necessarily be translated directly to another scale, hence questioning the applicability of results from numerous fine-scale studies in ecology to the broad-scale problems that were so pressing. Scale theory mandated that the understanding of landscape-level dynamics should be obtained from direct study of the landscape. Finer-scale processes could be considered *mechanisms* that explain the landscape dynamics. Broader-scale patterns could be viewed as *constraints* that limit the potential range of rate processes. The critical factor was, and remains, identifying the proper scale at which to address the problem.

Thus, land-management problems and hierarchy, or scale, theory encouraged ecologists to address the landscape as a distinct area of study. Landscape ecology recognizes that ecological systems are arrayed in space in response to gradients of topography, temperature, moisture, and soils. Additional pattern is imposed by disturbances, biotic interactions, and human use of the land. Spatial arrangement, in turn, influences many ecological processes, such as the movement patterns of organisms, the spread of disturbances, and the movement of matter or energy. Landscape ecology, focusing on spatial pattern and the ecological responses to this pattern, leads to a new set of principles, distinct from the principles that govern ecosystem and population dynamics at finer scales.

Technological Advances

Technological developments have also contributed to the emergence of landscape ecology. These developments include rapid advances in desktop computing power, availability of remotely sensed data such as satellite images, and development of powerful computer software packages called geographic information systems (GIS) for storing, manipulating, and displaying spatial data. New research techniques are required in landscape ecology because of the focus on spatial pattern and dynamics and on large areas that simply cannot be thoroughly sampled or easily manipulated. For example, laboratory and plot experiments are appropriate at fine scales, but broad-scale experiments are logistically difficult, and replication is often impossible. Landscape ecologists have needed to incorporate new sources of data into their studies and creatively study *natural experiments.* For example, large disturbance events (e.g., hurricanes, forest fires, and volcanic eruptions) as well as land-management practices (e.g., timber harvest and land-use change) create opportunities for studying ecological phenomena at the landscape scale. The availability of remote imagery has made it possible to study spatial pattern over large areas and its change through time, opening new horizons for landscape analysis. With the development of powerful GIS software, scientists can work with spatial data in ways that were not even imagined two decades ago. In addition, quantitative approaches such as spatial statistics and neutral modeling (discussed in Chapters 5 and 6) offer new possibilities for statistical analysis of spatial pattern and associated processes.

THE INTELLECTUAL ROOTS OF LANDSCAPE ECOLOGY

Although landscape ecology became more prominent within ecology in North America beginning in about 1980, it did not begin de novo at that time, but drew upon a rich history. Landscape ecology had its roots in Central and Eastern Europe. European biogeographers viewed the landscape as the total spatial and visual entity of human living space, thereby integrating the environment, the biota, and the human-created components of an area (Naveh and Lieberman, 1984). Troll, who coined the term *landscape ecology,* studied biology and then became a geographer. He was impressed by the ecosystem concept as defined by Tansley (1935) and fascinated by the comprehensive view of landscape units depicted on aerial photographs (Zonneveld, 1990; Schreiber, 1990). He viewed landscape ecology not as a new science, but as a special viewpoint for understanding complex natural phenomena (Schreiber, 1990). Troll (1968) wrote (as translated by Schreiber, 1990), "Aerial photo research is to a great extent landscape ecology, even if it is used, for instance, for archaeology or soil science. In reality, it is the consideration of the geographical landscape and of the ecological cause–effect network in the landscape." At about the same time, the Russian scientist Sukachef (1944, 1945) also developed a very similar concept of a biogeocenology.

Landscape ecology gained wider acceptance and appreciation in the German-speaking countries of Europe throughout the 1950s and 1960s, and it became closely linked with land planning and landscape architecture (Haber, 1990; Ruzicka and Miklos, 1990; Schreiber, 1990; Zonneveld, 1995). There was a strong emphasis on land evaluation, classification, and mapping as the basis from which land-use recommendations could be developed (Figure 1.3). A Society of Landscape Ecology was founded in The Netherlands in 1972; its members included a wide variety of scientists and practitioners whose concerns ranged from conservation to planning (Zonneveld, 1982, 1995). The major literature of landscape ecology from its inception until the early 1980s was predominantly in German and Dutch.

Despite the development of landscape ecology in Europe, the term was virtually absent from North American literature in the mid 1970s (Naveh and Lieberman, 1984). A handful of scientists from North America began attending European symposia and workshops on landscape ecology in the early 1980s (Forman, 1990) and disseminating these new ideas. Several influential publications in the early 1980s helped to introduce the developing field of landscape ecology to English-speaking scientists. Forman and Godron's (1981) article in *BioScience* asked whether the land-

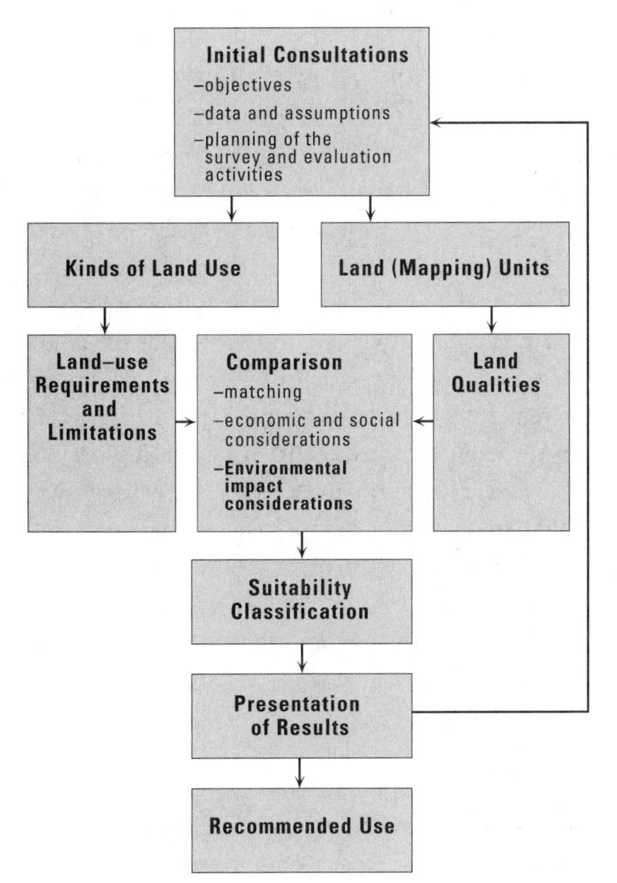

FIGURE 1.3.
Landscape classification and mapping approach developed by Dutch landscape ecologists. Note that the objective was to develop recommended uses of the land.

ADAPTED FROM ZONNEVELD, 1995.

scape was a recognizable and useful unit in ecology and provided a set of terms, such as patch, corridor, and matrix, that remain within the common parlance of landscape ecology. Naveh, an ecologist who focused on vegetation science, fire ecology, and landscape restoration, largely in Mediterranean climates, published a review that laid out a conceptual basis for landscape ecology (Naveh, 1982); his writing emphasized the integral relationship between humans and the landscape and the importance of a systems approach. These ideas were developed further as a book (Naveh and Lieberman, 1984) that delved into both concepts and applications of landscape ecology and stimulated much discussion among ecologists. Forman's (1983) editorial in *BioScience*, from which we quoted earlier, identified landscape ecology as the candidate idea for the decade, with a richness of empirical study, emergent theory, and applications lying ahead. And although not part of the infusion of ideas from Europe to North America, Romme's study of fire history in Yellowstone National Park (Romme, 1982; Romme and Knight, 1982) offered a breakthrough in the development of new metrics to quantify changes in the landscape through time.

Two pivotal meetings in the early 1980s helped to define the current scope of landscape ecology. A 1983 workshop held at Allerton Park, Illinois, brought together a group of North American ecologists to explore the ideas and potential of landscape ecology concepts (Risser et al., 1984). This meeting came soon after an influential meeting in The Netherlands that drew together landscape ecologists in Europe (Tjallingii and de Veer, 1982), and it represented the coalescence of several independent lines of research in the United States. The report that emerged (Risser et al., 1984) still makes for good reading. In many respects, the organized search for principles governing the interaction of pattern and process at the landscape scale began at these two meetings. The emphasis of landscape ecology in North American is somewhat different from Europe, where the association with land planning is so much closer and where the landscape itself has been more intensively managed for a much longer time. However, landscape ecology has grown out of intellectual developments that extended back many decades. The questions addressed by landscape ecologists typically couple the observation that landscape mosaics have spatial structure with topics that have interested ecologists for a long time (Wiens, 1995). Next we highlight several of the important precursors to the concepts of landscape ecology.

Phytosociology and Biogeography

Phytosociologists in Europe and the United States had long studied the spatial distribution of major plant associations (Braun-Blanquet, 1932), even going back to the observations of von Humboldt (1807) and Warming (1925). For example, it was well known that vegetation distributions in space responded to the north–south gradient of temperature combined with an east–west gradient of moisture. Vegetation pattern was further determined by topographic gradients in moisture, temperature, soils, and exposure. Thus, at broad scales, it was well established that ecological systems interacted with spatially distributed environmental factors to form distinct patterns.

Gradient analysis, an approach similar to the European phytosociology methods, developed in the United States as a means for explaining vegetation patterns; Robert Whittaker's analysis of communities in the Great Smoky Mountains provides an excellent example (e.g., Whittaker, 1952, 1956). In these eastern mountains, distinct patterns have formed with elevation, due to temperature, and with exposure, due to moisture. In a classic analysis, Whittaker was able to decipher the environmental signals creating the pattern. The complex vegetation system was arrayed on a vertical axis of elevation and a horizontal axis representing exposure from moist sites (mesic) to dry, exposed sites (xeric) (Figure 1.4). This simple two-dimensional diagram permits us to predict the vegetation type at any spatial location on the landscape based on its elevation and exposure.

One line of theory was particularly influential in the development of landscape ecology: island biogeography, the analogy between patches of natural vegetation and oceanic islands. The British biogeographer Lack (1942) had early observed that smaller and more remote offshore islands had fewer bird species. From this and similar observations, MacArthur and Wilson (1963, 1967) developed a general theory of island biogeography. The theory has two basic parts: (1) the probability of a species reaching an island is inversely proportional to the distance between the island and the source (mainland or source patch) and directly proportional to island size, and (2) the probability of extinction of a species on the invaded island is a function of island size.

The original theory of island biogeography has been subjected to considerable criticism (e.g., Simberloff, 1974) because of its simplifying assumptions. Nevertheless, it has proved useful as a heuristic construct in designing nature reserves (e.g., Burkey, 1989), and dozens of empirical studies have validated at least some general features of the model. Current efforts in landscape biogeography, dealing with population and community responses to fragmented landscapes, owe much to this body of theory. Nonetheless, metapopulation models (Hanski, 1998) have largely replaced island biogeography models as the theoretical framework within which issues of habitat fragmentation are considered.

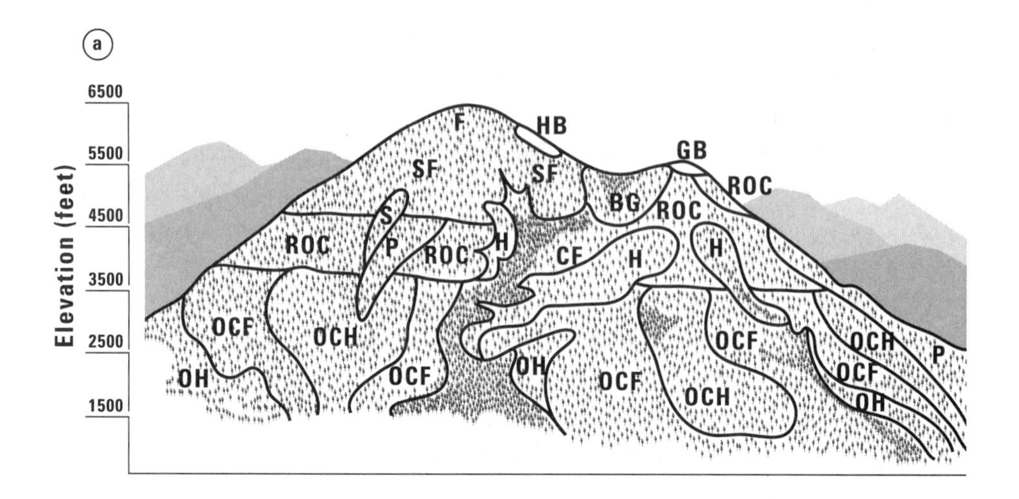

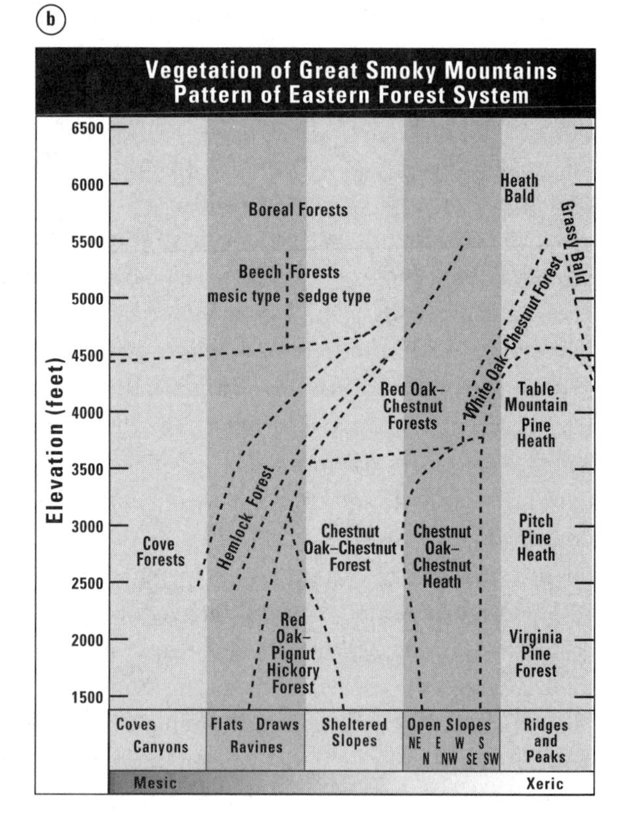

FIGURE 1.4.
(a) Topographic distribution of vegetation types on an idealized west-facing mountain and valley in the Great Smoky Mountains. Vegetation types are: BG, beech gap; CF, cove forest; F, fraser fir; G, grassy bald; H, hemlock forest; HB, heath bald; OCF, chestnut oak–chestnut forest; OCH, chestnut–oak–chestnut heath; OH, oak–hickory forest; P, pine forest and pine heath; ROC, red oak–chestnut oak; S, spruce; SF, spruce–fir; WOC, white oak–chestnut forest. (b) Vegetation of the Great Smoky Mountains, below the subalpine conifer forests, with respect to gradients of elevation and topography.

ADAPTED FROM WHITTAKER, 1956.

Landscape Planning, Design, and Management

The relationship between human societies and landscape change has been a fundamental concern of ecologists in Europe for many years (see Naveh, 1982). Indeed, the history of human-induced change is clearly apparent throughout Europe, with roads and viaducts constructed during the Roman Empire still having a visible effect in many regions (Marc Antrop, personal communication). The emphasis of ecological studies in North American has been on relatively undisturbed systems (Risser et al., 1984), but an awareness of human effects on landscapes has been evident for more than 140 years (Marsh, 1864, as cited by Turner and Meyer, 1993). The writings of a number of authors have provided an important context for integrating ecological effects with landscape planning, including the development of map overlay techniques (a precursor to current GIS methods) by McHarg (1969), the studies of Watt (1947) that focused on patch structure as fundamental to understanding vegetation pattern, an overview of the effects of ecosystem fragmentation in human-dominated landscapes edited by Burgess and Sharpe (1981), and the development of concepts of adaptive management by Holling (1978).

The goals of landscape planning, design, and management include the identification and protection of ecological resources and control of their use through plans that ensure the sustainability of these resources (Fabos, 1985). The result is that landscape planning is a primary basis for collaboration and knowledge exchange between landscape planners and landscape ecologists (Ahern, 1999). Perhaps the best examples of the integration of landscape planning, design, and management can be found in The Netherlands, where a national plan for a sustainable landscape is being implemented (Vos and Opdam, 1993). In North America, the best examples include the current plans for ecosystem management of national forests (Bartuska, 1999) and studies aimed at conservation design (Diamond and May, 1976; Mladenoff et al., 1994; Ando et al., 1998).

Multidisciplinary Studies and Regional Modeling

The geographic sciences have made important contributions to the methodology of landscape ecology. Satellite imagery, classified to cover type, has been an invaluable resource. Software developments (e.g., GIS and image analysis programs, spatial statistics) provide computer capabilities for displaying, superimposing, and analyzing spatial patterns. These analytical tools and the geographer's experience in handling large spatial databases have been a stimulus and critical resource for landscape ecologists.

During the 1960s, a number of diverse projects resulted in the development of large regional models. The result was the application of systems analysis and computer modeling at landscape scales that clearly established the broad-scale impact of human development. These models were often associated with urban development programs (Lowry, 1967) concerned with the interaction of spatial patterns and socioeconomic processes, a topic that remains important today. Transportation models were developed to link human activities at different positions on the landscape. Large-scale urban renewal programs (Pittsburgh Community Renewal Project, 1962) theorized about the optimal spatial arrangement of economic activities on the urban landscape. The central theme of these studies was the interactions by which socioeconomic processes produced spatial pattern and the patterns, in turn, encouraged or constrained human activities (Hemens, 1970). By the end of the 1960s, studies on pattern–process interactions had resulted in a considerable body of theory. Much of the development was synthesized under the titles of *urban dynamics* (Forester, 1969) and *regional science* (Isard, 1960, 1972, 1975). Considerable effort was expended toward linking spatial activities on the landscape with socioeconomic theory (Smith, 1976).

An important set of studies considered the spatial allocation of processes from the perspective of central place theory (Herbert and Stevens, 1960; Steger, 1964), which predicts that human activities will radiate outward from a center of economic activity, such as a city, transportation center, or highway intersection. This theory was later applied to the spatial pattern of foraging by animals (Aronson and Givnish, 1983), including ants (Harkness and Maroudas, 1985) and birds (Andersson, 1981) that forage outward from a nest. Central place theory was also used to predict land-use change following the installation of a sawmill in a rural area (Hett, 1971).

In subsequent years, regional modeling became concerned with predicting the effects of socioeconomic activities on the environment. Example applications with a strong spatial component included studies of the impact of large-scale energy developments (Basta and Bower 1982; Krummel et al., 1984) and the planning of river-basin systems (Hamilton et al., 1969). These studies resulted in new theoretical constructs to link socioeconomic and ecological variables in the same model (Klopatek et al., 1983), which were later applied to such diverse problems as modeling oil and gas extraction in the western United States (Mankin et al., 1981) and cattle herding societies in Africa (Krummel et al., 1986). All these studies focused on the interaction between landscape pattern and ecological processes and emphasized the need to include socioeconomic processes in landscape analyses, long before the principles of landscape ecology were articulated in Europe or North America.

Spatial Pattern and Theoretical Ecology

A number of theoretical population studies have considered the interaction between spatial patterning and ecological dynamics for terrestrial (Clark et al., 1978, 1979; Johnson et al., 1992a) and aquatic (Steele, 1974a; Harris, 1998) ecosystems. These studies demonstrated that unstable population interactions can sometimes be stabilized by spreading the interaction across a heterogeneous landscape (e.g., Reddingius and Den Boer, 1970; Roff, 1974a; Hastings, 1977; Scheffer and de Boer, 1995). At the same time, ecological processes alone can generate complex patterns in an otherwise homogeneous landscape (Dubois, 1975; McLaughlin and Roughgarden, 1991; Molofsky, 1994). Clark (1980) has pointed out that management practices that reduce heterogeneity to produce more stable dynamics are often counterproductive, because destruction of pattern can interfere with ecological mechanisms for persistence.

Many developments in population theory can be traced to the classic experiments of Huffaker (Huffaker, 1958; Huffaker et al., 1963), who studied the interactions of fructiverous and predatory mites in experimentally manipulated arrays of oranges. The oranges provided food for the fructivorous mites, which, in turn, were consumed by predatory mites. Spatial manipulation of the oranges could shift dynamics between unstable (oranges placed close together, allowing predators to locate and eliminate all prey) and stable (oranges formed into patches, preventing predators from locating and eliminating all prey). These experiments helped to define the importance of the spatial relationships among local populations that had been previously pointed out by Andrewartha and Birch (1954).

The interplay between spatial heterogeneity with species-specific patterns of dispersal has been extensively studied (e.g., Bradford and Philip, 1970; Caswell and Cohen, 1995; Cohen and Levin, 1991; Epperson, 1994; Hastings, 1996a; Kareiva, 1990; Levins, 1970; Levins and Culver, 1971). Spatial pattern of resources provides refuges (Comins and Blatt, 1974) that permit individuals to escape unfavorable conditions. The degree to which heterogeneity stabilizes relationships depends on the relative dispersal ability of predator and prey (Vandermeer, 1973; Taylor, 1990) and differences in their reproductive rates (Hilborn, 1975). Ziegler (1977) has also shown that dispersal or migration at discrete times can lead to a stable system even if continuous dispersal does not. The ability to disperse over a gradually changing environment could enable a population to survive extreme conditions (Roff, 1974b; Hamilton and May, 1977). It seemed clear that spatial pattern could affect both the stability of populations (Jones, 1975) and the total population size that could be supported (Steele, 1974b). Importantly, spatial pattern and the ability to disperse could spread the utilization of a resource over space so that it is not exhausted (Myers, 1976).

Another series of theoretical studies has been concerned with the biotic (Sprugel, 1976) and abiotic (Levin and Paine, 1974a, b) factors that cause the observed patterns on the landscape. Levin (1976a) provided an excellent general treatment of the subject, identifying three factors: (1) local uniqueness of sites on the landscape caused by variations in microhabitat, soils, and the likes, (2) phase differences, such that different points on the landscape are at different stages of development or different stages of recovery from localized disturbances, and (3) dispersal effects in which differential movement by organisms across landscapes leads to patchiness (e.g., Criminale and Winter, 1974).

Theoretical studies have suggested a number of specific mechanisms to explain landscape patterns. In areas where two species overlap, competitive interactions may produce sharp boundaries (Yamamura, 1976). Several workers (e.g., Kierstead and Slobodkin, 1953) have empirically demonstrated that spatial patterning in biota may reflect spatial patterning in abiotic factors such as water turbulence or topography. If a system has multiple stable states, a distinct spatial pattern may result simply by differences in microhabitat, which may be sufficient to structure phytoplankton communities (Powell et al., 1975). Even without microhabitat heterogeneity, Okubo (1974) has shown that the combination of competitive interaction and dispersal can result in patchiness. Segal and Levin (1976) reach a similar conclusion, particularly if there are mutualistic relationships among the prey.

Two important conclusions can be drawn from this brief survey of theoretical studies on spatial patterning: (1) it is clear that patterning is an important ecological phenomenon, with disruption of the pattern possibly resulting in the eruption of pests and subsequent population extinction events; (2) spatial patterns are the result of complex interplay between abiotic constraints, biotic interactions, and disturbances. The pattern is not simply a constraint imposed on the ecological system by topography and soils. Instead, there is an intimate tie between pattern and process that forms an important core for the understanding of landscape ecology.

Recent Theoretical Developments

New developments in theory are continuing to provide a stimulus for landscape ecology. We illustrate this with examples taken from fractal geometry, percolation theory, and self-organized criticality.

Fractal geometry (Burrough, 1981; Mandelbrot, 1983), which has identified classes of pattern that remain similar over a wide range of scales, has had intriguing applications in ecology (Sugihara and May, 1990). If the assumptions of the fractal theory are satisfied, extrapolation of spatial pattern across scales becomes possible, allowing broad-scale patterns to be predicted from fine-scale measurements. An early application of fractal geometry for landscape studies was the use of the fractal dimension as an index of human interference with landscape pattern (Krummel et al., 1987). Other applications include studies of insect movement (Johnson et al., 1992b; Wiens et al., 1995), measures of landscape texture (Plotnick et al., 1993), species perception of landscape structure (With, 1994a), generation of artificial landscapes (Palmer, 1992; With et al., 1997), characterizing landscape pattern (Milne, 1988; Overpeck et al., 1990), and using fractal theory for landscape design (Milne, 1991a).

Percolation theory (Stauffer and Aharony, 1992) deals with spatial patterns in randomly assembled systems. The application of percolation theory to landscape studies has addressed a series of questions dealing with the size, shape, and connectivity of habitats as a function of the percentage of a landscape occupied by that habitat type. Because percolation theory generates pattern in the absence of specific processes, the comparison of random maps with actual landscapes provides a neutral model capable of defining significant departures from randomness (Gardner et al., 1987a; With and King, 1997) of patterned landscapes. This theory has offered important insights into the nature of connectivity (or its inverse, fragmentation) on landscapes (Gardner et al., 1992a; Fonseca et al., 1996; Milne et al., 1996).

Descriptions of landscape pattern and process are beginning to benefit from insights provided by the theory of self-organized criticality (Bak et al., 1988). This theory states that open, complex systems (that is, systems with many independent components) may be described by power-law statistics over many orders of magnitude. Because these systems are self-similar (Grumbacher et al., 1993), a fundamental understanding of scale-dependent phenomena can emerge from studies of self-organized criticality. Well-studied examples of physical systems that display the properties of self-organized criticality include avalanches in sandpiles (Grumbacher et al., 1993), earthquakes (Ceva, 1998) and ferromagnetic systems (Tadic, 1998). Recently, these concepts have been applied to ecosystems (Milne, 1998), with examples that include canopy gaps in rain forests (Katori et al., 1998), river flows (Pandey et al., 1998), and coevolution in multispecies communities (Caldarelli et al., 1998). The importance of these results has recently been confirmed by using power-law statistics to estimate the risk of large fires from measurements taken from small to medium fires (Malamud et al., 1998).

OBJECTIVES OF THIS BOOK

It is clear that landscape ecology has a rich intellectual history and that it draws on a wide range of natural and social sciences. The remainder of this book will deal with the concepts, questions, methods, and applications of landscape ecology, with an emphasis on the ecological approach. This in no way diminishes the importance of the social sciences in the interdisciplinary study of landscapes; however, this text is written by ecologists, and our biases and expertise fall within the science of ecology. We hope that the book will be useful not only to students in ecology, but also to students in disciplines such as conservation biology, resource management, landscape architecture, land planning, geography, and regional studies who wish to delve more deeply into landscape ecology as an ecological science. In addition, we hope that this volume will complement other recent landscape ecology books that have somewhat different emphases (e.g., Haines-Young et al., 1993; Forman, 1995; Hansson et al., 1995; Bissonette, 1997; Farina, 1998; Klopatek and Gardner, 1999).

Landscape ecology may also serve as a source of new ideas for other disciplines within ecology. For example, aquatic ecologists have applied a landscape ecological approach to the study of riffle, cobble, and sandy substrates within streams (e.g., Wohl et al., 1995), patch distributions of fishes as measured by echolocation (e.g, Magnuson et al., 1991; Nero and Magnuson, 1992), patterns and processes of rocky benthic communities (e.g., Garrabou et al., 1998), and spatial variation in coral bleaching (Rowan et al., 1997). Thus, landscape ecology benefits from and contributes toward intellectual developments in other disciplines.

The development and application of models has emerged as an important component of landscape ecology, as in other areas of science. In particular, spatially explicit models of ecological dynamics have become widely used in landscape-level studies. There remains a strong need for enhanced integration of models with appropriate field or empirical studies. The combination of models, which provide a rigorous representation of our hypotheses or best understanding of the dynamics of a systems, and empirical data, which keep us firmly rooted in the ecological systems that we seek to understand, offers a powerful approach likely to result in greater insight than either approach applied alone. Nevertheless, we include a chapter on modeling to familiarize readers with the fundamental concepts of this important topic.

This is also not a textbook for geographic information systems (GIS) or remote sensing, although landscape ecology makes extensive use of these technologies. Often, landscape ecologists use the final products of GIS manipulations or the interpretation of spectral data, but many are not technically proficient in all the intricacies of the processes involved. Many fine texts are excellent resources for the landscape ecologist who needs a more thorough introduction to these subjects. For GIS, we suggest Burrough (1986), Bonham-Carter (1994), Fotheringham and Rogerson (1994), and Burrough and McDonnell (1998); for remote sensing, we suggest Lillesand and Kiefer (1994) or Jensen (1996).

We have organized the book in a sequence comparable to what we teach in a landscape ecology course. The first three chapters provide an introduction to the subject and its development (Chapter 1), a treatment of scale (Chapter 2), which influences everything that follows, and an introduction to basic modeling concepts (Chapter 3). We then examine the causes of landscape pattern (Chapter 4), including both biotic and abiotic factors, and consider observed changes over extended temporal scales. The quantification of landscape pattern, which is a necessary component of understanding the interaction between pattern and process, is presented in detail in Chapter 5. The use of neutral models in landscape ecology, which is closely related to quantification of pattern and to linkages of pattern with process, is considered in Chapter 6. The next three chapters deal with particular phenomena that have received considerable attention in landscape studies during the past two decades: disturbance dynamics (Chapter 7), the responses of organisms to spatial heterogeneity (Chapter 8), and ecosystem processes at landscape scales (Chapter 9). We then deal explicitly with the many applications of landscape ecology (Chapter 10) and, finally, suggest conclusions and future directions for the field (Chapter 11).

SUMMARY

Landscape ecology has come to the forefront of ecology and land management in recent decades, and it is still expanding very rapidly. Landscape ecology emphasizes the interaction between spatial pattern and ecological process, that is, the causes and consequences of spatial heterogeneity across a range of scales. Two important aspects of landscape ecology distinguish it from other subdisciplines within ecology. First, landscape ecology explicitly addresses the importance of spatial configuration for ecological processes. Second, landscape ecology often focuses on spatial extents that are much larger than those traditionally studied in ecology. These two aspects, explicit treatment of spatial heterogeneity and a focus on broad spatial scales, are complementary and encompass much of the breadth of landscape ecology.

The recent emergence of landscapes as an appropriate scale for ecological study resulted from (1) broad-scale environmental issues and land-management problems, (2) the development of new scale-related concepts in ecology, and (3) technological advances, including the widespread availability of spatial data, the software to manipulate these data, and the rapid rise in computational power. However, landscape ecology has a history, with its roots in Central and Eastern Europe. The major literature of landscape ecology from its inception in the late 1930s through the early 1980s was predominantly in German and Dutch; the term *landscape ecology* was virtually absent from North American literature in the mid 1970s. The recent search for principles governing the interaction of pattern and process at the landscape scale began with two influential workshops in the early 1980s in Europe and North America. The questions addressed by landscape ecologists typically couple the observation that landscape mosaics have spatial structure with topics that have interested ecologists for a long time. Landscape ecology has grown out of intellectual developments that extended back many decades and include phytosociology and biogeography, landscape design and management, geography, regional modeling, theoretical ecology, island biogeography, and mathematical theory.

DISCUSSION QUESTIONS

1. Reconcile the two different ways in which ecologists use the concept of landscape: as a relatively large area composed of elements that we recognize and as a theoretical construct for considering spatial heterogeneity at any scale (see Pickett and Cadenasso, 1995). Are these notions mutually exclusive or complementary? Do they confuse or enhance our understanding of landscape ecology?
2. Describe three current environmental issues that require consideration of the landscape, either as a causal factor or a response. What information or understanding is lost if a landscape perspective is not taken?
3. How has landscape ecology been influenced by the historical development of ideas in ecology? In landscape design and management?

4. Is landscape ecology defined by its questions or by its techniques? Do you consider it to be a broad or narrow avenue of inquiry within ecology?

RECOMMENDED READINGS

PICKETT, S. T. A., AND M. L. CADENASSO. 1995. Landscape ecology: spatial heterogeneity in ecological systems. *Science* 269:331–334.

TURNER, M. G. 1989. Landscape ecology: the effect of pattern on process. *Annual Review of Ecology and Systematics* 20:171–197.

URBAN, D. L., R. V. O'NEILL, AND H. H. SHUGART. 1987. Landscape ecology. *BioScience* 37:119–127.

CHAPTER 2

The Critical Concept of Scale

Nearly all ecologists now recognize that scale is a critical concept in the physical and natural sciences. In his MacArthur Award Address to the Ecological Society of America, Simon Levin noted that "the problem of relating phenomena across scales is the central problem in biology and in all of science" (Levin, 1992). Elsewhere in his address, Levin (1992) stated that

> we must find ways to quantify patterns of variability in space and time, to understand how patterns change with scale, and to understand the causes and consequences of pattern. This is a daunting task that must involve remote sensing, spatial statistics, and other methods to quantify pattern at broad scales; theoretical work to suggest mechanisms and explore relationships; and experimental work, carried out both at fine scales and through whole system manipulations, to test hypotheses.

Indeed, scale is a prominent topic in landscape ecology and rightfully so; it influences the conclusions drawn by an observer and whether the results can be extrapolated to other times or locations. Issues associated with scale will be touched on in many places throughout the remainder of this book, and entire books recently have been devoted to the topic (e.g., Schneider, 1994; Peterson and Parker, 1998). In this chapter, we provide an introduction to scale terminology and concepts.

Why is it that scale has taken on such a prominent role in ecology in general and landscape ecology in particular? There are several reasons, some practical and some theoretical, that brought scale issues to the forefront in the early 1980s. Pressing issues relating to the environment and the biosphere were manifested over larger and larger areas. For example, acid rain, global climate change, habitat fragmentation, and conservation of biodiversity all required understanding patterns and processes at very broad scales. Thus, ecologists found themselves challenged to use data and understanding usually obtained from fine scales (e.g., the square meter quadrat so common in field studies) and then infer or project consequences that would occur at broad scales. Consider a desert community that grades into a grassland community on a steep hillside. At one scale, an investigator might ask questions about the relative efficiency of the two communities and take samples from randomly placed quadrats in each community to calculate mean biomass and productivity. At a broader scale, however, we might ask where on the hillside's moisture–temperature gradient one community changes into the other. The finer-scale quadrat data within each community often cannot answer the question at the broader scale. Thus, practical questions about extrapolating from fine to broad scales became very important as scientists and managers struggled to find satisfactory methods to accomplish this.

In addressing broad-scale questions, ecologists became much more aware of the implications of choosing scales for conducting their research. There was increasing recognition that the answers obtained for a particular ecological question depended strongly on the scale at which the study was conducted. That is, changing the size of the quadrat or the overall extent of the area studied often yielded a different numerical result or pattern, and seemingly disparate results from different studies might simply be due to differences in the scales at which they were conducted. While change in patterns with scale was long recognized in ecology (Grieg-Smith, 1952), the expanded application of the scale concept in the design and interpretation of surveys, comparative studies, and controlled experiments was new (Schneider, 1994). Ecologists also became very aware that the spatial and temporal scales important to humans were not necessarily the scales that were relevant to other organisms or a wide range of ecological processes, from microbes through global change. Recognition that biological interactions in the environment occur at multiple scales was also influential. The increasing focus on scale appears to be an enduring change in the way that ecological research is pursued (Schneider, 1998).

Hierarchical structure in nature and a positive correlation in spatial and temporal scales of varying processes became important topics of discourse in ecology during the early 1980s, and three publications were particularly influential in stimulating this increased attention to scale. Allen and Starr's (1982) book on hierarchy theory laid out novel ways of considering scale that emerged from general systems theory and instigated tremendous discussion in all branches of ecology. The Delcourt et al. (1983) article on scales of vegetation dynamics in space and time synthesized paleoecological changes in landscapes and graphically presented the positive correlation of spatial and temporal scales as a time–space state space that has been widely used (Figure 2.1). The O'Neill et al. (1986) book reviewed the concept of an ecosystem and proposed a hierarchical framework for ecosystems. These three contributions laid an important foundation for the treatment of scale in ecology. In addition, new mathematical theory such as fractals (Mandelbrot, 1983) seemed to explain some complicated patterns in nature, while offering potential for the development of rules that would allow observations to be transferred from one scale to another. Collectively, these events and insights engendered an appreciation for concepts of scale and a mandate to better understand its effects in ecology.

SCALE TERMINOLOGY AND ITS PRACTICAL APPLICATION

The terminology related to scale is not used consistently, and this often results in confusion in the literature (Allen and Hoekstra, 1992; Allen, 1998). It is important for landscape ecologists to be unambiguous in their use of scale-related terms. **Scale** refers to the spatial or temporal dimension of an object or a process.* This is distinguished from **level of organization**, which is used to identify a place within a biotic hierarchy. For example, a sequence of differing levels of organization might be organism, deme, population, community, and biome. Each level of organization is characterized by a variety of processes that have their own scales of space and time. A population of a particular species may occupy a given amount of space, move or disperse a set distance, and reproduce within a characteristic time period. However, the community to which that population belongs will be char-

*For the reader's later convenience the terms in boldface are gathered together in Table 2.1.

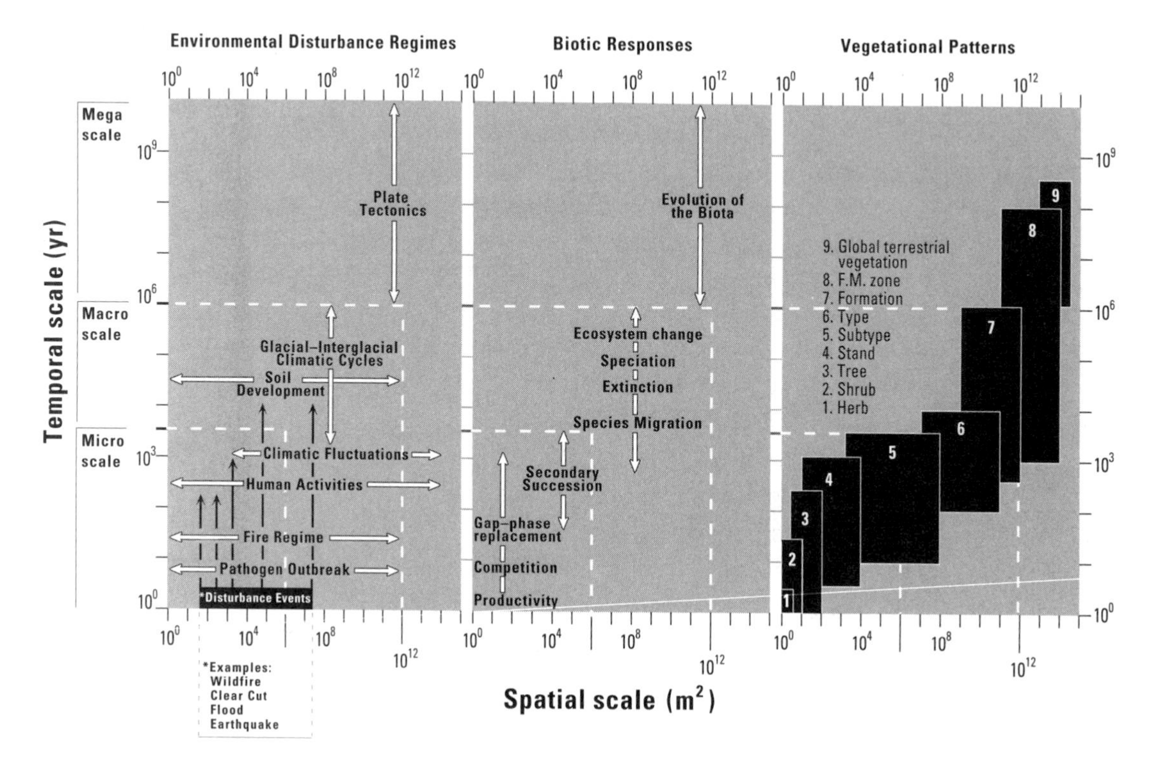

FIGURE 2.1.

Space–time hierarchy diagram proposed by Delcourt et al. (1983). Environmental disturbance regimes, biotic responses, and vegetational patterns are depicted in the context of space–time domains in which the scale for each process or pattern reflects the sampling intervals required to observe it. The time scale for the vegetation patterns is the time interval required to record their dynamics.

MODIFIED FROM DELCOURT ET AL., 1983.

acterized by spatial and temporal scales associated with the collection of populations composing the community.

Scale is characterized by grain and extent (Figure 2.2). **Grain** refers to the finest spatial resolution within a given data set; for example, grain refers to the cell size for gridded maps or the minimum mapping unit of maps drawn with polygons. **Extent** refers to the size of the overall study area. Grain and extent are easy to think of when considering remote imagery. Different satellite sensors have different cell sizes, or grain; for example, there is a cell size of 10 m by 10 m for SPOT panchromatic imagery, 30 m by 30 m for Landsat Thematic Mapper imagery, and

Table 2.1.
Definitions of scale-related terminology and concepts.

Term	Definition
Absolute scale	Actual distance, direction, shape, and geometry.
Cartographic scale	Degree of spatial reduction indicating the length used to represent a larger unit of measure; ratio of distance on the map to distance on Earth's surface represented by the map, usually expressed in terms such as 1:10,000. In cartography, large scale means fine resolution and small scale means coarse resolution.
Critical threshold	Point at which there is an abrupt change in a quality, property, or phenomenon.
Extent	Size of the study area or the duration of time under consideration.
Extrapolate	To infer from known values; to estimate a value from conditions of the argument not used in the process of estimation; to transform information (1) from one scale to another (either grain size or extent) or (2) from one system (or data set) to another system at the same scale.
Grain	Finest level of spatial resolution possible within a given data set.
Hierarchy	System of interconnections or organization wherein the higher levels constrain and control the lower levels to various degrees depending on the time constraints of the behavior.
Holon	Representation of an entity as a two-way window through which the environment influences the parts and parts communicate as a unit to the rest of the universe (Koestler, 1967).
Level of organization	Place within a biotic hierarchy (e.g., organism, deme, population).
Relative scale	Transformation of absolute scale to a scale that describes the relative distance, direction, or geometry based on some functional relationship.
Resolution	Precision of measurement; grain size, if spatial.
Scale	Spatial or temporal dimension of an object of process, characterized by both grain and extent.

90 m by 90 m for the earlier Landsat Multispectral Scanner imagery. The detail that can be gleaned from these different sensors varies, in part because of the differences in grain. Extent can vary independently of grain, although there is some degree of correlation; for example, a small extent will require a small grain size. When we say that a pattern, process, or phenomenon is *scale dependent*, we mean that it changes with the grain or the extent of the measurement. Schneider (1994) defines scale-dependent processes as those in which the ratio of one rate to another varies with either resolution (grain) or range (extent) or measurement.

One source of confusion is that ecologists and geographers usually mean the opposite when they say *large* or *small* scale. The long-standing use in geography of **cartographic scale** refers to the degree of spatial reduction indicating the length used to represent a larger unit of measure. Cartographic scale is typically expressed as the ratio or representative fraction (RF) of distance on the map to distance on the surface of Earth that is represented on a map or aerial photograph, for example, 1:10,000 or 1:100,000. When geographers and cartographers say large scale, they mean very fine resolution (e.g., 1:500), which in practice means a very

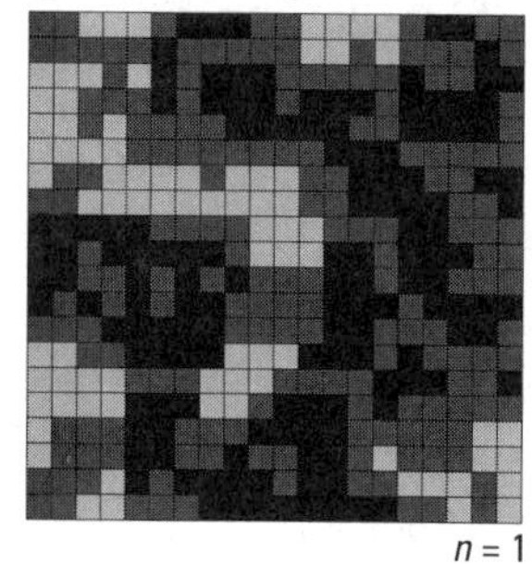

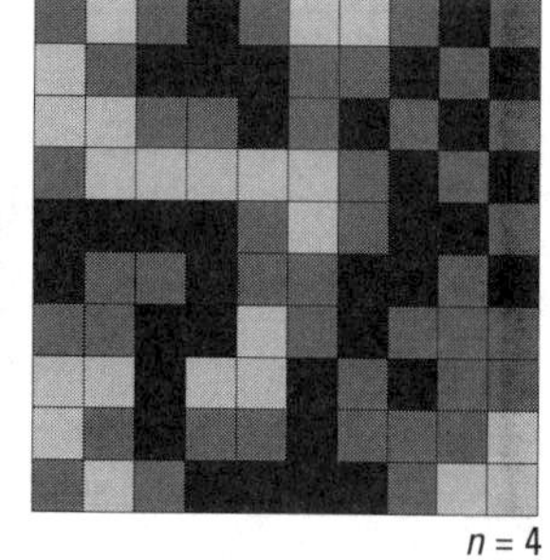

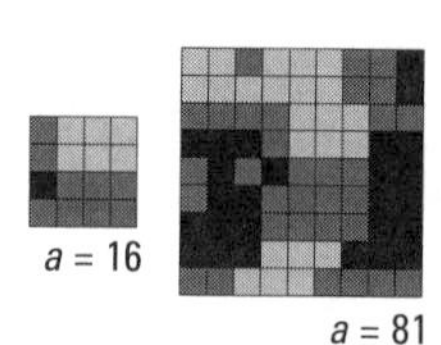

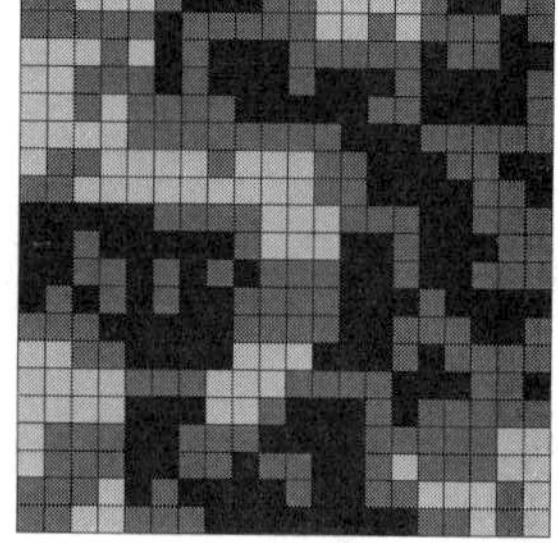

FIGURE 2.2.
Schematic of two components of spatial scale: (a) grain and (b) extent. The number of cells aggregated to form the new data unit (i.e., new grain size) are indicated by n; total area, or extent, is indicated by a.

MODIFIED FROM TURNER ET AL., 1989B.

large map of a small spatial extent; similarly, when they say small scale, they mean very coarse resolution, or maps of large areas that do not contain much detail (e.g., 1:250,000). This use of small and large is opposite to what ecologists usually mean by these terms! To avoid confusion, we recommend and use here the terms *fine* and *broad* to modify scale such that fine-scale refers to small areas, greater resolution, and more detail and broad-scale refers to larger areas, lower resolution, and less detail.

In practical terms, the scale at which you make any measurement, for example, the size of the quadrat, length of transect, area of census, or size of the grid cell in remotely sensed data, influences the numerical answer obtained. The species–area curve (Figure 2.3) is an early example of this phenomenon. In trying to determine the species richness for a particular community, ecologists soon realized that the number of species detected would increase asymptotically with the size of the area that was censused. Thus, to avoid erroneous inferences due to mismatches in scale, the study area size had to be accounted for before estimates of species richness among sites or through time could be compared.

Another consideration in scale terminology is the distinction between absolute and relative scale. We generally talk about **absolute scale**, that is, the actual distance, time, area, or the like. On the other hand, we might consider distance relative to the energy that an animal would need to expend to travel between different points on a landscape. With this **relative scale**, two points that are closest to each other as the crow flies may be far apart if they are separated by a large peak or ravine that would require much energy to traverse; two points farther apart as the crow flies, but connected by level ground that is easily traversed, may have a closer relative distance.

When we seek to **extrapolate**, we attempt to infer from known values, that is, to estimate a value from conditions beyond the range of the data used in the process of estimation. For example, we extrapolate when we use a regression line to predict values of y based on a value of x that is beyond our original data. We also extrapolate when we transfer information from one scale to another (either grain size or extent) or from one system (or data set) to another system at the same scale. Searching for techniques or algorithms to extrapolate across scales or among landscapes remains an important research topic in landscape ecology. In practical terms, this occurs because scientists never have all the data they need at all the right scales. Extrapolating may be straightforward in some cases when the relationship of a variable with changes in scale is linear, or additive; however, if the relationship is nonlinear and there are **critical thresholds** at which there is an abrupt change in some quality, then extrapolation is problematic.

SCALE PROBLEMS

Landscape ecologists often refer to scale as a problem or challenge. This occurs for several reasons, including the inherent difficulty of understanding or predicting an ecological attribute over a large area, the logistical problems associated with sampling or experimentation over large areas, and the associated statistical problem of pseudo-replication (Hurlbert, 1984; Hargrove and Pickering, 1992)

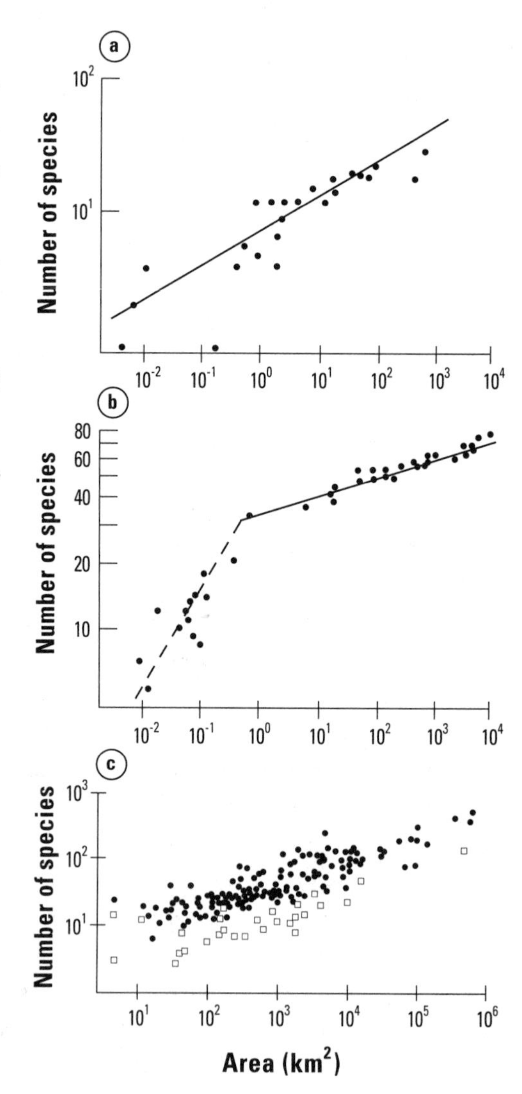

FIGURE 2.3.

Examples of species–area relationships in which the number of species observed in locations that vary in area typically increases with increasing area. (a) Species–area plot for 24 islands in the Sea of Cortez (from Cody, 1983; copyright University of California Press). (b) Species–area plot for birds in the Solomon Islands, illustrating nonlinearity. (Data from Diamond and Mayr, 1976, and Williamson, 1981.) (c) Species–area plots for landbirds of islands in tropical and subtropical oceans, illustrating interaction between island size and distance from mainland. Open squares are islands ≥300 km from the next largest land mass or in the Hawaiian or Galapagos archipelagos. Closed circles are <300 km from next largest landmass.

MODIFIED FROM WIENS, 1989B, BASED ON SLUD, 1976, AND WILLIAMSON, 1981.

(Table 2.2). However, these scale issues have been well recognized for a long time in the field of geography, which focuses on the spatial distribution of many natural and human phenomena. Haggett (1963) identified three scale problems, circumstances in which scale was perceived to hinder geographical research.

1. *The scale coverage problem.* Essentially, Haggett's first problem recognized one identified by the ancient Greeks: the surface of Earth is so large that this vastness poses difficulties in mapping and understanding its spatial variability. If the purpose of geography is "to provide accurate, orderly, and rational description and interpretation of the variable character of the earth surface," then the magnitude of the task is enormous.

2. *The scale linkage problem.* A direct consequence of the scale coverage problem is that field work is often restricted to relatively small areas, raising the problem of relating fine-scale data to broader spatial scales. McCarty et al. (1956)

TABLE 2.2.
COMPARISON OF THE ATTRIBUTES OF FINE- AND BROAD-SCALE STUDIES.

	Scale	
Attribute	Fine	Broad
Detail resolution	High	Low
Effects of sampling error	Large	Small
Experimental manipulation	Possible	Difficult
Generalizable	Low	High
Model form	Mechanistic	Correlative
Replication	Possible	Difficult
Rigor	High	Low
Sampling adequacy	Good	Poor
Study length	Short	Long
Survey type	Qualitative	Quantitative
Testability of hypotheses	High	Low

ADAPTED FROM WIENS 1989A AND BISONETTE, 1997.

wrote, "In geographic investigation, it is apparent that conclusions derived from studies made at one scale should not be expected to apply to problems whose data are expressed at other scales. Every change in scale will bring about the statement of a new problem, and there is no basis for assuming that associations existing at one scale will also exist at another." This nicely describes the practical problem of comparing data obtained at different scales.

3. *The scale standardization problem.* The ability to compare different locations, extrapolate from one place to another, or assemble different types of data for the same place is influenced by differences in how data are collected and reported. Any landscape ecologist who has built or worked with a multilayer GIS database is probably painfully aware of this problem. For geographers, Hagget (1963) noted that most social data were reported for areas rather than points (e.g., census tracts, counties, states, countries) and that these areas vary wildly in size and shape, both between and within countries. This issue is very relevant for ecologists when we consider the political–ownership boundaries that have been superimposed on ecological units such as watersheds. Developing methods for combining different types of data (e.g., point- and area-based measures) remains an active research area today.

SCALE CONCEPTS AND HIERARCHY THEORY

Concepts of scale and hierarchy are inextricably linked. Hierarchy is usually identified with the concept of levels of organization in the ecological literature (O'Neill et al., 1986). In the simplest series (cell, organism, population, community), each level is composed of subsystems on the next lower level and is constrained by the level above it. Ecological organizations do indeed show hierarchical structure (Rowe, 1961), but this simple view of hierarchy is not adequate to fully characterize the range of processes and scales in ecology. We clearly distinguish between scale and level of organization, but hierarchy theory offers considerably more insight into scale in ecology. Here, we briefly highlight some of the important messages that are especially significant to landscape ecologists.

A **hierarchy** is defined as a system of interconnections wherein the higher levels constrain the lower levels to various degrees, depending on the time constraints of the behavior. The concept of hierarchy has a long history in science, but Koestler's (1967) *The Ghost in the Machine* was a landmark publication. Koestler identified entities that were at the same time composed of parts, yet were also a whole that fits within its environment. At every level in a hierarchy there are these elements, termed **holons**, that are both wholes and parts.

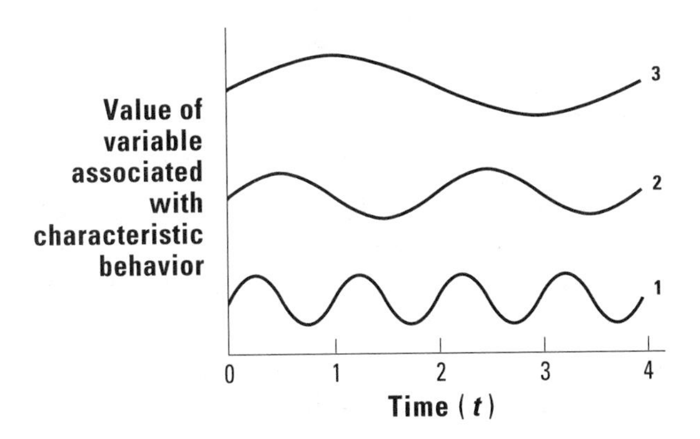

FIGURE 2.4.
Value of variables associated with a level of an ecological hierarchy as they change through time. The top line (line 3) is a slow variable, one that would serve as a constraint to the lower levels; this may change so slowly that it is perceived as a constant by an observer. The middle line (line 2) might be the scale at which an observer measures change in the system. The lower line (line1) is a fast variable, one that might change so quickly that it could be perceived as a constant.

REDRAWN FROM ALLEN AND STARR, 1982:12.

Within a hierarchical system, the levels are distinguished by differences in the rates, or frequencies, of their characteristic processes (Figure 2.4). Holons have characteristic rates of behavior, and these rates place them at certain levels in the hierarchy of holons. For example, an individual organism, as a holon, can interact with other individual organisms because both operate at the same space–time scale. But an individual organism cannot interact with a biome; they are orders of magnitude different in scale. To the individual organism, the biome is a relatively constant background or context within which it operates. Thus, temporal scales serve as important criteria for identifying levels within a hierarchy, and there are different scales of space and time over which controls operate.

An important concept from hierarchy theory is the importance of considering at least three hierarchical levels in any study (Figure 2.5). The focal level or level of interest is identified as a function of the question or objective of the study. For example, answering the question, "What is the effect of insect herbivory on tree growth rate?" would require focusing on individual trees, whereas "What is the effect of insect herbivory on the distribution of live and dead trees across the landscape?" would require focusing on the forest as a whole. Two additional levels must then be considered. The level above the focal level constrains and controls the lower levels, providing context for the focal level. The level below the focal level provides the details needed to explain the behavior observed at the focal level. Returning to our example of the individual organism as a holon, we explain how it is able to function as a predator by examining its structure and physiology (e.g., sensory organs, teeth, and/or claws). However, the availability of prey species (and ultimately the success of the predator) will be constrained by the broader-scale system in which it is located.

A second important message is that although the variables that influence a process may or may not change with scale, a shift in the relative importance of the variables or the perceived direction of a relationship often occurs when spatial or temporal scales are changed. There are numerous examples of this. For example, predicting the rate of decomposition of plant material at a very local scale requires detailed knowledge of the microclimate, variability in the environment, and characteristics of the litter, such as its lignin content; however, effectively predicting rates of decomposition at regional to global scales can be done based solely on temperature and precipitation (Meentemeyer, 1984). Studies of oak seedling mortality at local scales in the western United States showed that mortality decreased as precipitation increased, whereas studies at regional scales demonstrated that mortality decreased in the drier latitudes (Neilson and Wullstein, 1983).

An effect of changing temporal scales is exemplified by how sampling frequency in lakes influences the relationship observed between phytoplankton and zooplankton abundance. Carpenter and Kitchell (1987) used an aquatic ecosystem model to study the correlation structure among ecosystem components. When the relationship between algal production and zooplankton biomass was examined at 3-day intervals, a negative correlation was observed. However, if a 6-day interval was used, the

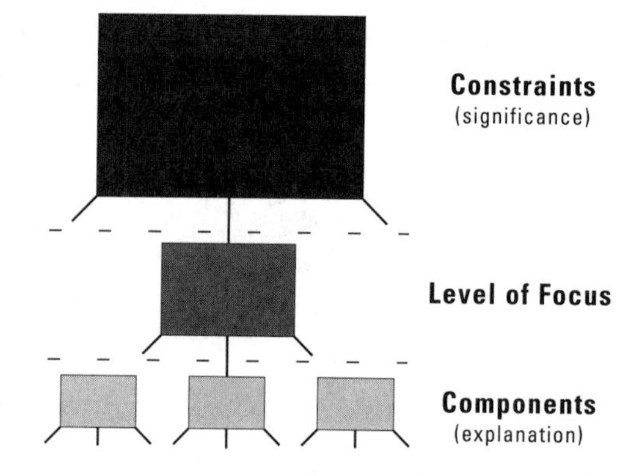

FIGURE 2.5.
Three levels in a hierarchy. Upper levels constrain the focal level and provide significance; lower levels provide details required to explain response of focal level.

ADAPTED FROM O'NEILL ET AL., 1986.

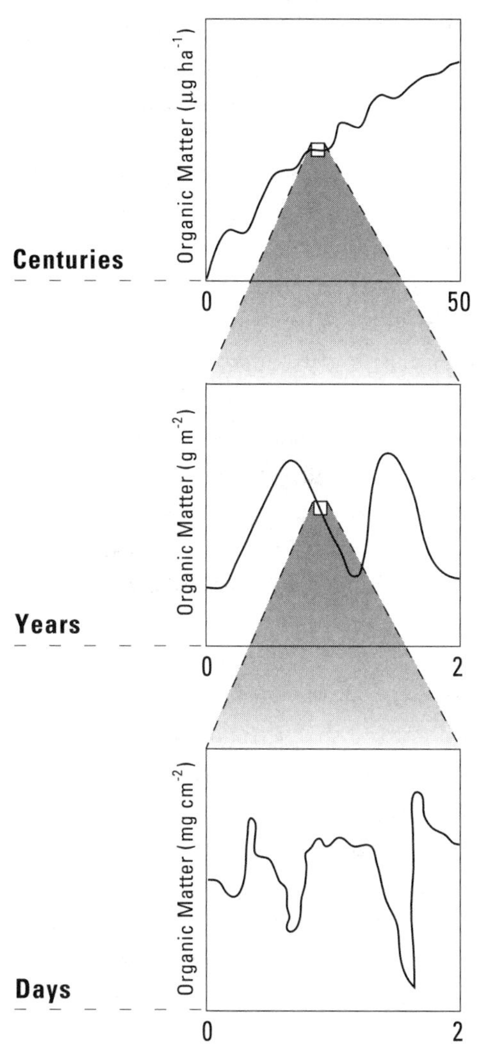

FIGURE 2.6.

Changes in the apparent dynamics of organic matter in the soil when the temporal scale of observation changes. An observation window of days (lower panel) reveals rapid fluctuations in litter due to wind and arthropod activity. Over a scale of years (middle panel), seasonal patterns of decomposition are apparent. Over a scale of centuries (upper panel), the accumulation of organic matter is observed with oscillations that relate to succession.

ADAPTED FROM SOLLINS ET AL., 1983.

seasonal dynamics of nutrients became most important, and the correlation between algal production and zooplankton biomass was positive. These seemingly contradictory results from relationships examined at different spatial or temporal scales demonstrate that a change in scale results in a change in the processes that are important.

Extending the time frame of observation of a system may lead to different observations and conclusions about the function of the system (e.g., Sollins et al., 1983; Magnuson, 1990). The results of different processes become apparent at different temporal scales (Figure 2.6), and conclusions about the directional change may change qualitatively as observation time increases. The constraint of having a limited observation window has been characterized as "the invisible present" (Magnuson, 1990), and this has been illustrated numerous times with long-term ecological data.

Hierarchy theory also suggests that multiple scales of pattern will exist in landscapes because of the multiple scales at which processes are acting. Consider the processes that may give rise to pattern in a hypothetical forest landscape. Over broad scales of space and time, geomorphological processes result in distributions of substrate and soil that influence the tree species might occur at different positions. Within the forest that develops, the pattern and frequency of large disturbances, such as fire or pathogen epidemics, may generate a coarse-grained pattern of different successional stages across the landscape. Local processes of individual tree death may result in small canopy gaps distributed throughout the forest landscape. Collectively, then, the spatial pattern of forest communities at any given time in this landscape may reflect these three processes operating over different scales of space and time. An analysis of the spatial pattern of the forest communities across the landscape may well detect these multiple scales of pattern (e.g., Kuuluvainen et al., 1998).

Thus, hierarchy theory tells us that attention should be focused directly on the scales at which phenomena of interest occur, that there is no single correct scale for studying landscapes or any other ecological system, and that if we change the scales, the relevant processes or even the direction of relationships that we observe may well change. The scale of interest must be dictated by the question or phenonenon of interest. Finer-scale processes may be viewed as the details required to explain the phenomena at the focal scale, while broader-scale patterns are the constraints that limit the potential range of rate processes. These concepts are rich and provocative, but still leave us with the critical challenge of identifying the proper scale at which to address a given problem.

IDENTIFYING THE "RIGHT" SCALE(S)

Returning to Levin's (1992) MacArthur address, we read "That there is no single correct scale or level at which to describe a system does not mean that all scales serve equally well or that there are not scaling laws." Developing the rules for identifying the correct scale for a particular question continues to be an important topic of current research. However, ecologists must recognize that identifying the "right" scale still requires combining art with science, because satisfactory algorithms do not currently exist.

The developing field of spatial statistics offers a variety of techniques that is useful in identifying scales (see Chapter 5). For example, O'Neill et al. (1991b) analyzed vegetation transect data from three locations to determine if multiple scales of pattern could be detected. Data were obtained from a semiarid grassland in New Mexico, calcareous openings in a deciduous forest in Tennessee, and a shrub–steppe system in Washington. Analyses revealed that between three and five scales of pattern could be identified on all three sites.

Spatial statistics techniques may also be used to identify the spatial scales over which successive sampling points are correlated. Data compared at scales where the correlation is near zero can be regarded as statistically independent, an important assumption for many statistical tests. Pearson et al. (1995) used this approach to identify the range of distances over which measurements of ungulate foraging intensity were autocorrelated. After identifying these relationships, the data were resampled to select observations greater than the correlation length. The resampled data then met the independence criterion of regression analyses relating grazing intensity to environmental heterogeneity.

Concepts such as ecological neighborhoods (Addicott et al., 1987, discussed in Chapter 8) use frequency distributions for the space and time components of the behavior of organisms to determine the correct scale of analysis. They suggest that a process, such as foraging or reproduction, be selected first and the activity of the organism then be monitored to determine the spatial scale relevant to the process. The spatial extent that encompasses most of the organism's activity relevant to the process (and over a relevant amount of time) is then the appropriate scale for this process.

Multiple regression can be used to quantify the explanatory power of a set of variables at different scales (e.g., Pearson, 1993; Pearson et al., 1995; Gergel et al., 1999). The approach used by Pearson (1993) to examine the relationship between the presence and abundance of wintering birds in the Georgia, USA, piedmont and environmental variation at different spatial scales has been widely applied (e.g., Rescia et al., 1997; Sisk et al., 1997; Elliot et al., 1999; Estades and Temple, 1999). This approach has confirmed that there is no single appropriate scale at which ecologists may expect to analyze their data. Rather, the identification of a suite of appropriate scales, or *multiscale analysis*, must continue to be employed as the science of scale identification and interpretation continues to mature.

REASONING ABOUT SCALE

In the introductory chapter to his book *Quantitative Ecology, Spatial and Temporal Scaling*, Schneider (1994) provides an insightful discussion of the ways in which scale is used as a routine part of research and reasoning in disciplines other than ecology. The explicit treatment of scale is found in a very high percentage of research articles in the physical sciences (e.g., geophysical fluid dynamics), engineering, and measurement science. Scale issues are also familiar to biologists, who have routinely used allometric scaling of form and function to body size throughout the 20th century. Indeed, most ecologists are familiar with relationships among life history characteristics such as longevity, generation time, fecundity and body size in animals. Schneider (1994) observes that "The most important characteristic of quantitative reasoning . . . is directed at scaled quantities obtained by measurement or by calculations from measurement." Reasoning about scaled quantities includes questions such as these: What are the algebraic rules for rescaling quantities? What is the best way to visualize and verify these operations? How are these operations used in solving ecological problems?

SCALING UP

Issues of extrapolating to broader scales in space and time, or scaling up, continue to be on the cutting edge of research in landscape ecology. Ecologists remain vexed by the need to make predictions at broad scales when most measurements have been made at relatively fine scales. How can it be done and what is the current state of knowledge? The challenges of scaling up lie in (1) correctly defining the spatial and temporal heterogeneity of the fine-scale information and (2) correctly integrating or aggregating this heterogeneity to the broader scale (King, 1991).

The simplest approach to scaling across space is to multiply a measurement made at one scale (e.g., unit of area) to predict at a broader scale. For example, a standing biomass for a 10,000-ha forest might be predicted by multiplying the amount of biomass measured in a 1-ha stand by the factor of 10,000. Termed "lumping" by King (1991), this approach assumes that the properties of the system do not change with scale and that the broader-scale system behaves like the average finer-scale system. From a modeling perspective (see Chapter 3), this assumption holds only if the equations describing the system are linear. Lumping is known to lead to considerable bias, because it does not account for variability (spatial or temporal) in the scaling process and ignores nonlinear changes in the variable of interest with changes in scale (Rastetter et al., 1992). The nature of the bias depends on the specifics of the spatial dependencies and/or nonlinearities in the system (O'Neill, 1979a, b). This approach should only be used with careful consideration of the potential errors and biases that may result.

An improvement on the simple multiplicative approach to scaling is an additive approach that accounts for spatial variability within an area of relatively large extent. King (1991) identifies two general methods for this type of scaling. The first, *direct extrapolation*, uses data or model simulations from a tractable number of discrete elements within a landscape. For example, instead of assuming that biomass is the same throughout the 10,000-ha forest, we recognize that biomass varies with stand age and composition and that we can account for this spatial variability by mapping or using a remote image. In this case, we use empirical measurements to estimate the biomass in each type defined and then multiply by the area of each type within the 10,000-ha forested landscape and sum the results. Within a modeling framework, King (1991) notes that this approach is probably the most widely used. The direct extrapolation approach works reasonably well for quantities such as biomass or net primary production, which themselves do not interact spatially and which can be related to attributes that can be measured remotely over large areas (e.g., color of the ocean, vegetation composition).

For translating models across spatial scales, a variation on this theme is King's (1991) second method, *extrapolation by expected value*. The general algorithm for this approach entails (1) a model simulating local behavior of a system, (2) a larger landscape over which the model is to be extrapolated, (3) the frequency distributions of variables that describe landscape heterogeneity, and (4) calculation of expected values of the system behavior as a function of the variables describing the heterogeneity of the landscape. The principal source of error in this approach lies in the estimation of the probability or frequency distributions of the landscape variables. However, like direct extrapolation, the more simple approaches to extrapolation by expected value also do not account for spatial interactions or feedbacks.

It is important to note that the error, or variance, associated with the original measurements should also be scaled accordingly to estimate confidence in predictions at the broader scale. However, this problem is easier to recognize than to resolve; confidence intervals around a measurement made at one scale may not translate directly to another scale (Schneider, 1994). Some quantities increase in variance as scale increases, the so-called pink or red noise identified by spectral analysis (Platt and Denman, 1975; Ripley, 1978; Caswell and Cohen, 1995; Cohen et al., 1998). This may occur when the extent of observations is increased and greater environmental heterogeneity is encountered. For example, the air temperature of a small area might be characterized by a mean temperature, say 11°C, with a small range of values, say ±1°C. As extent increases in topographically rough terrain, sites with temperatures ranging from 6° and 24° may be included, resulting in an increase in the variance even if the mean value remains constant. Thus, as the grain becomes coarser the cell characterized by a single temperature may also have greater variability.

Rastetter et al. (1992) and Wiens et al. (1993) suggest that a combination of approaches is needed to scale up. The multiplicative approach can be implemented as a first approximation with the recognition that it may work in some cases, but be widely off the mark in others. Additional detail can be added as needed, either through improving the computations at the finer scales (Wiens et al., 1993) to reduce the error that would be translated to the broader scale or by identifying the spatial subunits across which extrapolations can be summed (Rastetter et al., 1992).

Mandelbrot (1967) was the first to point out that simple geometric relationships can be used to quantify changes in the measured properties of objects with changes in scale. This new concept was illustrated with a power-law relationship between the scale of a ruler and the measured length of the coast of Britain: $L(\lambda) = K\lambda^{1-D}$, where L is the length of the coast, K is a constant, λ the scale of the ruler, and D the dimension of the object. The range of scales over which this power-law holds defines the range of self-similarity of the object. Unlike smooth geometric objects (squares, circles, etc.), the value of D is not always an integer, but may take on fractional values. Objects with fractional values of D are therefore referred to as fractals (also see Chapter 5). Since introduced by Mandelbrot in 1967, fractals have had an immense appeal and impact for addressing problems of scale and hierarchy (Sugihara and May, 1990). The fractal nature of many objects has been confirmed (see Hastings and Sugihara, 1993), and departures from the power-law relationship have been used to identify scales where processes may alter patterns (see Krummel et al., 1987). A fractal dimension may be estimated for almost any object, but this alone does not guarantee a self-similar or scaling relationship (the range of the power-law relationship may be uselessly small). The use of fractals for extrapolating across scales requires two things: estimation of the fractal dimension, D, and verification of the range of the power-law relationship.

Although our discussion has emphasized the processes associated with scaling up, that is, moving from fine-scale measurements to predictions at broad scales, the inverse process of scaling down is also important. For example, the temperature and precipitation patterns predicted by the general circulation models used to simulate potential changes in global climate typically have a coarse resolution of 1° or 2° latitude and longitude. However, we know that precipitation and temperature vary considerably within areas that are 100 km by 100 km in size and that this variability is important for local ecological processes (Lynn et al., 1995; Kennedy, 1997; Russo and Zack, 1997). Using a rather different example, tabulations of population density or housing units for a census tract do not account for the spatial variability within the census tract, yet this variation may be most important for predicting future patterns of land-use change or the movement of nutrients from land to water. Recognition of the importance of developing methods for extrapolating information from coarse to fine grains has increased greatly with the wide use of GIS technologies. However, progress in this arena has been slowed by the immense data requirements for verifying these extrapolations.

SUMMARY

Scale is a prominent topic in landscape ecology because it influences the conclusions drawn by an observer and whether inferences can be extrapolated to other places, times, or scales. Scale refers to the spatial or temporal dimension of an object or a process, and this is distinct from level of organization, which is used to identify a place within a biotic hierarchy. Scale is characterized by grain, the finest level of spatial resolution possible within a given data set, and extent, the size of the overall study area. The related concept of cartographic scale refers to the degree of spatial reduction indicating the length used to represent a larger unit of measure. In practical terms, the scale at which you make any measurement influences the numerical answer obtained.

Scale issues are problematic for several reasons. The magnitude of the task of describing and understanding patterns and processes over large areas is enormous. Because observations are influenced by the scales at which they are made, assembling and comparing data from studies conducted at different scales is tedious and time consuming. Although considerable progress is being made, general methods for extrapolating information across scales have not yet been established.

Hierarchy theory is closely related to scale and provides a framework for organizing the complexity of ecological systems. A hierarchy is defined as a system of interconnections wherein the higher levels constrain and control the lower levels to various degrees, depending on the time constraints of the behavior. The levels within a hierarchy are differentiated by their rates of behavior. Ecological studies should consider three levels within a hierarchy: the focal level, the level above, which provides constraint and context, and the level below, which provides mechanism. When scales change, a shift in the relative importance of variables or the perceived direction of a relationship may also change.

Scale issues often arise in attempting to extrapolate ecological studies to larger or smaller scales. The current understanding has led to a few useful rules of thumb for extrapolation. First, scale changes may be ignored in homogeneous space, but not under conditions of spatial heterogeneity. Average dynamics can be applied to a larger area only when the area is homogeneous for the characteristic of interest. If the spatial heterogeneity is present, but is random rather than occurring with a structured pattern, then the average plus the variance can be used to apply local measurements to the broader area. Second, as long as major processes and constraints do not change, the theory of fractals shows that under certain conditions quantities may be extrapolated across scales. Third, when spatial heterogeneity combines with nonlinear dynamics and the possibility of major changes in constraints, extrapolation becomes a very difficult problem that does not, at present, have any simple solution.

There is no right scale for landscape ecological studies. Scales must be selected based on the question or objective of a study. However, identifying the appropriate scale remains challenging, and developing methods for doing so remains a topic of current research. Indeed, most of the topics covered in this chapter have many unknowns associated with them. Ecologists are still learning how to take the knowledge we have gleaned about patterns and processes at multiple scales into consideration when developing field studies and models and the techniques for extrapolating across scales and landscapes.

DISCUSSION QUESTIONS

1. Select a landscape of your choice and list the important ecological processes that occur in the landscape. Next, estimate the temporal and spatial scales over which these processes operate and plot these in a time–space state space (see Figure 2.1). How might such a diagram assist you in selecting appropriate scales for a field study or model? What scales are appropriate for different hypotheses that you might test?
2. Do you think that scale issues will be of passing or enduring interest to ecologists? Provide a rationale for your opinion.
3. Describe how scale may be considered as a problem as well as an opportunity.

~ RECOMMENDED READINGS

Delcourt, H. R., P. A. Delcourt, and T. Webb. 1983. Dynamic plant ecology: the spectrum of vegetation change in space and time. *Quaternary Science Review* 1: 153–175.

King, A. W. 1997. Hierarchy theory: a guide to system structure for wildlife biologists. In J. A. Bissonette, ed. *Wildlife and Landscape Ecology. Effects of Pattern and Scale,* pp. 185–212. Springer-Verlag, New York.

Levin, S. A. 1992. The problem of pattern and scale in ecology. *Ecology* 73:1943–1983.

O'Neill, R. V., D. L. DeAngelis, J. B. Waide, and T. F. H. Allen. 1986. *A Hierarchical Concept of Ecosystems.* Princeton University Press, Princeton, New Jersey.

Schneider, D. C. 1994. *Quantitative Ecology. Spatial and Temporal Scaling.* Academic Press, San Diego, California.

Wiens, J. A. 1989. Spatial scaling in ecology. *Functional Ecology* 3:385–397.

CHAPTER 3

INTRODUCTION TO MODELS

MODELS ARE IMPORTANT TOOLS IN LANDSCAPE ECOLOGY, as they are in many scientific disciplines. Spatial models, in particular, are playing a more and more prominent role for investigating the consequences of heterogeneous distributions of ecological resources. Because we refer to models throughout this book and because we are aware that many students have not had any formal training in modeling or systems ecology, this chapter presents an elementary set of concepts and terms for students to understand what models are, why they are used, and how models are constructed and evaluated. Our purpose is to introduce models and their development, define what we mean by a spatial model, and indicate when spatial models are important. Students interested in the modeling process in greater depth are referred to the recommended readings at the end of the chapter.

WHAT ARE MODELS AND WHY DO WE USE THEM?

What Is a Model?

A model is an abstract representation of a system or process. Models can be formulated in many different ways. Physical models are material replicas of the object or system under study, but at a reduced size; for example, model ships and airplanes are developed to better understand the forces that act on them, and architectural models allow the space and structure of a building to be visualized. Physical models are used in many branches of engineering, but ecologists also build physical models of streams, ponds, and even whole ecosystems, such as the Bio-sphere 2 (Macilwain, 1996). In contrast, abstract models use symbols rather than physical devices to represent the system being studied. For example, verbal models are constructed out of words, graphical models are pictorial representations, and mathematical models use symbolic notation to define relationships describing the system of interest. We focus here primarily on mathematical models, which have played an important role in ecology since the beginning of the 20th century (Figure 3.1).

Why Do Landscape Ecologists Build Models?

Models serve a variety of useful purposes in the sciences. They help to define problems more precisely and concepts more clearly. They provide a means of analyz-

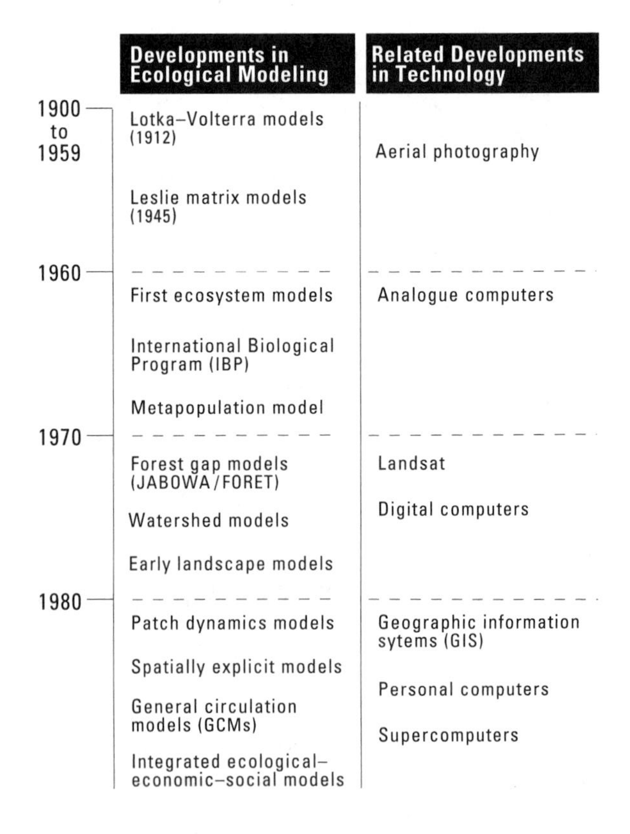

FIGURE 3.1.
Time line of the development of models in ecology, with important technological and programmatic developments that influenced ecological modeling highlighted. Developments shown are not comprehensive but selected for illustration.

ing data and communicating results. Finally, and perhaps most importantly, models allow us to make predictions. However, landscape models should be regarded as tools or methods to achieve an end and should not be considered as goals unto themselves. Because knowledge is incomplete, assumptions are always required within models to fill in the blanks. Consequently, most models are employed to explore the consequences of our hypotheses regarding system structure and dynamics.

Landscape ecologists have found models to be an important part of their tool box for several reasons. When faced with studying a large and complex landscape, it is difficult, and sometimes impossible, to conduct experiments at the appropriate scale. Experimental manipulation of large landscapes with the appropriate number of independent replicates is not very common because of the high cost and logistical difficulty involved, although landscape-management actions framework (e.g., agricultural systems and forest harvesting) can be studied experimentally. Experimental manipulations of microlandscapes avoid this difficulty, and this approach is being used to gain insights into the response of insects, small mammals, and some plants to alternative landscape patterns (e.g., Johnson et al., 1992a; Glenn and Collins, 1993; Imes et al., 1993; Wiens et al., 1995, 1999). However, extrapolation of results from small experimental landscapes to large regions remains a perplexing problem. More common are field studies that provide correlative relationships, for example, by comparing locations that vary in a variable of interest, say the abundance of land cover or connectivity of a specific habitat. Landscape ecologists also take advantage of natural, uncontrolled events (especially natural disturbances, such as fires, storms, and floods) and study their effects from an experimental viewpoint (see Chapter 7). However, these serendipitous opportunities result in a limited range of conditions being studied and do not allow either replication or controlled manipulation. Under these circumstances, the unique features of each landscape and disturbance event may dominate results. Models relax these empirical restraints, providing a means of systematic comparison across a broad range of conditions.

Landscape models help to formalize our understanding and develop theory about how spatial patterns and ecological processes interact, producing general insights into landscape dynamics. Models may also generate testable hypotheses that can be used to guide field studies by exploring conditions that cannot be manipulated in the field. For example, scientists do not have the option to implement large, severe disturbances, yet it may be very important for both basic and applied science to have expectations of what the effects of large disturbances may be (Dale et al., 1998). Models may also be used to explore ecological responses to a broader range or combination of conditions than could be established in a field experiment. For example, a landscape simulation model was used to explore the effects of a wide range of fire sizes (10% to 90% of the landscape) and spatial arrangements (random to clumped) on wintering ungulates in northern Yellowstone National Park (Turner et al., 1993a, 1994a). These simulation experiments clearly could not be implemented in the field.

Classification of Models

Models may be described and classified in various ways, and it is helpful to understand some commonly used terms. We review the terms often used to describe ecological models; similar distinctions are also presented by Grant et al. (1997).

Deterministic versus Stochastic

A model is deterministic if the outcome is always the same once inputs, parameters, and variables have been specified. In other words, deterministic models have no uncertainty or variability, producing identical results for repeated simulations of a particular set of conditions. However, if the model contains an element of uncertainty (chance), such that repeated simulations produce somewhat different results, then the model is regarded as stochastic. In practice, the heart of a stochastic simulation is the selection of random numbers from a suitable generator. For example, suppose that periodic movements of an organism are being simulated within a specified time interval. It may be likely that the organism will move, but it is not certain when this event will occur. One solution is to represent the movement event as a probability, say 0.75, and the probability of not moving as $(1.0 - 0.75) = 0.25$. Selection of a random number between 0.0 and 1.0 is done to decide randomly if movement occurs during a specific time interval. If the simulation is repeated, the time-dependent pattern of movement will be different, although the statistics of many movement events will be very similar. Inclusion of stochastic events within a model produces variable responses across repeated simulations, a result that is very similar to our experience of repeated experiments.

Analytical versus Simulation

These terms refer to two broad categories of models that either have a closed-form mathematical solution (an analytical model) or lack a closed-form solution and therefore must rely on computer methods (a simulation model) to obtain model solutions. For analytical models, mathematical analysis reveals general solutions that apply to a broad class of model behaviors. For instance, the equation that describes exponential growth in a population is an example of an analytical

TABLE 3.1.
COMPARISON OF DIFFERENTIAL AND DIFFERENCE EQUATION FORMS OF SOME SIMPLE EQUATIONS FOR POPULATION GROWTH.

Type of population growth	Differential form	Difference form
Linear growth	$dN/dt = r$	$N_{t+\Delta t} = N_t + r\,\Delta t$
Exponential growth	$dN/dt = rN$	$N_{t+\Delta t} = N_t + rN_t\,\Delta t$
Logistic growth	$dN/dt = rN[(K - N)/K]$	$N_{t+\Delta t} = N_t + rN_t\,\Delta t[1 - (N_t/K)]$

SEE FIGURE 3.2 FOR THE GRAPHICAL REPRESENTATION OF THESE FORMS.

model (Table 3.1), as are many of the model formulations used in population ecology (May, 1973; Hastings, 1996b).

In contrast, the complexity of most simulation models means that these general solutions may be difficult or impossible to obtain. In these cases, model developers rely on computer methods for system solution. Simulation is the use of a model to mimic, step by step, the behavior of the system that we are studying (Grant et al., 1997). Thus, simulation models are often composed of a series of complex mathematical and logical operations that represent the structure (state) and behavior (change of state) of the system of interest. Many ecological models, especially those used in ecosystem and landscape ecology, are simulation models.

DYNAMIC VERSUS STATIC

Dynamic models represent systems or phenomena that change through time, whereas static models describe relationships that are constant (or at equilibrium) and often lack a temporal dimension. For example, a model that uses soil characteristics to predict vegetation type depicts a relationship that remains the same through time. A model that predicts vegetation changes through time as a function of disturbance and succession is a dynamic model. Simulation models are dynamic.

CONTINUOUS VERSUS DISCRETE TIME

If the model is dynamic, then change with time may be represented in many different ways. If differential equations are used (and numerical methods are available for the solution), then change with time can be estimated at arbitrarily small time steps (Figure 3.2). Often models are written with discrete time steps or intervals. For instance, models of insects may follow transitions between life stages, vegetation succession may look at annual changes, and so on. Models with discrete time steps evaluate current conditions and then jump forward to the next time, while assuming that conditions remain static between time steps. Time steps may be constant (a solution every week, month, or year) or event driven, resulting in irregular intervals between events. For example, disturbance models (e.g., hurricane or fire effects on vegetation) may be represented as a discrete time-step, event-driven model.

Mechanistic, Process-Based, Empirical Models

These three terms are frequently confusing. A *mechanism* is "the arrangement of parts in an instrument." When used as an adjective to describe models (i.e., a *mech-*

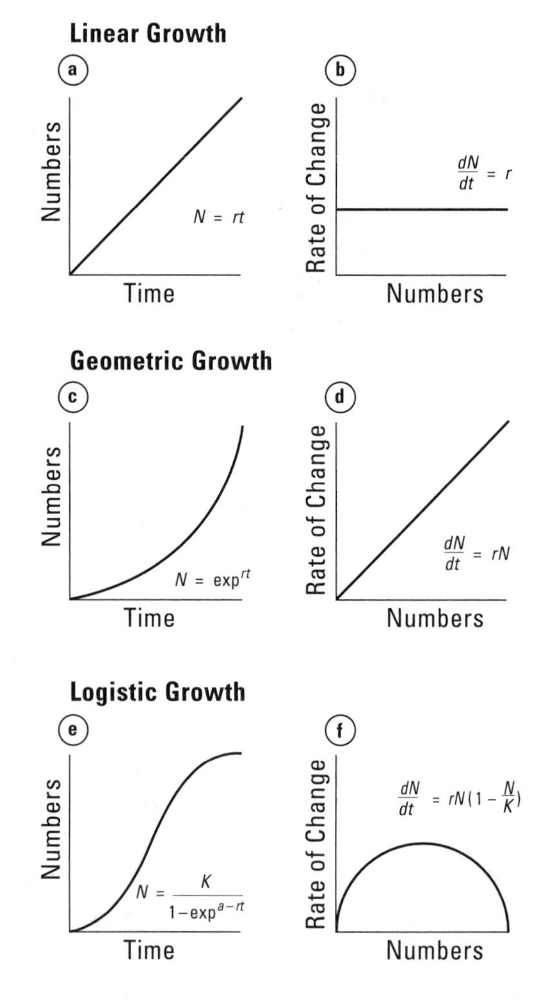

FIGURE 3.2.
Some examples of the integral curves (a, c, and e, on the left) for particular forms of population growth, with the corresponding graphs of the differential curves (b, d, and f, on the right).

ADAPTED FROM KITCHING, 1983.

anistic model), the term implies a model with parts arranged to explain the whole. In the best sense of the term, a mechanistic model attempts to represent dynamics in a manner consistent with real-world phenomena (e.g., mass and energy conservation laws or the laws of chemistry). Although there has been waning support for mechanistic approaches to ecological modeling (Breckling and Muller, 1994), the use of mechanistic in the strictest sense distinguishes these models from black-box models that grasp at any formulation that might satisfactorily represent system dynamics. Confusion arises when the term *mechanistic* is loosely applied to distinguish less detailed models from more detailed ones. Often the implication is that mechanistic models are more desirable than less mechanistic (less detailed) models. Unfortunately, the assertion that additional detail produces a more reliable model must be demonstrated on a case by case basis (Gardner et al., 1982).

A *process-based model* implies that model components were specifically developed to represent specific ecological processes; for example, equations for birth, death, growth, photosynthesis, and respiration are used to estimate biomass yields, rather than simpler, more direct estimates of yields from the driving variables of temperature, precipitation, and sunlight. Although this concept seems clear, there is no a priori criterion defining formulations that qualify (or conversely do not qualify) as process models. Thus, depending on the level of detail, it is possible to have a mechanistic process-based model or an empirical process-based model.

An *empirical model* usually refers to a model with formulations based on simple, or correlative, relationships. This term also implies that model parameters may have been derived from data (the usual case for most ecological models). Regression models (as well as a variety of other statistical models) are typically empirical, because the equation was fitted to the data.

The problem of distinguishing between types of model is illustrated by the simulation of diffusive processes based on well-defined theoretical constructs (Okubo, 1980). These formulations of diffusion allow simple empirical measurements to define the coefficients estimating diffusive spread. Thus, there is a strong theoretical base along with empirically based parameters. Is such a model considered empirical or theoretical? Should complex formulations always be considered more theoretical or simply harder to parameterize?

The essential quarrel with each of these three terms is that most ecological models are a continuum of parts, processes, and empirical estimations. Separating models into these arbitrary and ill-defined classifications lacks rigor and repeatability. One person's mechanistic model is the next person's process-based model, and so on. There does not appear to be a compelling reason to use these vague and often confusing terms to distinguish between alternative model formulations.

Spatial Models

A model is spatial when the variables, inputs, or processes have explicit spatial locations. Spatial models are useful when the heterogeneity of resources and processes is required to represent and predict system dynamics properly. Although such models have received considerable attention in recent years, the complexity and difficulty of formulating a spatial model may not be necessary to address every landscape question. *A spatial model is only needed when explicit space—what is present and how it is arranged—is an important determinant of the process being studied.* Although this condition is easily stated, it is often difficult to determine a priori. There are three general conditions for which spatial models are important.

1. Spatial pattern (abundance and configuration of elements) may be one of the independent variables in the analysis. That is, we are particularly interested in how some ecological response variable changes as a function of the configuration of landscape elements. Questions that illustrate this requirement include these: What is the effect of different arrangements of the same amount of habitat on species diversity? How does input of nitrogen to an aquatic ecosystem vary with the positioning of vegetation or land-cover types in a watershed? How does the spatial patterning of resources influence movement or foraging dynamics of a species? How is the spread of disturbance influenced by the spatial pattern of areas that are or are not susceptible to a disturbance? A spatial model developed to address these types of questions would use spatial pattern as one of the input variables. Most often, using a map for one of the driving variables does this; this map may or may not change through time.
2. A spatial model is needed when predicting spatial variation of an attribute of interest and how it changes through time (Figure 3.3). General questions include these: How will disturbance, land-use activity, or land management affect land-use change? How does habitat pattern affect the distribution and abundance of species within the landscape? These questions require the initial spatial patterns as input to a model that then simulates pattern change through time (Turner, 1987a; Costanza et al., 1990; Sklar and Costanza, 1990).
3. A spatial model is required when the question involves sets of processes or biotic interactions that generate pattern. This situation is often explored in theoretical population ecology (Holmes et al., 1994; Ives et al., 1998). Questions might include these: If two or more species compete for space, how do their interactions lead to heterogeneous distributions of the species across the landscape? What happens to the patterns of species distributions if re-

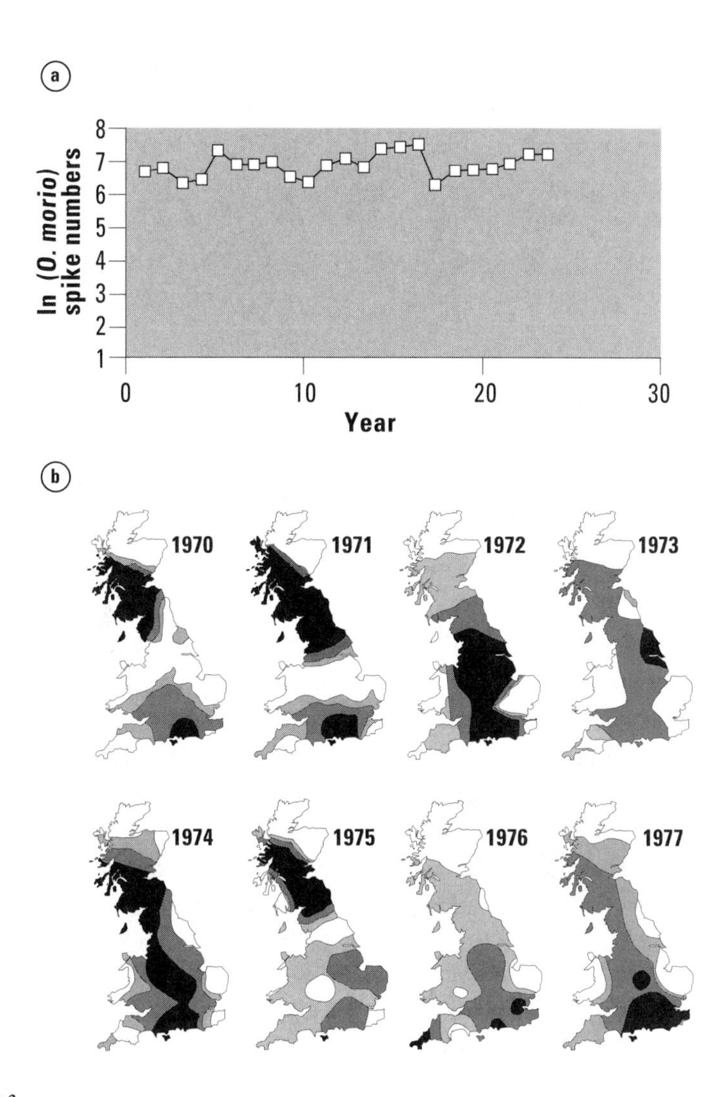

FIGURE 3.3.

Comparison of temporal versus temporal plus spatial population dynamics. (a) Change through time in a population of the green-winged orchid (*Orchis morio*) at one locality. In this case, changes are temporal only. (b) Change in the spatial distribution in Great Britain of elder aphid (*Aphis sambuci*) between 1970 and 1977, where the density of the shading represents local population density. This illustrates changes through both space and time; predicting such dynamics would require a spatial model.

ADAPTED FROM GILLMAN AND HAILS, 1997.

source gradients are changed? These models generally begin with a homogeneous space, and pattern is allowed to develop through time. Population ecologists often used analytic models (May, 1974) to study these situations, but cellular automata models (systems of cells interacting in a simple way, but generating complex overall behavior (e.g., Wolfram, 1984; Hogeweg, 1988) are being increasingly used to include specific effects of landscape heterogeneity on system response.

STEPS IN BUILDING A MODEL

The steps that we describe for the modeling process are derived from the systems ecology approach to ecological modeling (Figure 3.4). Systems ecology uses the techniques of analysis developed primarily by engineers for studying, characterizing, and making predictions about the dynamics of complex entities. Common among the many definitions of a system is the notion that sets of objects or components that interact together in space and/or time produce a unique set of measurable outcomes (Kitching, 1983). In an engineering sense, a system might be a set of electrical components that constitutes a radio or a set of machines that form an assembly line. In a biological sense, a system might be the set of organs that constitutes an organism, populations that constitute a community, or components and processes that produce measurable ecosystem dynamics. The success of the systems approach has had a strong influence in the early development of landscape ecology in both Europe (Naveh and Leiberman, 1990; Zonneveld, 1995) and North America (Johnson et al., 1981; Gardner et al., 1987; Opdam, 1987; Sklar and Costanza, 1990).

The basic principles of the systems approach go back to the philosophy of holism formulated by Smuts (1926) and developed more rigorously by Von Bertalanffy (1968, 1969). Numerous works written more than 30 years ago describing the principles of general systems theory and their application to ecological systems still provide insightful reading today (e.g., Watt, 1968; VanDyne, 1969; Patten, 1971). Here, we draw from the sequence of modeling steps outlined in Kitching's (1983) text on systems ecology. We also provide a reference table for terms commonly used in modeling (Table 3.2).

Step 1: Define the Problem

The initial step in model development is to specify the purpose of the model. What are the questions being addressed and the objectives for which the model is to be used? The more specific the purposes are, the more tractable the solution! Simply

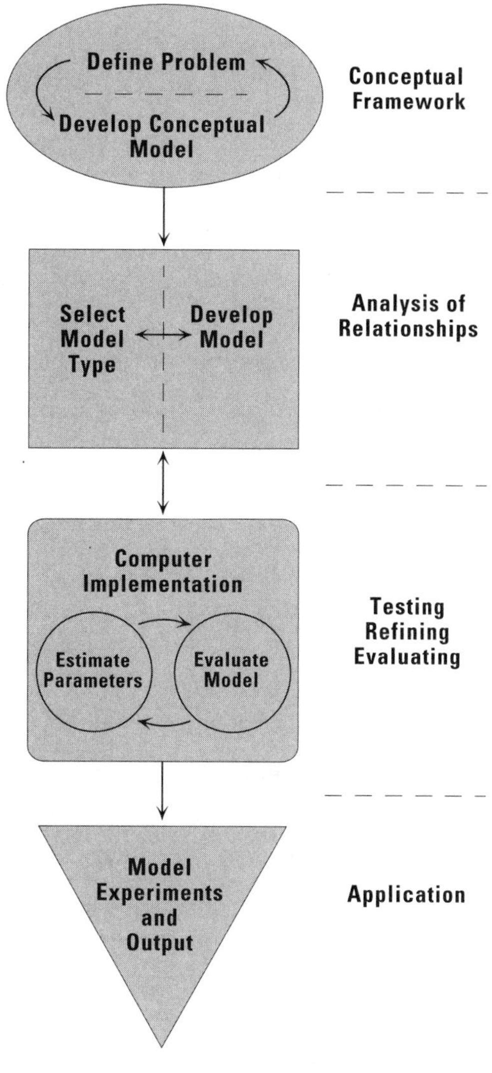

FIGURE 3.4.
Flow chart of the major steps in building a model.

having a lot of data that may appear useful is *not sufficient grounds* for developing a model; there must be a clear and compelling reason for the model. Model objectives provide the framework for model development, the standard for model evaluation, and the context within which simulation results must be interpreted. Thus, a clear statement of objectives is arguably the most crucial step in the entire modeling process (Grant et al., 1997). Definition of the problem will influ-

Table 3.2.
Terminology for model components and common procedures.

Term	Definition
Parameter	Constant or coefficient that does not change in the model.
Variable	Quantity that assumes different values in the model.
State variable	Major elements of the model whose rates of change are given by differential equations.
Initial conditions	Values of the state variables at the beginning of a simulation.
Forcing function, external variable, or driving variable	Function or variable of an external nature that influences the state of the system, but is not influenced by the system.
Output variables	Variables that are computed within the model and produced as results.
Sink	Compartment in the model into which material or flow goes, but from which it does not return.
Source	Compartment from which the material flowing in the model flows, but to which it does not return.
Dimensional analysis	Process in which the units in a model are checked for consistency.
Calibration	Process of changing model parameters to obtain an improved fit of the model output to empirical data.
Corroboration	Process of determining whether a model agrees with the available data about the system being studied.
Sensitivity analysis	Methods for examining the sensitivity of model behavior to variation in parameters.
Validation	Commonly, the process of evaluating model behavior by comparing it with empirical data; we prefer *corroboration* because it does not imply "truth."
Verification	Process of checking the model code for consistency and accuracy in its representation of model equations or relationships.

ence the form of the model, the degree of complexity needed, and the spatial and temporal scales at which it will operate. Surprisingly, however, this step often receives far less attention than its importance warrants. The danger is that ignorance of a model's purpose ultimately results in the misapplication and/or misinterpretation of the results.

Step 2: Develop the Conceptual Model

Based on a clear statement of the model purpose or objectives, a conceptual or qualitative model of the system follows. This phase includes identification of system boundaries, categorizing model components and identifying the relationships among them, and describing the expected patterns of model behavior. Kitching (1983) notes that "The ecological modeler, having decided that the problem being faced demands the construction of a model, is immediately confronted with the necessity of defining the system which he or she is going to study." What is to be included within the scope of the model? What are the important variables and parameters (coefficients that control model processes, but do not themselves change)? What drives the system? [*Driving variables* (Table 3.2) influence model behavior, but are external to the model; that is, they are not affected by variables within the model.] What outputs will be generated? What is the appropriate level of spatial and temporal resolution for the model? Specifying the scale of the model is an important step, because the same model may produce different results if spatial or temporal scales are changed (Figure 3.5). What are the initial conditions (state of variables at the outset of the model run) going to be? Conceptual development of a model is often the most intellectually challenging phase and frequently results in a valuable refining and/or redefinition of the model problem (Figure 3.4).

There are two different approaches that modelers frequently take in defining their model system (Kitching, 1983). The first is to make the initial selection of system components as simple as possible, including only those variables that the modeler is confident are necessary. Alternatively, the initial model might include all variables that might conceivably have an effect on the problem of interest, with the assumption that the modeling process will prune unnecessary variables and model components.

Once the initial selection of variables and processes is made, the modeler must consider how the components interact with each other. Flow diagrams or flow charts are common techniques used in this phase of modeling as precursors to writing equations for these interactions. Essentially, flow diagrams are a formal representation of the qualitative relationships of the conceptual model.

Step 3: Select the Model Type

Once the problem is defined and the system is identified, along with the nature of the interactions or processes to be modeled, the type of model to be developed must be selected. For example, will the model be a simulation or analytical model? Stochastic or deterministic? Spatial or nonspatial? This decision depends strongly on the expected use of the model. If the question and system are relatively simple, then building an analytical model may be both feasible and desirable, because results from these models are often elegant in their simplicity and generality. If the question is more complex, then messier numerical solutions provided by computer simulations may be required. The way in which the model will be compared to data may influence the choice between the deterministic or stochastic modeling approach. One advantage to using a stochastic model is that multiple runs of the same model can be summarized statistically, as we would do with empirical data, thereby allowing the mean, median, and ranges of output variables to be examined and compared. This variation in model output allows a statistical comparison with measurements from experimental or field observations (Kitching, 1983).

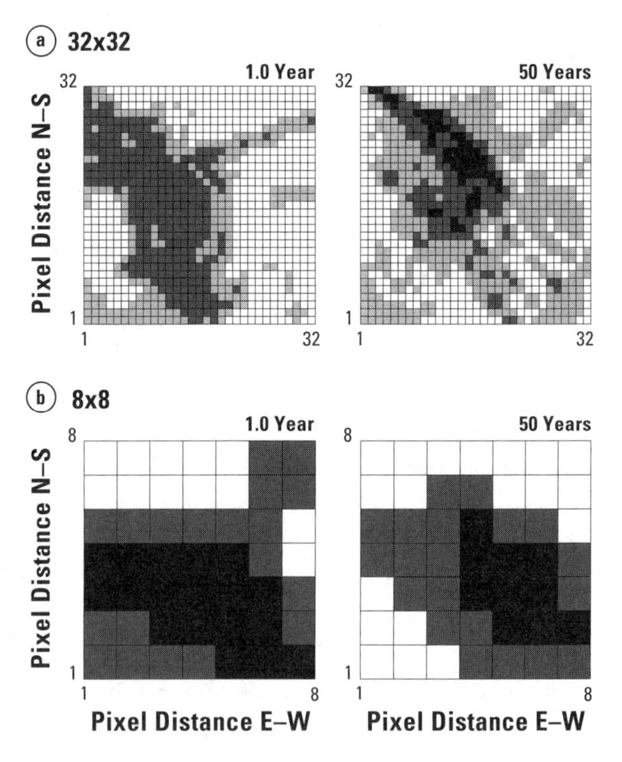

FIGURE 3.5.
Modeled concentrations of nitrogen in space and time at two different spatial scales for a section of the Walker Branch Watershed in eastern Tennessee, USA. Note the qualitative differences in the predicted patterns between the two scales.

ADAPTED FROM BARTELL AND BRENKERT, 1991.

Step 4: Model Development

When the structure of the model has been decided (step 3), then it is time to actually write the equations and/or logical operations to be performed by the model. A wide variety of mathematical formulations may be used, and it may not be clear at the outset which will be superior for a given problem. Problems encountered during this stage often require adjustments in model type, making steps 3 and 4 a tightly linked process, as illustrated in Figure 3.4. A wealth of model types is available, including techniques from graph theory, diffusion theory, game theory, percolation theory, fractal geometry, chaos theory, optimization theory, and aspects of probability theory such as Markov chains or Bayesian models. It is beyond our scope to review and describe these techniques here. Readers should refer to Swartzman and Kaluzny (1987), who provide an excellent introduction to specifying and coding both linear and nonlinear simulation models.

Step 5: Computer Implementation

Model developers are often faced with a number of technical challenges in the implementation of a model. Questions include the available computer resources and usefulness of different programming language. Will existing modeling packages such as STELLA suffice (Costanza et al., 1998), or will technical complexities require new program development? If your model is spatial, how will the input and output of data from GIS software be handled?

The coding of models into a computer language (or a modeling package) requires careful checking for the accuracy of equations and relationships. This phase is both challenging and demanding, often requiring much more time than any other step represented in Figure 3.4. Logical errors, or even simple typographic errors, may change the form or function of the model in subtle ways. Thus, *model verification*, the tedious process of checking and testing the internal logic and consistency of the model, is an essential component of model development. It has been our experience that a simple set of test data, along with a pencil, sheet of paper, and calculator, provides the best means of identifying many errors. However, all model developers (and users) should be aware that it is probably impossible to prove that any given model is truly bug free.

Another unsung but important aspect of model development is the production of adequate *model documentation*. The documentation may appear within the computer program by the liberal use of comment statements or in a separate text describing the model and its application. Although this sounds trivial, it is important to realize that parameter names, or logical constructs within the model, that seem so obvious at the time that the model is being developed may be much less clear at a later date or to another scientist. Documentation of the model should include the objectives and conceptual basis of the model; space and time scales and other units associated with model parameters, variables, inputs, and outputs; the essential underlying model assumptions; rationale for equations controlling program logic and flow; data sources and estimation methods for all model parameters; and tests performed to ensure model accuracy and reliability. Adequate documentation is the best measure to ensure the useful and productive application of the model.

Step 6: Parameter Estimation

This step refers to the selection of values of model parameters, model inputs, and initial values of the state variables within the model (Table 3.2). All values must be consistent with the model's purpose and equations used to represent the system of interest. These values are typically estimated from data or obtained from a variety of published values. Modelers must be especially attentive to the consistency of units of all parameters and variables, as well as the degree of certainty associated with how well each value is known. For example, if the model assumes that dynamics are linear and always near equilibrium, parameters that are estimated from systems that have been severely perturbed (that is, are far from equilibrium) will cause a strong bias in system dynamics and errors that may accumulate with time. Note that the process of parameter estimation differs from model *calibration*, which refers to the iterative adjustments to inputs and parameters to improve model fit to measured output variables. This iterative fitting is shown by the feedback loop of model implementation, evaluation, and parameter estimation in Figure 3.4. Calibration is most successful when direct estimates of parameters are not available, but net changes in system dynamics have been carefully measured. In such cases, unmeasured parameters may be iteratively adjusted until the difference (error) between predictions and observations has been minimized. The ideal situation is to have available a second data set to test the validity of these calibrated (but unmeasured) parameters.

Step 7: Model Evaluation

Once the model is operational, you need to know how well it works. Does its behavior agree with empirical observations? How well does the model meet its objectives? Are the underlying assumptions reasonable? Is the behavior resulting from these assumptions realistic? How sensitive is model behavior to changes in these assumptions? Model evaluation includes both comparisons of model results with data and understanding the sensitivity of the model, as structured, to the parameters within it.

Comparing models and data is sometimes referred to as model validation, but this terminology remains problematic (Rykiel, 1996). To validate means to assess the truth of; given that models are never "true," many modelers reject this terminology and prefer to use more meaningful and quantifiable terms such as model testing and reliability (Mankin et al., 1975). Objective testing requires that data used to estimate model parameters must be independent from data used to test model behavior. Lack of independence may merely confirm the goodness of fit to a particular data set, but does not indicate the range of conditions over which predications may be regarded as reliable.

Comparing models with data may be done graphically (note that the variance in the data must be included), statistically (variance estimates for model output also required), or in tabular form (model prediction, data, and ranges should be available). All comparisons should be based on the model objectives. If the purpose of the model is to assess the direction of change (e.g., urbanization will increase at the expense of agriculture), only the direction of change predicted by the model needs to be assessed. However, if the estimation of the amount and location of agricultural land-use change is the model's objective (a more difficult task), spatial statistics may be required to verify model response. A key point here is that verification of one set of model objectives does not automatically qualify a model for other comparisons, no matter how similar these may appear to be.

Sensitivity analysis is the evaluation of the relative importance of particular parameters within the model (Gardner et al., 1981; Caswell and Trevisan, 1994). A small change in a sensitive parameter will result in large changes in model output. Conversely, even large changes of insensitive parameters will not significantly affect model output. A collection of methods is available for evaluating how sensitive model output is to changes in parameter values (Metzger et al., 1998). The most straightforward approach changes the parameter values, either singly or in concert, by the same fixed percentage and observes changes in model output. (Note that parameter perturbations should be $\leq 1\%$ to qualify as sensitivity analysis. Larger perturbations produce results that are unreliable and difficult to interpret; see Gardner et al., 1981, for further discussion of this topic.) One may also vary the parameter values based on their expected range of variation, or uncertainty; this approach, referred to as *uncertainty analysis* (Gardner and Trabalka, 1985; Gardner et al., 1990; Wallach and Genard, 1998; Ricotti and Zio, 1999), uses the statistical distributions of parameter values (and correlations among parameters) to estimate uncertainties of model output.

The Final Step: Experimentation and Prediction

When the steps illustrated in Figure 3.4 have been successfully completed, the ecologist has a tool with which to conduct experiments and address the problems initially defined. Model predictions, and associated comparison with data, are usually the most desired endpoint of model development. The verification of predictions across a range of conditions confirms the model hypotheses and assumptions and increases our confidence that we have new insight and understanding of system behavior. As our confidence builds, model applications will move from hypothesis testing to more serious applications, such as conservation planning (Baker, 1989a; Bender et al., 1998), landscape management and design (Baskent, 1997), and assessment of potential changes due to land development or climate change (Baker et al., 1991; Neilson and Koerper, 1992; He et al., 1999). Care must be taken at each stage of model development to assure the accuracy and adequacy of the model. The widely available software now available for model development can also make model development very easy; however, the danger remains that jumping directly to computer implementation (step 5) without using steps 1 through 4 is extremely dangerous. However, no amount of care will guarantee that a model is a perfect representation of the ecological system that it was intended to mimic. Therefore, wise use and application requires an awareness of the problems and pitfalls common to the use of all models.

LANDSCAPE MODELS

It has been more than 10 years since Baker (1989b) and Sklar and Costanza (1990) reviewed landscape models. Focused reviews on specific topics have been published since then (e.g., Turner and Romme, 1994; Roberts and Gilliam, 1995; Lambin, 1997; Fries et al., 1998; Mladenoff and Baker, 1999), but no comprehensive overview has been provided. To assess current models and modeling methods in landscape ecology quickly, we performed an informal survey. ISI's Web-of-Science (1999) was searched for papers published in the last 5 years (1994–1999) referencing "landscapes" and "model." This search located 177 papers from 34 different journals (books, proceedings, and technical reports were not included in this search), and the abstracts and keywords were downloaded and reviewed. Only original papers discussing simulation models were retained, and multiple publications of the same model were eliminated, giving a final database of 101 manuscripts. The journals most frequently cited were *Ecological Modelling* (23), *Landscape Ecology* (11), *Ecology* (9), *Ecological Applications* (9), and *Oikos* (6).

The review of these papers showed a broad diversity of approaches and subject matter currently being considered by landscape models and also illustrated the difficulty of placing these models into discrete categories. The most frequent subject of these models (37%) concerned single-species issues, including metapopulation dynamics (12 of 37), effects of habitat fragmentation (15 of 37), corridors (5 of 37), and dispersal or invasion processes (17 of 37). Models of disturbance effects (17%) and vegetation dynamics (15%) were the next two largest categories. Fire was the most frequent disturbance type (11 of 17) and forests the most frequently modeled vegetation type (11 of 15). Nine papers concerned management issues, including forestry (6 of 9), cultivation (1 of 9), and socioeconomic factors affecting landscapes (2 of 9). Surprisingly, only 6 papers were found that dealt explicitly with human effects through economic change and its impact on landscape change. This extremely important topic received far less attention than expected. However, modeling human impacts on landscapes requires an immense amount of information and the consideration of multiple impacts on landscapes, making the construction of these models a daunting task. Physical factors including hydrology and meteorology composed 11% of the papers, while 6% of the manuscripts dealt with simulation tools (languages, analysis methods, and the like) required to compare models among themselves and against landscape data.

Not surprisingly, the simulation methods used to tackle these problems are as diverse as the subjects being studied. Simple sets of categories to describe these models are probably impossible to construct and would not adequately convey the diversity of approaches being employed. Dispersal models are continuing to increase, with multiple approaches being used, including reaction–diffusion methods (e.g., Bevers and Flather, 1999), individual-based random-walk models (e.g., Gustafson and Gardner, 1996; Liu and Ashton, 1998), and probabilistic-based cellular automata models (e.g., Zhou and Liebhold, 1995; Darwen and Green, 1996). Approaches to modeling disturbance effects are also varied, ranging from partial differential equations (e.g., Emanuel, 1996; Jin and Wu, 1997), probability distributions and stochastic simulations (e.g., Boychuk and Perera, 1997; He and Mladenoff, 1999), to specialized methods for unique disturbance effects (e.g., hurricanes studied by Boose et al., 1994). Models of vegetation change include Markov chain models (e.g., Li, 1995; Thornton and Jones, 1998), continued extensions of the familiar JABOWA-FORET forest succession models (e.g., Acevedo et al., 1996; Malanson and Armstrong, 1996; Pausas et al., 1997), as well as fuzzy set theory combined with vital attributes modeling (Roberts, 1996). Linkages to GIS data layers is an active area of research, with new models and methods being developed (e.g., Schippers et al., 1996; DeAngelis et al., 1998; Baker, 1995; Butcher, 1999). The diversity of approaches is a healthy sign of a growing field of research. Modeling approaches to landscape issues continues to reflect the diversity of subjects of interest to landscape ecologists, and this will become apparent throughout the remainder of this book. Clear and simple paradigms for solving similar sets of problems have yet to emerge.

CAVEATS IN THE USE OF MODELS

Models are extremely important tools in landscape ecology. Indeed, it may well be that all research of spatially extensive systems requires the use of a variety of models and associated theory to understand the dynamics of change (Hartway et al., 1998). Wise application of these tools should recognize the pitfalls and problems of model development and interpretation. We review here, in concise form, what we consider to be the most important caveats for modeling in landscape ecology:

1. *Know thy model.* The performance of each model is the logical consequence of the hypotheses and assumptions on which the model is based. Alternative assumptions regarding systems behavior might be equally viable, but produce dramatically different results. Comparison among alternative model formulations is extremely desirable and should be attempted where possible (e.g., Rose et al., 1991a, b; VEMAP, 1995; Pan et al., 1998).
2. *Errors propagate.* Small errors in sensitive parameters can lead to large errors in outputs (Rose et al., 1991c). Techniques for analysis of the effects of parameter errors are available (Metzger et al., 1998) and should always be employed before predictions are made. Assessment of errors of spatially explicit models remains a challenge (Cherril and McClean, 1995; Henebry, 1995; Heuvelink, 1998), largely because of the added complexity of evaluating qualitative and quantitative spatial predictions.
3. *All models are simplifications of reality.* This is not a casual philosophical statement. It simply means that no single model will ever be a completely adequate description of reality. Therefore, the goal of model studies should be to define the applications for which a given model provides reliable and useful results. New applications of old models are not released from this requirement.
4. *There are never enough data.* The incomplete nature of data often requires parameter values to be estimated from a diversity of sources. Inconsistency

in the methods of data collection and parameter estimation may result in model biases that are difficult to identify. Gaps in empirical information that do not allow adequate estimation of key parameters are often the greatest source of uncertainty in model predictions.

5. *High-tech methods do not guarantee a good model.* Technologically advanced methodologies, including the availability of higher-level programming languages that facilitate model coding, do not assure the accuracy or reliability of results. When developing or interpreting models, it is critical for the user to understand fully the structure of the model, the assumptions that went into its development, and the constraints (such as spatial or temporal scales) on its appropriate use.
6. *Keep an open mind.* There is no single paradigm for spatial modeling of landscapes. Model development and testing require a broad perspective of landscape ecology and systems analysis techniques.

SUMMARY

A model is an abstraction or representation of a system or process. There are many different kinds of models, and mathematical models are commonly used in ecology. In landscape ecology, model development is an important tool that complements empirical techniques. Models permit the landscape ecologist to explore a broader range of conditions than can usually be set forth experimentally. Landscape models help to formalize our understanding and develop theory about how spatial patterns and processes interact, producing general insights into landscape dynamics.

Models are characterized in various ways: for example, models may be deterministic or stochastic; analytical or simulation; dynamics or static; and represent time as continuous or discrete. A model is spatial when the variables, inputs, or processes have explicit spatial locations represented in the model. A spatial model is only needed when explicit space, that is, what is present and how it is arranged, is an important determinant of the process being studied.

The process of building a model is multifaceted and includes the following steps: (1) Define the problem. (2) Develop the conceptual model. (3) Select the model type. (4) Develop the model by writing out the mathematical equations and relationships. (5) Computer implementation, including verification and documentation of the code. (6) Estimate the parameters, and calibrate if necessary. (7) Evaluate the model by comparison with empirical observation and perform a sensitivity or uncertainty analysis. (8) Use the model for experiments and prediction.

A broad diversity of approaches and subjects are represented in landscape models. Many landscape models address metapopulation dynamics, habitat fragmentation, and dispersal and invasion processes. Disturbance and vegetation dynamics are also well represented in the landscape modeling literature. Integrated models of ecological and socioeconomic processes are becoming a presence in the literature, but less attention is given to this important topic than it deserves. Approaches to implementing models of landscape patterns and processes are very diverse, although linking models to GIS data is common.

Models are and will remain extremely important tools in landscape ecology. Wise application of these models requires care, however, particularly with the following points: (1) Performance of any model results from the hypotheses and assumptions on which it is built. Comparing alternative model formulations is extremely valuable. (2) Understanding the sensitivity of models to error in estimating parameters is critical; however, assessing error propagation in spatial models remains challenging. (3) All models are simplifications of reality, and the domain of applicability for each model must be defined. (4) Gaps in empirical data for estimating key parameters are often a great source of uncertainty in model predictions. The empirical database that contributes to a model must be understood. (5) Technologically advanced methodologies do not assure the accuracy or reliability of model results!

DISCUSSION QUESTIONS

1. What are the distinguishing characteristics of landscape models? Is any spatial model also a landscape model? Must all landscape models have a spatial component?
2. Can parameter values estimated for simple models be directly used in models with more complicated formulations? Refer to Table 3.1 to illustrate your answer.
3. The survey of recent models presented in this chapter provides an overview of current modeling activities. Are models being applied in a balanced manner to the broad spectrum of landscape issues? What areas of landscape ecology are missing from the list of topics reviewed? Why?
4. Technological advances may eventually allow complex spatial simulations to be easily performed within GIS software. What should be the key concerns of landscape ecologists for the development and application of these methods?

RECOMMENDED READINGS

Ecological Modeling: General Reference

GRANT, W. E., E. K. PEDERSEN, AND S. L. MARIN. 1997. *Ecology and Natural Resource Management. Systems Analysis and Simulation.* John Wiley & Sons, Inc., New York.

KITCHING, R. L. 1983. *Systems Ecology.* University of Queensland Press, St. Lucia, Queensland, Australia.

STARFIELD, A. M., AND A. L. BLELOCH. 1986. *Building Models for Conservation and Management.* Macmillan Publishing Company, New York.

SWARTZMAN, G. L., AND S. P. KALUZNY. 1987. *Ecological Simulation Primer.* Macmillan Publishing Company, New York.

Spatial Modeling

BAKER, W. L. 1989. A review of models of landscape change. *Landscape Ecology* 2:111–133.

KAREIVA, P. 1990. Population dynamics in spatially complex environments: theory and data. *Philosophical Transactions of the Royal Society of London, B* 33:175–1990.

SKLAR, F. H., AND R. COSTANZA. 1990. The development of spatial simulation modeling for landscape ecology. In M. G. Turner and R. H. Gardner, eds. *Quantitative Methods in Landscape Ecology*, pp. 239–288. Springer-Verlag, New York.

TILMAN, D., AND P. KAREIVA. 1997. *Spatial ecology: The role of space in population dynamics and interspecific interactions. Monographs in Population Biology.* Princeton University Press, Princeton, New Jersey, 367 p.

CHAPTER 4

Causes of Landscape Pattern

When we view a landscape, we look at its composition and spatial configuration: the elements present and how these elements are arranged. In an agricultural landscape, we may observe forests occurring along streams and on steep ridges, whereas croplands and pastures occupy upland areas of gentler slope. In a fire-dominated boreal forest landscape, we may observe large contiguous areas of old forest, young forest, and early successional vegetation. In a deciduous forest, we may observe small gaps in an otherwise continuous canopy of trees, and we may detect transitions between forest communities dominated by different species of trees. In a coastal landscape, we may observe long narrow bands of similar vegetation as one moves from the land–water margin further inland. In landscapes of small extent (e.g., 100 m by 100 m), we may observe complex patterns of vegetated and unvegetated surfaces. How do all these different patterns develop? How do they change through time?

Today's landscapes result from many causes, including variability in abiotic conditions such as climate, topography, and soils; biotic interactions that generate spatial patterning even under homogeneous environmental conditions; past and present patterns of human settlement and land use; and the dynamics of natural disturbance and succession. Broad-scale variability in the abiotic environment sets the constraints within which biotic interactions and disturbances act. In this chapter, we discuss a variety of ways in which patterns develop on landscapes and provide a longer temporal context for understanding present-day patterns.

Much of what we as humans observe as landscape pattern is actually the spatial distribution of dominant vegetation types: for example, forest versus grasslands versus desert. The dominant vegetation establishes the resource base for the rest of the ecosystem. The pattern in the dominant vegetation, therefore, affects the spatial patterning of all components of the system. The patterning of the dominant vegetation may be defined by *ecotones*, the spatial divisions between vegetation types used to identify *patches* of similar vegetation or land cover. The ecotone forms the demarcation line that divides the dominant vegetation types and structures the basic spatial pattern on the landscape. In general, conditions of steep environmental gradients or recent disturbance lead to sharper boundaries between communities.

Levin (1976a) identified three general categories of causes of spatial pattern. The first category, *local uniqueness*, deals with unique features of a point in space, such as abiotic variability or unique land uses imposed by society. In addition to unique constraints at a local point, there are also the vagaries of colonization. In a sort of founder's effect, the seeds of a long-lived plant can become established and determine unique local features for decades. Chance alone may determine which of several different long-lived species arrives first at a site and becomes established. Finally, local uniqueness may depend on the existence of multiple stable states that may result from competition. That is, competition among interacting populations at a particular site may result in different relative abundances of these populations.

Levin's second category, *phase difference*, deals with spatial pattern resulting from disturbances (also see Chapter 7). The ecosystem responds to a local disturbance by going through succession. When viewed at any point in time, the landscape will have a number of disturbance sites of different age and in different stages of succession, that is, different phases. The individual sites will be in different phases of recovery, and the result will be a patchy pattern of vegetation.

Levin's third category, *dispersal*, prevents the landscape from becoming uniformly covered with a single, dominant population. The mechanism is a simple "fugitive" strategy (Platt and Weis, 1985). Prairie plants found in small patches of disturbed ground provide an example of this strategy. By producing many seeds that disperse far and wide, a fugitive species can establish itself whenever an opportunity arises, such as when ground squirrels or badgers have dug holes and displaced the prior vegetation. The fugitive species reach adulthood and produce seeds their first year, and plants from the surrounding undisturbed prairie spread slowly over the disturbed area. Given spatial heterogeneity of the landscape, frequent small disturbances, and limited dispersal ability of the dominant species, a fugitive species can maintain itself at isolated places throughout the landscape. Finally, interacting populations with differential dispersal abilities can also impose a quasi-periodic pattern on the landscape.

The following sections introduce a variety of factors that contribute to the patterns observed in landscapes, including abiotic factors, biotic interactions, human land-use patterns, and disturbance and succession. Climate, physiography, and soils establish the template for biotic interactions characteristic of each landscape. Because these abiotic factors are spatially and temporally variable, spatial patterning in soil formation and vegetation growth naturally occurs. Other processes, including disturbance and recovery from disturbance, as well as variability in human land use, may amplify this heterogeneity over a broad range of spatial and temporal scales.

ABIOTIC CAUSES OF LANDSCAPE PATTERN

Landscape patterns result, in part, from variability in climate and landform. *Climate* refers to the composite, long-term, or generally prevailing weather of a region (Bailey, 1996), and climate acts as a strong control on biogeographic patterns through the distribution of energy and water. Climate effects are modified by *landform*, the characteristic geomorphic features of the landscape, which result from geologic processes producing patterns of physical relief and soil development. Together, climate and landform establish the template on which the soils and biota of a region develop.

All landscapes have a history; understanding landscape pattern and process requires an understanding of landscape history. *Paleoecology* is the study of individuals, populations, and communities of plants and animals that lived in the past and their interactions with and dynamic responses to changing environments. This field offers a wealth of insight into the long-term development of today's landscapes. Although we do not attempt to review this rich field, we draw on paleoecological studies to discuss the role of climate in the spatial structuring of the biota and the role of prehistoric humans in influencing landscapes.

Climate

General climatic patterns will be familiar to all ecologists from introductory classes in biology or geography. At the broadest scale, climate varies with latitude, which influences both temperature and the distribution of moisture, and with continental position. Because of differential heating of land and water, coastal regions at a given latitude differ from inland regions. The distributions of biomes on Earth result from these broad-scale climate patterns. However, the effects of both latitude and continental position are then modified locally by topography, leading to finer-scale heterogeneity in climate patterns (Bailey, 1996). Temperatures generally decrease with increasing elevation, and north- and south-facing slopes experience different levels of solar radiation and hence different temperatures and evaporation rates.

Long-term Climate Change

The distribution of plant and animal communities, and indeed of entire biomes, has varied tremendously with past changes in climate, even in the absence of human activities. The spatial distribution of life forms today as a function of latitude and longitude look very different compared to those of 5000 or 10,000 years (yr) before present (BP). Furthermore, present assemblages of plants and animals represent only a portion of the ecosystems that have existed during Earth's history.

Climatic changes on Earth during the past 500,000 yr have been dramatic (Figure 4.1). Each glacial–interglacial cycle is about 100,000 yr in duration, with 90,000 yr of gradual climatic cooling, followed by rapid warming and 10,000 yr of interglacial warmth. The peak of the last glacial period, or ice age, was about 18,000 yr BP and ended approximately 10,000 yr BP. These long climate cycles may be produced by cyclic changes in solar irradiance resulting from long-term and complex variation in Earth's orbital pattern (the Milankovitch cycles) as Earth wobbles on its rotational axis (Crowley and Kim, 1994). This orbital eccentricity results in approximately 3.5% variation in the total amount of solar radiation received by Earth and changes its latitudinal distribution.

Looking more closely at the most recent climate cycle, we can examine changes in mean global temperature for the past 150,000 yr. Mean global temperature is the only reliable expression of global surface air temperature because climatologists want to remove the spatial variability in climate to detect trends in the entire global climate system; thus, small changes in mean global temperature may reflect very large fluctuations in temperature at many locations on Earth. During the past 150,000 yr, there was a 5°C shift in average global temperature between the glacial and interglacial periods (Figure 4.2). Peak warming, about 1° to 2°C warmer than today, occurred between 9000 and 4000 yr ago. This seemingly small increase led to a 70-km shift eastward in the prairie–forest boundary in the upper Midwest compared to its present location. Since the end of the last ice age, mean global temperature has fluctuated by little more than 1°C; indeed, the Little Ice Age, which lasted for >500 yr, was a 1°C fluctuation. If past patterns continue, the Milankovitch cycle indicates a decrease in global temperatures with the onset of another glaciation during the next 25,000 yr. Alternatively, a major climatic warming of at least 2°C is proposed as a superinterglacial that will last for at least 1000 yr because of the anticipated build up of carbon dioxide (CO_2) and other greenhouse gases that trap infrared radiation within the atmosphere and warm Earth.

Earth's biota obviously must respond to these large fluctuations in climate. In general, organisms may respond in three ways (Cronin and Schneider, 1990), which contribute to long-term changes in their distribution: (1) they may evolve and speciate; (2) they may migrate long distances, each according to its limits of tolerance and movement capability; or (3) they may become extinct. Considerable work has been done to describe and understand the vegetation changes that accompanied past changes in climate. For example, range limits of tree species in eastern North America changed dramatically during the past 13,000 yr (Figure

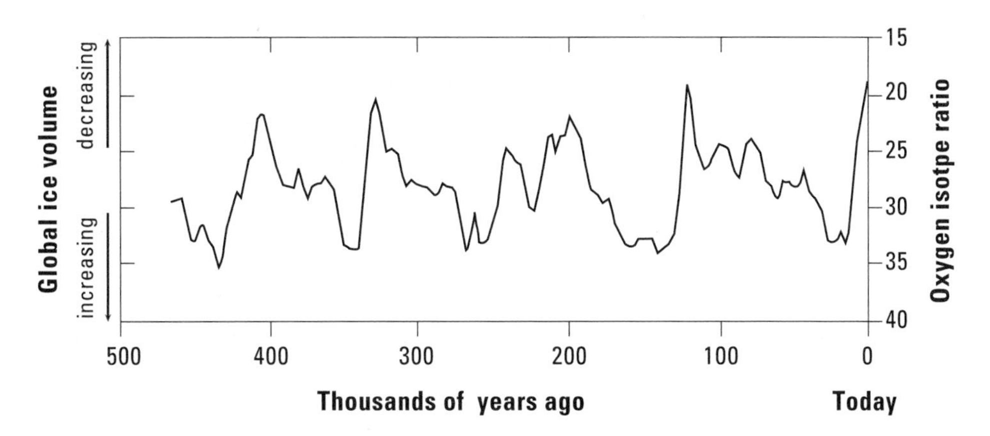

FIGURE 4.1.
Record of climatic changes over the past 500,000 years as measured by oxygen isotope ratios from cores of deep-sea sediments obtained from the Indian Ocean. Note the cycles of rapid warming followed by gradual cooling.

ADAPTED FROM DELCOURT AND DELCOURT, 1991, BASED ON IMBRIE AND IMBRIE, 1979.

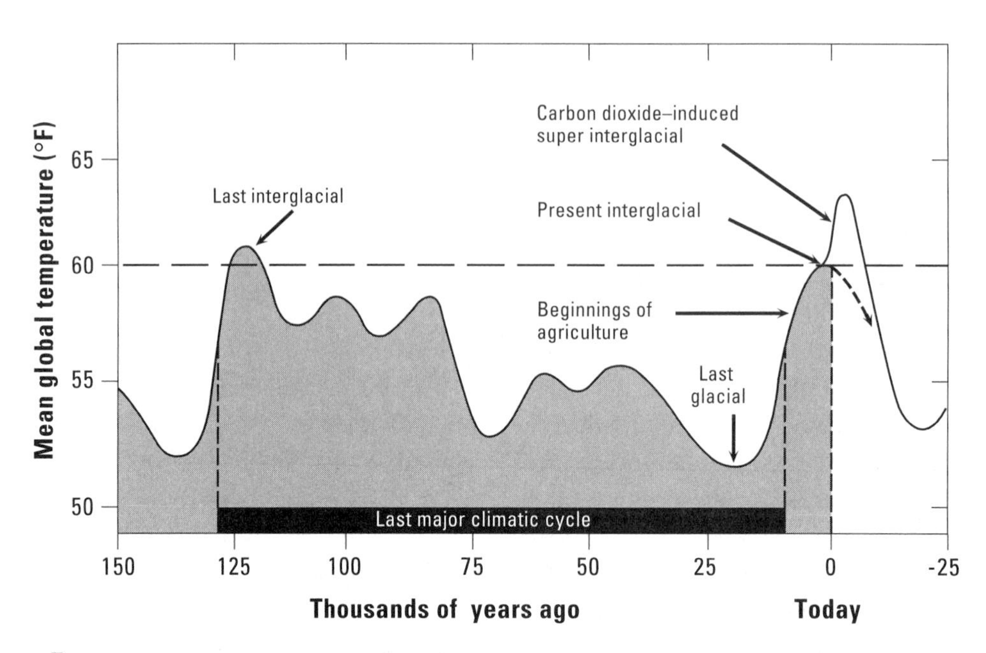

FIGURE 4.2.
Global climate changes over the past 150,000 years and projected for the next 25,000 years. A future cooling trend is projected based on the Milankovitch cycles, but this may be delayed by a warming period induced by elevated concentrations of carbon dioxide and other greenhouse gases in the atmosphere.

ADAPTED FROM DELCOURT AND DELCOURT, 1991, BASED ON IMBRIE AND IMBRIE, 1979.

4.3) (Davis, 1983). Not only have species varied in their ranges, but also the local abundances, and thus relative dominance, of taxa have changed. The range of oak (*Quercus*) in eastern North America has expanded northward during the past 20,000 yr, and the population centers where oak dominated also varied spatially (Delcourt and Delcourt, 1987).

Several points important for providing a context for interpreting patterns on today's landscapes emerge from the studies of vegetation response to climate. First, the glacial–interglacial cycles trigger the disassembly of communities, followed by reassembly that is unpredictable in terms of either species composition or abundance. Compared to present-day communities, the past communities at many sites feature mixtures of species that are absent or very rare on the modern landscape (e.g., Barnosky et al., 1987). Second, the characterization of past plant communities suggests that the displacement of entire vegetation zones or communities was the exception rather than the rule. That is, species responded individualistically to climatic change, each according to its limits of tolerance, dispersal capability, and interactions with the surrounding biota. Third, disturbance regimes (see Chapter 7) have been very sensitive to past changes in climate. For example, the fire regime in northwestern Minnesota, USA, shifted from a 44-yr fire cycle during the warm, dry 15th and 16th centuries to an 88-yr fire cycle after the onset of cooler, moister conditions after AD 1700 and throughout the Little Ice Age (Clark, 1990). In summary, it is important for the landscape ecologist to recognize the dynamic responses of the biota to variability in climate in space and time.

The implications of potential climate change for the distribution of Earth's biota and the patterns observed across landscapes are profound. Current climate exerts a very strong effect on landscape patterns (see Bailey, 1996, for an excellent treatment of this), and the most conspicuous effect of climate change may be shifts in landscape pattern (Neilson, 1995). Teams of mathematical modelers have pro-

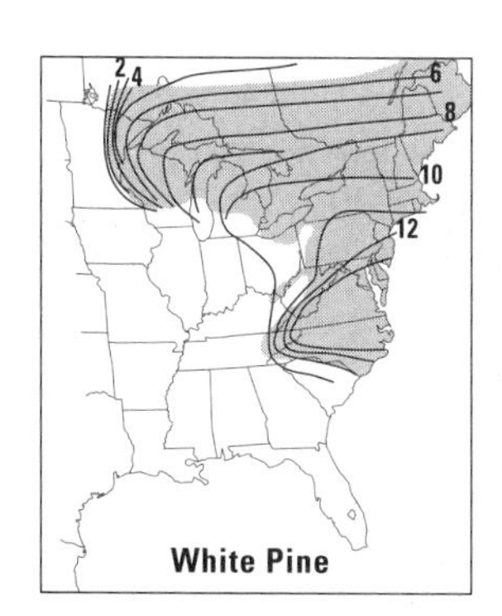

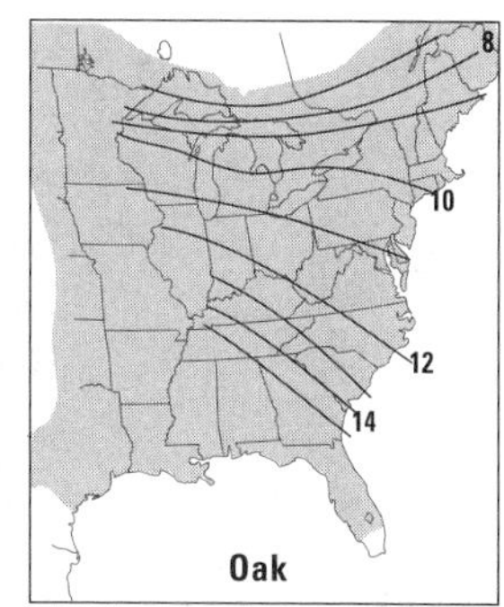

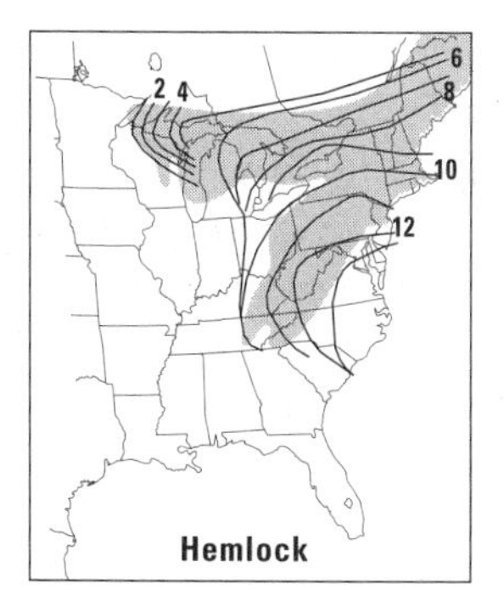

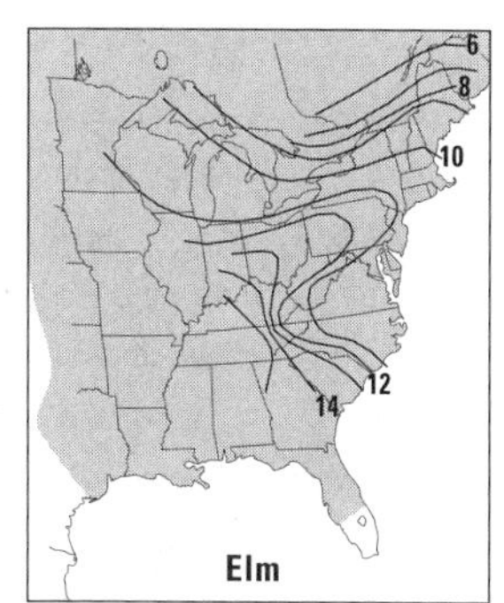

FIGURE 4.3.
Changes in northern and western range limits for four eastern North America tree taxa during the late Quaternary based on pollen records. Numbers indicate the time (in thousands of years before the present) at which pollen from each species was recorded at a given site. Shading indicates current geographic range.

ADAPTED FROM DAVIS, 1983.

BOX 4.1
LONG-TERM VEGETATION CHANGES AT GRAYS LAKE, IDAHO, USA

Many examples could be used to illustrate vegetational changes during the Pleistocene and Holocene. Beiswenger's (1991) study of the Grays Lake Basin in southeastern Idaho, USA, offers one fine case study. Grays Lake sits within the central Rocky Mountains at relatively low elevation (1950 m) and is ideal for studying late-Quaternary vegetation dynamics because it was not glaciated. The current vegetation includes marshes dominated by *Scirpus americanus*, sagebrush (*Artemisia*) steppe, coniferous forests (including *Pseudotsuga menziessii, Pinus contorta, P. flexilis, Picea engelmannii,* and *Abies lasiocarpa*), and aspen (*Populus tremuloides*) forest. Fossil pollen were identified and dated from sediment cores obtained from the lake at the snow–ice surface; the cores ranged in length from 14 to 21 m. Results demonstrate a dominance of *Artemisia* ~70,000 to 30,000 yr before present (BP) prior to the last major glacial advance in the Rocky Mountains. This indicates an arid climate in which trees were limited to the adjacent mountains. *Pinus* pollen then dominates between ~30,000 and 11,500 yr BP, and the increase in pine pollen indicated that a forest occupied the basin during the full glacial period. At the transition from the late-glacial to Holocene ~11,500 to 10,000 yr BP, there is an increase in pollen from both *Picea* and *Artemisia* along with a tenfold increase in total pollen influx, suggesting a vegetation response to increased moisture accompanying climatic warming. Initial climatic change produced cool, moist conditions suitable for *Picea*, which had been limited by a cold and dry glacial climate. However, this transitional period of increased moisture reversed before 10,000 yr BP, and the percentages of *Picea* and *Pinus* pollen both decline near the end of the period. The conifers moved to higher elevations, while *Artemisia* and other species (e.g., Compositae) became more abundant at lower elevations. As the Holocene began, the percentage of pollen from steppe plants increased, with a peak in Gramineae pollen ~8500 yr BP. Warm, dry conditions occurred from ~10,000 to at least 7100 BP, with a xeric maximum suggested ~8200 yr BP. Around 7300 to 2000 yr BP, *Pinus, Artemisia,* and *Juniperus* pollen all increased, reflecting moderate cooling, increased precipitation, or both. The most recent 2000 yr were characterized by increases in *Pinus, Picea, Abies, Pseudotsuga,* and *Populus* pollen percentages and declines in *Juniperus, Artemisia,* Compositae, and Chenopodiaceae pollen. Further cooling and/or increased precipitation has continued since ~2000 yr ago.

The Grays Lake study reveals a strong relationship between vegetation and climate in the Central Rocky Mountains over the past 70,000 yr (Beiswenger, 1991). The data indicated that the vegetation around Grays Lake has shifted from a cold, dry, *Artemisia* steppe to a conifer woodland during the last glacial period. Rapid expansion of spruce and sagebrush followed with the cool, moist conditions produced by climatic warming. A dry steppe developed next with the rising temperatures and increased aridity of the early Holocene, but conifer forest established with a subsequent cooling. This work demonstrates the wide range of vegetation types that occupied a particular landscape through time and emphasizes that the landscapes that we observe today are by no means static. Landscape ecologists must strive to understand the history of the landscapes that they study.

duced maps of how the dominant ecotones will move across the United States in response to temperature and moisture shifts caused by a doubling of atmospheric CO_2 (VEMAP, 1995). Bartlein et al. (1997) projected the potential distributions of selected tree taxa in the region of Yellowstone National Park, Wyoming, USA. They used a coarse-resolution climate model that incorporated a doubling of atmospheric CO_2 and interpolated the projections onto a 5-min grid of topographically adjusted climate data. Simulated vegetational changes included elevational and directional range adjustments. That is, taxa could move up or down elevational gradients or latitudinally. The ranges of high-elevation species (e.g., *Pinus albicaulis*) diminished under the future climate scenario, and some species were

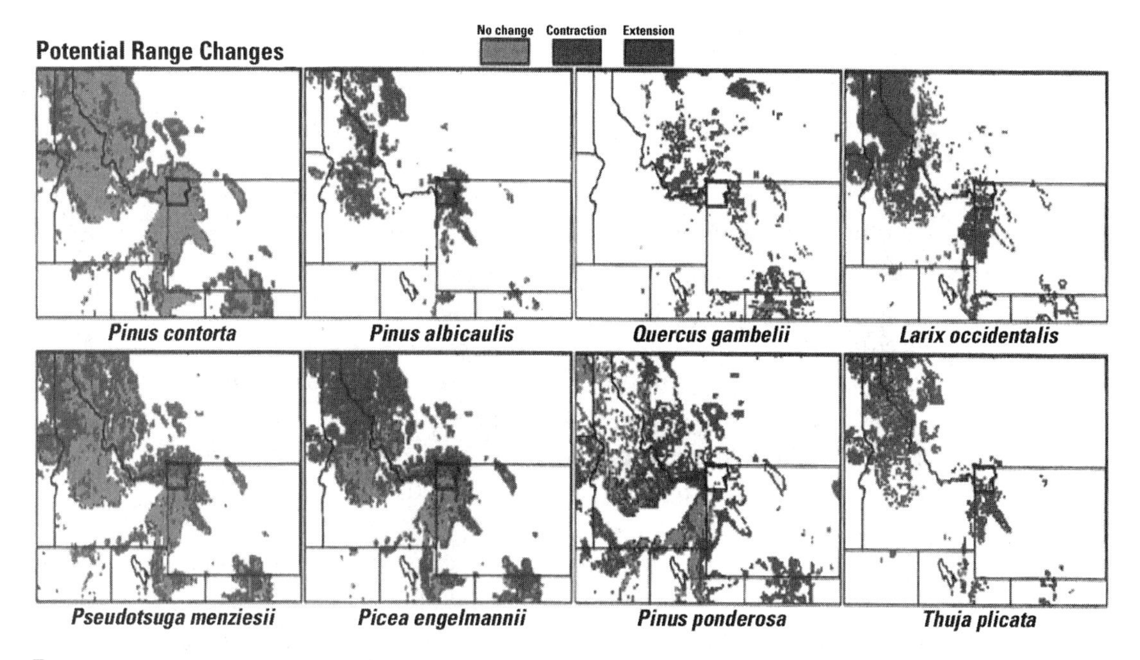

FIGURE 4.4.

Potential range changes of selected tree taxa in the Yellowstone National Park region of the Rocky Mountains under projections of a $2 \times CO_2$ climate. Green shading indicates grid points where the taxon occurs under both the current and $2 \times CO_2$ scenario. Red shading indicates grid points where the taxon occurs under current climate, but does not occur under the $2 \times CO_2$ climate. Blue shading indicates grid points where the taxon does not occur under current climate, but does occur under the $2 \times CO_2$ climate. (Refer to the CD-ROM for a four-color reproduction of this figure.)

REPRODUCED FROM BARTLEIN ET AL., 1997.

extirpated locally (Figure 4.4). Projected mild, wet winters also produced new areas of suitable habitat for other taxa (e.g., *Pinus ponderosa, Larix occidentalis,* and *Quercus gambelii*) (Figure 4.4). Of particular note was that the new communities had no analogue in the present-day vegetation, because low-elevation montane species currently in the region were mixed with species that might colonize from the northern and central Rocky Mountains and the Pacific Northwest. In addition, the potential range adjustments projected for different species equaled or exceeded the changes seen in the paleoecological record during previous warming intervals (Bartlein et al., 1997).

Landform

Landforms range from nearly flat plains to rolling, irregular plains, to hills, to low mountains, to high mountains (Bailey, 1996) and are identified on the basis of three major characteristics: (1) relative amount of gently sloping (<8%) land, (2) local relief, and (3) generalized profile, that is, where and how much of the gently sloping land is located in valley bottoms or in uplands (Bailey, 1996). Landforms may be described further by considering the topographic sequence of variation, or *soil catena*, of soils and associated vegetation types within each landform. For example, a mountainous landform may have a toposequence that includes ridgetops, steep slopes, shallow slopes, toe slopes, and protected coves. If different areas are composed of similar landforms with similar geology, then soil catenas and vegetation types may also be expected to be similar.

Four general effects of landform on ecosystem patterns and processes (Figure 4.5) were categorized by Swanson et al. (1988).

1. The elevation, aspect, parent materials, and slope of landforms affect air and ground temperature and the quantities of moisture, nutrients, and other materials available at sites within a landscape. For example, south-facing slopes receive more solar radiation than northward slopes, resulting in warmer, drier conditions. These topographic patterns are strongly related to the distribution of vegetation across a landscape (e.g., Whittaker, 1956).
2. Landforms affect the flow of many quantities, including organisms, propagules, energy, and matter through a landscape. The funneling of winds, for example, may lead to dispersal pathways for wind-blown seeds. The position of lakes relative to groundwater-flow pathways may strongly influence the chemical and biological characteristics of these lakes (Kratz et al., 1991; also see Chapter 9).

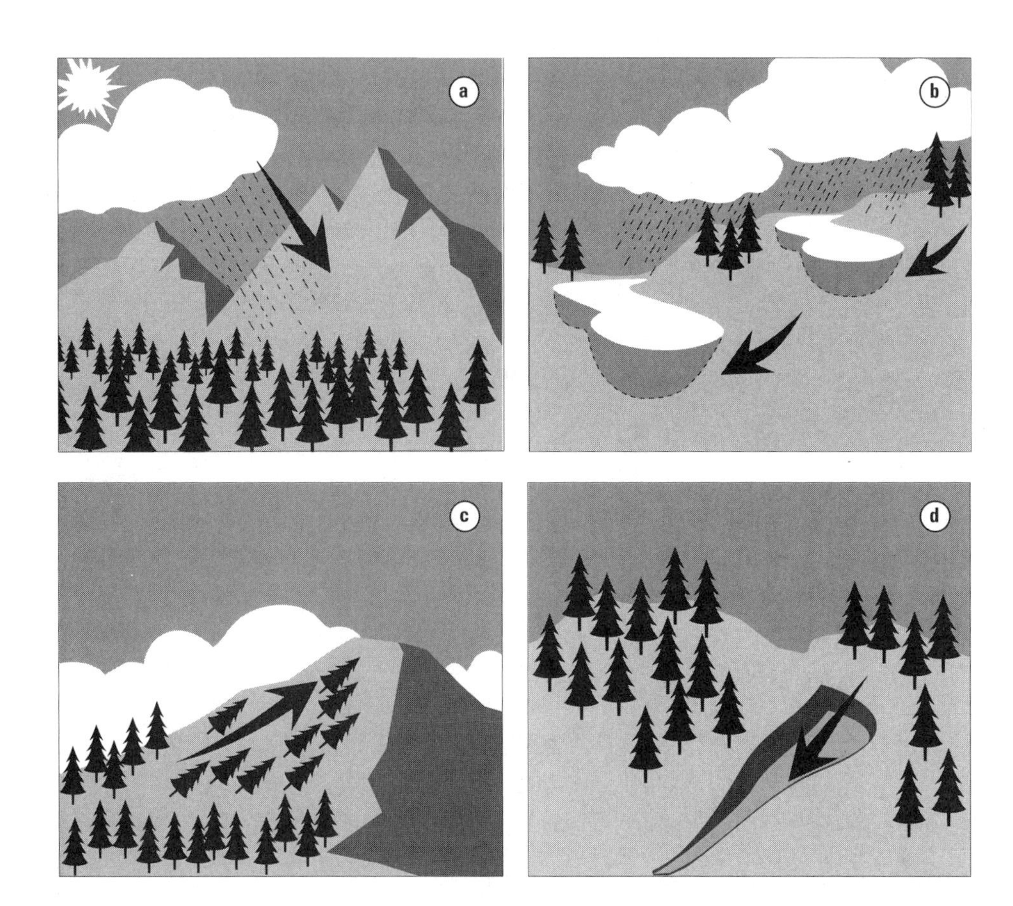

FIGURE 4.5.

Examples of four classes of landform effects on ecosystem patterns and processes. (a) Topographic influences on rain and radiation (arrow) shadows. (b) Topographic control of water input to lakes. Lakes high in the drainage system receive a greater proportion of water input by direct precipitation than lakes lower in the landscape, where groundwater (arrows) predominates; also see Chapter 9. (c) Landform-constrained disturbance by wind (arrow) may be more common in upper-slope locations; also see Chapter 7. (d) The axes of steep concave landforms are most susceptible to disturbance by small landslides (arrow).

MODIFIED FROM SWANSON ET AL., 1988.

3. Landforms affect the frequency and spatial pattern of natural disturbances such as fire, wind, or grazing. Across a New England landscape, susceptibility to damage from hurricanes varied with landscape position, with greater damage observed in more exposed topographic positions (Foster and Boose, 1992; Boose et al., 1994). In Labrador, fire and topography jointly influenced vegetation patterns (Foster and King, 1986) with nearly all patches of birch (*Betula*) forest occurring on steep slopes or ridges with high moisture (Figure 4.6). Lightning would ignite fires on ridge tops covered by spruce–fir (*Picea–Abies*) forest, sweep down the ridges, and stop at existing birch stands or wetter areas in the valley bottoms. These newly burned areas along the slopes provided opportunities for birch to colonize.

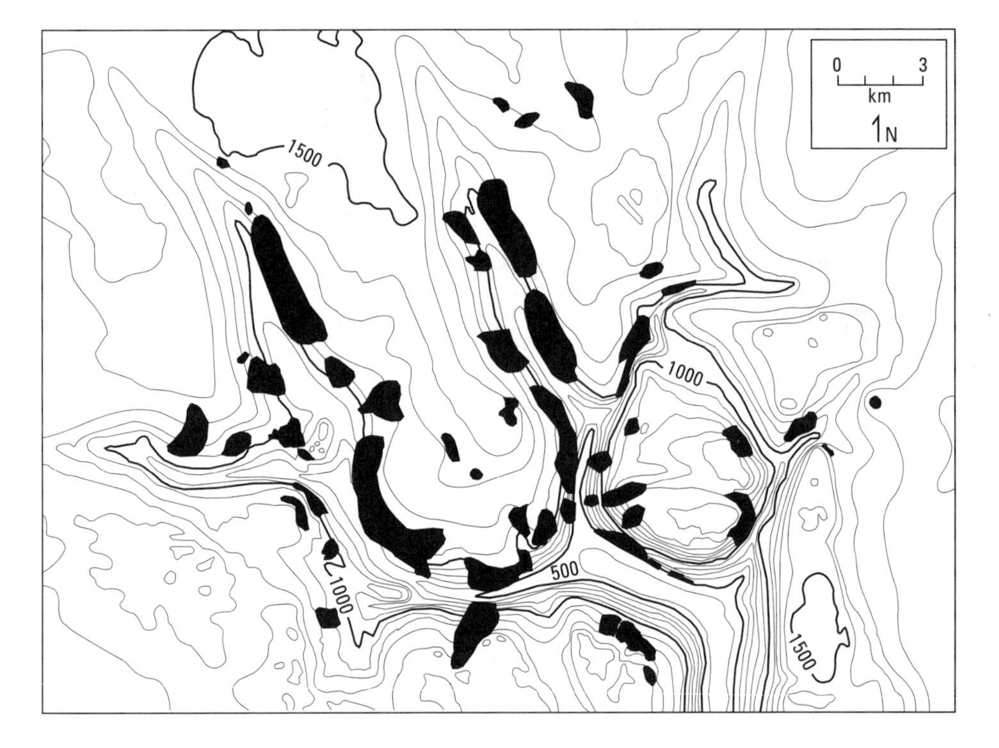

FIGURE 4.6.
Distribution of *Betula papyrifera* forests (black) on the hillslopes and canyon walls of the St. Augustin River Valley, southeast Labrador.

ADAPTED FROM FOSTER AND KING, 1986.

4. Landforms constrain the spatial pattern and rate or frequency of geomorphic processes, the mechanical transport of organic and inorganic material, that alter biotic characteristics and processes. Portions of a landscape may be more or less susceptible to landslides or to shifts in river channels. Taken together, landforms significantly contribute to the development and maintenance of spatial heterogeneity across a landscape through their multiple effects on soils, vegetation, and animals (Swanson et al., 1988). Even in areas of relatively little topographic relief, such as the glacial landforms of the upper Midwest of the United States, landform explains a great deal of the variability in successional pathways (Host et al., 1987) and biomass (Host et al., 1988) across the landscape.

BIOTIC INTERACTIONS

Interactions among organisms, such as competition and predation, may lead to spatial structuring even in a completely homogeneous space. Theoretical population ecology focuses much attention on these dynamics (Ives et al., 1998), with an emphasis on how interactions within and among populations can generate spatial patterns and how these patterns influence the outcome of interactions. The product of these theoretical approaches often is a map of species distributions.

Competition between two species in a landscape without any abiotic variation theoretically could result in homogeneous spatial distribution (i.e., one species remaining) through competitive exclusion (Gause, 1934). The best competitor would win out and establish itself throughout the landscape, resulting in a homogeneous distributional pattern. However, there are important exceptions to competitive exclusion.

Groups of competing organisms may interact in complex ways so that final distributions take on one of many alternative stable states. These *multiple stable states* (Sutherland, 1974) may often occur when several different species can potentially occupy and dominate a site. Which species actually occurs on a specific site is determined by very small stochastic changes in the initial conditions. Once in one of these states, the community may remain dominant in spite of minor disturbances. However, a major disruption may result in a new configuration that is different, but also stable. This type of shifting, stochastic pattern may be observed near ecotones between major community types. For example, small, stable stands of trees may extend out into grassland, and small stable patches of grasses may intrude into the forest. Along this ecotonal edge, both communities are stable, and there are very small differences in the competitive advantage of one community over the other. Chance plays a role in which community is established, and once established this community can maintain itself until a major disruption occurs.

Competition between vegetation types can also form ecotones, resulting in a sharp line between vegetation, even when differences in environmental conditions on either side of the ecotone are small. Along a north–south transect, for example, temperature and moisture may change gradually and continuously, with no sharp discontinuities. Conditions to the south may favor one species and conditions to the north, another. Somewhere along the transect, conditions will be suitable for the growth of both species. Competition for space may form a sharp ecotone between them, rather than a gradation or intermingling.

A different sort of pattern emerges from *reaction–diffusion models* of interacting populations (Okubo, 1975). In these models, the growing and competing populations are also dispersing across a uniform environment. In many cases (Levin, 1978), the expected uniform distribution is destabilized by the action of diffusion, and the system spontaneously assumes a patchy, periodic spatial distribution. For example, in *predator–prey models*, a patchy distribution results if the diffusion coefficient of the predator is sufficiently larger than the prey. A fixed spatial pattern with peaks and troughs in the density of both predators and prey can result. This mechanism of *diffusive instability* has been suggested as the cause of patchy distribution in plankton (Kierstead and Slobodkin, 1953; Steele, 1974a; Edelstein-Keshet, 1986; Murray, 1989). We might suspect this type of mechanism whenever a periodic or quasi-periodic pattern is detected on the landscape.

Pattern also results from the activities of a *keystone species*. Paine (1974, 1976) studied the interactions between the mussel *Mytilus californianus* and its starfish predator, *Pisaster ochraceous*, in the intertidal zone. The mussel is a superior competitor, but predation by the starfish keeps the mussel population in check. Higher up on the shoreline, the starfish has difficulty reaching the mussels. The mussels completely dominate the rock surfaces and eventually grow too large for the starfish to handle. Farther down the shoreline, the starfish consumes all young mussels. The result is a very distinct striped pattern on the rocks, mussel above, but not below this line. When Paine (1974) experimentally removed the starfish, the mussels moved down the surface of the rock, outcompeting and eliminating 23 other species of invertebrates. The starfish is clearly the keystone predator that creates and maintains the spatial pattern. Holling (1992) believes that keystone species and processes are a common cause of pattern, stating that "All ecosystems are controlled and organized by a small number of key plant, animal, and abiotic processes that structure the landscape at different scales."

Influence of Dominant Organisms

In many respects, it is the dominant organisms that define spatial pattern on the landscape. It is, for example, the patches of trees or natural vegetation that define the pattern on most natural terrestrial landscapes. Within the context of the abiotic template, the dominants alter the abiotic conditions and provide resource base and substrate for the other populations in the ecosystem. In these cases, the rest of the ecosystem is constrained to operate within the spatial pattern of the dominants. The interactions of the plants with the soil, climate, and topography produce the underlying spatial context. This is not only true in terrestrial ecosystems; for example, coral is a dominant organism along tropical shorelines. The coral forms the substrate and resource base for the entire food web, and its spatial distribution dictates the spatial pattern for the rest of the ecosystem.

A common example of a dominant consumer that may produce and maintain spatial pattern is a lethal pest. Insects such as the spruce budworm (*Choristoneura fumiferana*) and the balsam wooly adelgid (*Adeiges picea*) act very much like other disturbances in causing patches to revert to earlier successional stages. The bark beetle provides a simple example (Rykiel et al., 1988). Lightning strikes and kills a single tree and permits the beetle to invade. Once established, the beetle can attack adjacent trees and spread from this original point of attack. Eventually, a large patch is opened and reverts to early successional stages.

The beaver (*Castor canadensis*) provides a fascinating example of landscape pattern resulting from the activities of a dominant organism. The beaver uses sticks and mud to dam a second- to fifth-order stream, impounding water behind the dam (Johnston and Naiman, 1990a). Aerial photography (Johnston and Naiman, 1990b) shows that as much as 13% of the landscape can be altered in this way. The animals also affect the riparian vegetation and saturate the soils, forming wetlands (Naiman et al., 1986). When the dam breaks down and the pond is abandoned, a characteristic beaver meadow remains as a distinct spatial feature on the landscape (Remillard et al., 1987).

A similar story can be told of the American bison (*Bison bison*). At one time there were 75 million bison in North America (Roe, 1951). Huge herds migrated regularly and determined plant composition along these linear routes, both by preferential grazing and by recycling nutrients in dung. The animals also used dust baths to control skin parasites and formed characteristic circular patches on the landscape. In general, large mammals will act as a mechanism in pattern formation (Botkin et al., 1981). More generally, large mammals often directly alter vegetation and rates of nutrient recycling (Dyer et al., 1986). A moose, for example, consumes five to six metric tons of food a year (Pastor et al., 1988), increasing nutrient recycling and altering patterns of productivity. By selectively browsing hardwoods (Pastor et al., 1993), moose also directly affect species composition at landscape scales. An effect of excluding elephants from their native habitat is a change in the pattern of vegetation (Harton and Smart, 1984).

HUMAN LAND USE

Patterns of land use can alter both the rate and direction of natural processes, and land-use patterns interact with the abiotic template to create the environment in which organisms must live, reproduce, and disperse. *Land use* refers to the way in which and the purposes for which humans employ the land and its resources (Meyer, 1995). For example, humans may use land for food production, housing, industry, or recreation (Nir, 1983). A related term, *land cover*, refers to the habitat or vegetation type present, such as forest, agriculture, and grassland. Although they are related, it is important to note the distinction between these terms: an area of forest cover may be put to a variety of uses, including low-density housing, logging, or recreation. We use *land-use change* to encompass all the ways in which human uses of the land have varied through time. The ways in which humans use the land are important contributors to landscape pattern and process.

Prehistoric Influences

Prehistoric humans had a major role in influencing landscapes, and their past effects contribute to present-day landscape patterns. Using the pollen record, indications of human activities can be traced back thousands of years, and discrete episodes of human disturbance can be correlated with archeological data. Consider, for example, the historical expansion of human influences in Europe (Delcourt and Delcourt, 1991). In the early Holocene, there was broad-based foraging throughout the Mediterranean region. The switch from a nomadic to a more sedentary way of life was just beginning ~10,000 BP, and by ~8000 BP permanent settlements were established in Greece. These settlements included cultivation of crops and maintenance of livestock, and food production became more labor intensive. Cereal cultivation caused a major shift in patterns of land use because the permanent fields needed weeding and required nutrient replenishment, both of which were activities requiring considerable human labor. By about 6500 BP, farming expanded north of Greece as winters became warmer and precipitation increased. Development of more efficient technologies also contributed to the continued expansion of agriculture in Europe. Use of the *ard*, a tool that used the angle between the trunk and roots of a tree to break through the soil and that was pulled by an oxen, became prevalent ~5000 BP. Further human expansion became based on the maintenance of work animals, because the oxen-drawn *plow* that could both furrow and turn over the soil was developed and used by ~3000 BP. More efficient bronze sickles also replaced wooden sickles.

What were the effects of this expansion of human activities in Europe on native vegetation? The impact of the axe and spade on ecosystems began to transform natural landscapes into cultural ones through plowing, burning, and trampling. The ard, because it did not overturn the soil, left perennial roots intact. The plow, however, removed perennials from the soil and encouraged establishment of annual plants. The process of deforestation and conversion of land to pasture or crop cultivation changed the landscape from a natural to a cultural mosaic (Delcourt, 1987). This also occurred in North America, although early settlements of Native Americans were more restricted to floodplains; uplands were used much later than in Europe (Delcourt, 1987). However, Native Americans in North America profoundly influenced the landscape by establishing settlements, practicing agriculture, hunting, and using fire to induce vegetation changes (Denevan, 1992).

The influences of prehistoric humans on landscapes were characterized by Delcourt (1987) into five main types.

1. Humans changed the relative abundances of plants, especially the dominance structure in forest communities. In the pollen record from Crawford Lake, Ontario, land clearance and maize cultivation by the Iroquois is documented by pollen sequences spanning the 14th to 17th centuries. During this time, the dominance of tree species in the surrounding forest changed from late-successional species such as beech (*Fagus grandifolia*) and sugar maple (*Acer saccharum*) to forest of oak (primarily *Quercus rubra*) and white pine (*Pinus strobus*).
2. Humans extended or truncated the distributional ranges of plant species (woody and herbaceous). In Europe, for example, the range of olives (*Olea europaea*) after 3000 yr BP was extended through cultivation from the Mediterranean coast only to throughout southern Europe. Truncation of the

range of a native tree species by prehistoric humans has been documented for bald cyprus (*Taxodium distichum*) in the central Mississippi and lower Illinois valleys in eastern North America. Charcoal evidence suggests a preference for cyprus wood during the period from 2000 yr BP to AD 1450, with the species becoming locally extinct as human populations increased (Delcourt, 1987).

3. Opportunities were created for the invasion of weedy species into disturbed areas. In many places, weedy species assemblages associated with cultivated fields increase in abundance in the pollen record, and these increases are correlated with archeological evidence of human occupation (Delcourt, 1987).
4. The nutrient status of soils was altered through both depletion and fertilization.
5. The landscape mosaic was altered, especially the distribution of forest and nonforest. This last change is also easiest to detect in the paleoecological record by examining ratios of tree to herbaceous pollen.

A key point from this brief discussion of the long-term development of the cultural landscape is that the landscapes we may perceive to be natural today probably have a history of human influence that dates back a long time. Of course, there is variability in the degree to which humans influenced different ecosystems on different continents. However, humans have long been a presence in many landscapes, and their role in creating landscape pattern should not be discounted.

Historic and Present-Day Effects

Both worldwide and in the United States, land-cover patterns today are altered principally by direct human use: by agriculture, raising of livestock, forest harvesting, and construction (Meyer, 1995). Human society relies on natural habitats for a variety of services, including productivity, recycling of nutrients, breakdown of wastes, and maintenance of clean air, water, and soil. In North America, land-use changes have been particularly profound since Europeans settled the continent three centuries ago. Landscapes have become mosaics of natural and human-influenced patches, and once-continuous natural habitats are becoming increasingly fragmented (e.g., Burgess and Sharpe, 1981; Harris, 1984).

Land-use changes in the United States serve as a handy example. At the time of European settlement, forest covered about half the present lower 48 states. Most of the forestland was in the moister east and northwest regions, and it had already been altered by Native American land-use practices (Williams, 1989). Clearing of forests for fuel, timber, and other wood products and to open the land for crops led to a widespread loss of forest cover that lasted through the early 1900s. So extensive was this loss that by 1920 the area of virgin forest remaining in the conterminous United States was but a tiny fraction of that present in 1620 (Figure 4.7). Some originally cleared areas, for example, New England, the Southeast, and the upper Midwest, have become reforested due to lack of cultivation. In other regions, clearing for agriculture has been more permanent (e.g., the lower Midwest), or harvest of primary forest has continued until recent times (e.g., Pacific Northwest).

Developed land in the United States has expanded as the population has grown in number, with most of the population now living in cities, towns, and suburbs

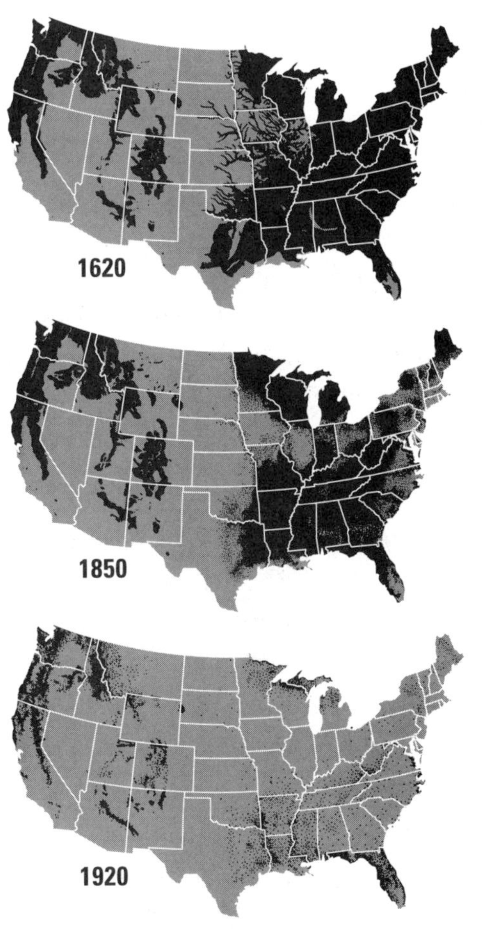

FIGURE 4.7.
Approximate area of virgin old-growth forest in the contiguous United States in 1620, 1850, and 1920. Note that this does not depict total forest area, because many forests, especially in the eastern United States, have regrown following clearing and the abandonment of agriculture.

ADAPTED FROM MEYER, 1995.

rather than on farms. Americans spread out more across the land as transportation technologies improved, especially as the automobile became the primary mode of transportation. Present-day patterns of settlement take up more land per person than in the past, and homes and subdivisions are more dispersed across the landscape. A frontier of rapid and sometimes chaotic land-use change surrounds urban areas (Meyer, 1995). Trends in developed land are unique because they run in only one direction; that is, developed land expands and does not revert to other categories. Thus, the distribution of developed land across the United States will leave a long-lasting footprint on the landscape (Turner et al., 1998a). The most remarkable aspect of the landscape of the United States since European settlement is its continual change. Effects of these vast changes are long lasting and crucial to our understanding of the present-day plants and animals that inhabit our landscapes (Foster, 1992; Dale et al., 2000).

DISTURBANCE AND SUCCESSION

Disturbance and the subsequent development of vegetation are key contributors to pattern on the landscape. By *disturbance*, we mean any relatively discrete event in time that disrupts ecosystem, community, or population structure and changes resource availability, substrate, or the physical environment (White and Pickett, 1985). Examples include fires, volcanic eruptions, floods, and storms. Disturbances are often described by a variety of attributes, including their spatial distribution, frequency, spatial extent, and magnitude. The spread of disturbance and spatial patterns of recovery have received considerable attention in landscape ecology, and we devote a chapter to exploring these dynamics (see Chapter 7). Here, we simply recognize disturbance as an important agent of pattern creation at a variety of spatial and temporal scales.

SUMMARY

Today's landscapes result from many causes, including variability in abiotic conditions such as climate, topography, and soils; biotic interactions that generate spatial patterning even under homogeneous conditions; past and present patterns of human settlement and land use; and the dynamics of natural disturbance and succession. Three general causes of spatial pattern were identified by Levin (1976a): (1) local uniqueness, that is, the unique features of a point in space, such as abiotic variability or unique land uses imposed by society; (2) phase differences, or variation in spatial pattern resulting from disturbances; and (3) dispersal, which prevents landscapes from becoming uniformly covered with a single, dominant population.

Landscape patterns result, in part, from variability in climate and landform; these broad-scale abiotic drivers constrain other causes of landscape change. Climate refers to the composite, long-term, or generally prevailing weather of a region (Bailey, 1996). Climate effects are modified by landform, which includes both geology and topography, or physical relief. The distribution of plant and animal communities and indeed of entire biomes has varied tremendously with past changes in climate, even in the absence of human activities. Not only have species varied in their ranges, but also the local abundances and thus the relative dominance of taxa have changed. Landforms are important influences on landscape pattern because they influence moisture, nutrients, and materials at sites within a landscape; they affect flows of many quantities; they may influence the disturbance regime; and they constrain the pattern and rate of geomorphic processes. It is important for the landscape ecologist to understand the influence of climate and landform on the biota and to recognize the dynamic responses of the biota to variability in climate in space and time.

Interactions among organisms, such as competition and predation, may lead to spatial structure, even in the absence of abiotic variation. Keystone species or dominant organisms may define spatial pattern on a landscape. Disturbance and succession (see Chapter 7) are key contributors to landscape pattern. Humans are also a strong driver of landscape patterns, because land-use patterns interact with the abiotic template to create the environment in which organisms must live, reproduce, and disperse. Nearly all landscapes, even those that we perceive as natural today, probably have a history of human influence that dates back a long time. Many landscapes today have become mosaics of natural and human-influenced patches, and once-continuous natural habitats have become increasingly fragmented. Effects of past land use are increasingly recognized as important determinants of the present-day biota that inhabit our landscapes.

The understanding of what causes landscape pattern and pattern change with time is often translated into models used to project future landscape scenarios (Baker, 1989b). These models simulate changes in the abundance and spatial arrangement of elements on the landscape, such as vegetation or cover classes. Developing predictive models of landscape pattern and how such patterns vary through time is an active, rapidly changing field.

DISCUSSION QUESTIONS

1. For a landscape of your choice, define its spatial extent and describe the dominant factors causing landscape pattern in each of the following categories: abiotic factors, biotic interactions, human land use, and disturbance and succession. Repeat this exercise after reducing the extent of the landscape to 10% of its original size. Does the importance of the factors shift when the scale is changed? Why or why not?
2. Consider the variety of factors that create landscape pattern. How would you rank their relative importance? Do you think this ranking has changed through time? Explain your answers.
3. How do abiotic factors provide the template for the development of landscape pattern?
4. Why is it important to understand the history of a landscape? What types of effects of events from the past may remain in present-day landscape patterns?

RECOMMENDED READINGS

Bailey, R. G. 1996. *Ecosystem Geography.* Springer-Verlag, New York.

Delcourt, H. R., and P. A. Delcourt. 1988. Quaternary landscape ecology: relevant scales in space and time. *Landscape Ecology* 2:23–44.

Johnston, C. A., and R. J. Naiman. 1990. Aquatic patch creation in relation to beaver population trends. *Ecology* 71:1617–1621.

Knapp, A. K., J. M. Blair, J. M. Briggs, S. L. Collins, D. C. Hartnett, L. C. Johnson, and E. G. Towne. 1999. The keystone role of bison in North American tallgrass prairie. *BioScience* 49:39–50.

Swanson, F. J., T. K. Kratz, N. Caine, and R. G. Woodmansee. 1988. Landform effects on ecosystem patterns and processes. *BioScience* 38:92–98.

CHAPTER 5

Quantifying Landscape Pattern

The quantification of landscape pattern is an area of broad practical interest. Interest in measuring landscape pattern has been driven by the premise that ecological processes are linked to and can be predicted from some (often unknown) broad-scale spatial pattern (Baskent and Jordan, 1995; Gustafson, 1998). We begin this chapter by considering why pattern is quantified, briefly discuss sources of landscape data, and highlight some caveats and cautions that are important before analysis and interpreting of landscape structure. We then deal with two major categories of quantification: landscape metrics and spatial statistics.

Why Quantify Pattern?

Because landscape ecology emphasizes the interaction between spatial pattern and ecological process, methods by which spatial patterning can be described and quantified are necessary. There are numerous practical examples of where knowledge

FIGURE 5.1.
Changes in forest cover (shaded green) since the time of European settlement for Cadiz Township in southeastern Wisconsin. This pattern can be observed in many areas and illustrates both the changes in the abundance and spatial arrangement of forest in the landscape.

ADAPTED FROM CURTIS, 1956.

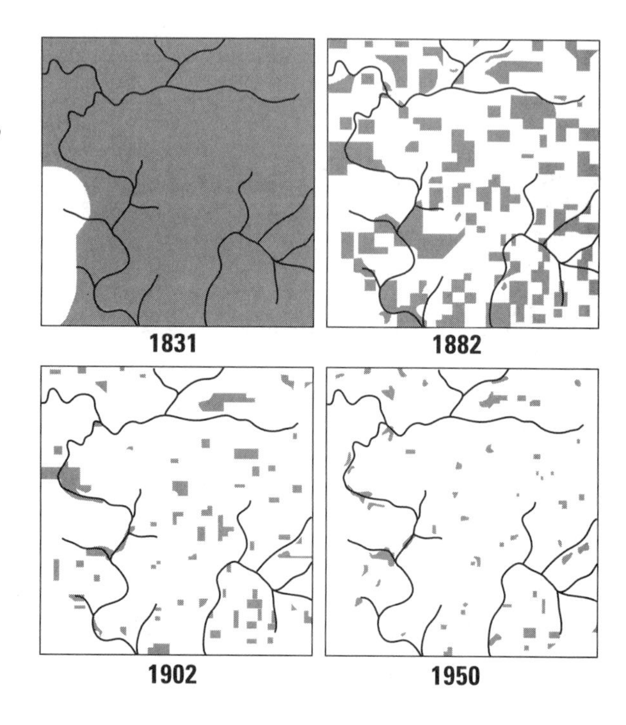

FIGURE 5.2.
Differences in landscape pattern are apparent along the western boundary of Yellowstone National Park in this false-color aerial photo. The National Park lands with relatively continuous forest cover (in red) can be seen to the right. To the left, areas with dispersed patches of clearcuts (white) on National Forest and private lands are evident. (Refer to the CD-ROM for a four-color reproduction of this figure.)

of the pattern is important. First, landscapes change through time, and we may be interested in knowing whether the pattern is different at time $t + 1$ than it was at time t. Furthermore, we may want to know specifically *how* landscape pattern has changed. Landscapes have undergone dramatic change during the past two centuries, as illustrated by the changes in forest cover in Cadiz Township, Wisconsin (Figure 5.1). Second, we may wish to compare two or more different landscapes or places within a given landscape and determine how different or similar they are. In some cases, a political boundary may result in dramatically different landscape configurations within close proximity, as seen along the western boundary of Yellowstone National Park, Wyoming (Figure 5.2). Third, when considering options for land management or development, we may need to evaluate quantitatively the different landscape patterns that result from the alternatives. Spatial analyses have been especially informative when comparing alternative forest harvest strategies (Figure 5.3). Finally, different aspects of spatial pattern in the landscape may be important for processes such as the movement patterns of organisms, the redistribution of nutrients, or the spread of a natural disturbance. Again, metrics are required to describe these patterns.

The quantification of pattern has received considerable attention in the past few years, but presently we can quantify more about pattern than we understand in terms of its ecological importance. The relationship between pattern and process remains a challenging and important area of research. Because pattern is fundamental to many of the relationships that we seek to understand, it is important to become familiar with the metrics that are used and, more importantly, to understand the factors that influence the interpretation of any landscape analysis.

DATA USED IN LANDSCAPE ANALYSES

Many analyses of landscape pattern are conducted on land use/land cover data that have been digitized and stored within a GIS. As summarized by Dunn et al. (1991), there are three main types of data. *Aerial photography* remains an important data source for landscape studies, particularly for detecting changes in a landscape during this century. Aerial photos are generally available back through the 1930s, although the quality may be uneven. *Digital remote sensing* is now widely used and is accessible to many researchers. The Landsat and SPOT satellites have provided frequent and spatially extensive coverage worldwide and are a very useful source of digital data. Airborne imaging scanners may also be used to provide fine-resolution data for a particular locale. *Published data and censuses* provide a valuable source of landscape data, particularly for temporal comparisons that extend back beyond the record of aerial photography. For example, the U.S. General Land Office Survey data have been used extensively to describe vegetation prior to European settlement (e.g., McIntosh, 1962; Lorimer, 1977; Whitney, 1986; White and Mladenoff, 1994; Delcourt and Delcourt, 1996; Silbernagel et al., 1997). In addition to these three sources, *field mapped data* may be used for landscapes of smaller extent in which the investigator might map the spatial

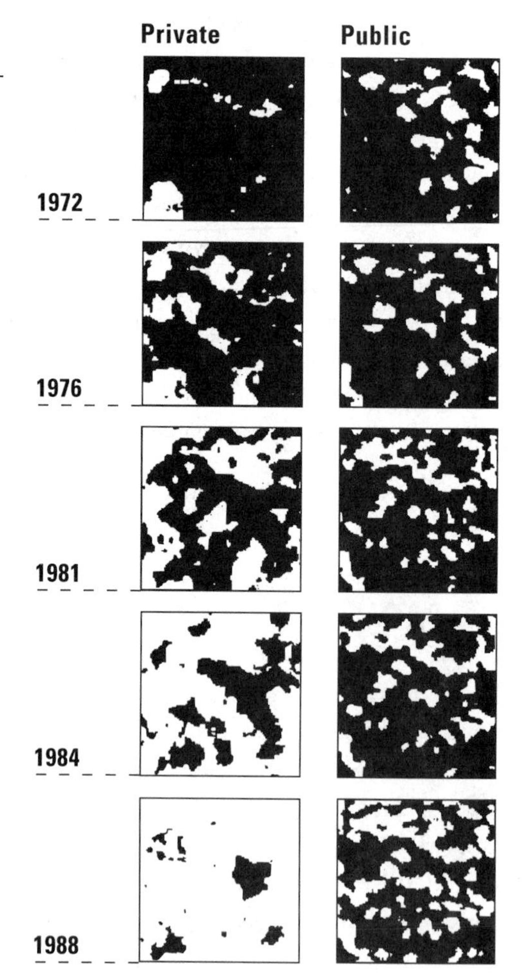

FIGURE 5.3.
Changes in conifer (green) and other forest types for a private and public landscape (2500 ha) with similar initial conditions and rates of change that are relatively high for the ownership types. Landscape metrics were used to quantify the differences in landscape pattern between ownerships.

REDRAWN FROM SPIES ET AL., 1994.

patterns of particular vegetation classes or landscape elements of interest in a relatively small area. Field mapping is not generally feasible for studies that cover a large area (e.g., hundreds to thousands of hectares).

Whatever the selected source, pattern analysis is generally conducted on a spatial data set in which the images or spectral data have been classified into some meaningful number of categories. In other words, a digital map, using map in a general sense, provides the baseline for the analysis. The land-use and land-cover scheme developed by the U.S. Geological Survey (Anderson et al., 1976) is an example of the types of categories that might be used for such a data set (Table 5.1). This scheme is a hierarchical arrangement of categories from general to specific. For example, forest (a Level I of the Anderson classification system) may be subdivided into deciduous, evergreen, or mixed forestland (Level II). Further divisions (Level II) would distinguish dominant species groups. Landscape data arranged by this classification are presumed to be homogeneous, an important assumption that must be recalled when interpreting landscape data. There are methods in remote sensing image analysis that do not require the user to determine landscape categories, but we will not cover these here. (Interested readers might consult a text such as Lillesand and Kiefer, 1994.) The remainder of this chapter will assume the availability of a spatial data set that is categorical rather than continuous; point data will be considered separately.

Most researchers store their landscape data in a GIS for ease of manipulation and display (Figure 5.4). However, ASCII text files may also be used. Most computer programs that have been written for landscape analyses were developed for use with raster, or grid cell, data, although vector-based versions are sometimes available. In raster format, a landscape is divided into a grid of square or hexagonal cells of equal size (Figure 5.4). The size of the grid cell determines the grain (resolution) of the mapped data. Irregularly shaped landscapes can be represented within a rectangular perimeter larger than the landscape itself. In vector format, lines are defined by ordered sets of coordinate pairs defining the boundaries of polygons (Figure 5.4). The polygons may be of variable size and shape, but the minimum mapping unit (grain size) corresponds to the minimum patch size that was mapped. Raster data are more commonly used in landscape analyses largely because the computer programming of the analyses is somewhat easier and most satellite imagery is in raster format.

It is very important to consider the accuracy of the spatial data or map on which the analysis of landscape pattern is to be performed. Often, an analyst may be using data provided from other sources, and the quality of these data should be known. Within GIS/remote sensing data, there are a number of recognized potential sources of error (Table 5.2). The old maxim of computer programming, "Garbage in, garbage out," also holds for landscape pattern analysis; the end prod-

Table 5.1.
USGS land-use, land-cover classification systems: an example of a hierarchical classification system that can be used in landscape analyses.

Level I	Level II
1. Urban or built-up land	11. Residential
	12. Commercial or services
	13. Industrial
	14. Transportation, communication, or utilities
	15. Industrial and commercial
	16. Mixed urban or built up
2. Agricultural land	21. Cropland and pasture
	22. Orchards, groves, vineyards, horticulture
	23. Confined feeding operations
	24. Other agricultural land
3. Rangelands	31. Herbaceous rangelands
	32. Shrub and brush rangelands
	33. Mixed rangelands
4. Forest land	41. Deciduous forest land
	42. Evergreen forest land
	43. Mixed forest land
5. Water	51. Streams and canals
	52. Lakes
	53. Reservoirs
	54. Bays and estuaries
6. Wetland	61. Forested wetlands
	62. Nonforested wetlands
7. Barren land	71. Dry salt flats
	72. Beaches
	73. Sandy areas, except beaches
	74. Bare exposed rock
	75. Strip mines, quarries, gravel pits
	76. Transitional areas
	77. Mixed barren land
8. Tundra	
9. Perennial snow or ice	

From Anderson et al., 1976

uct is only as good as the data on which the analysis is based. Understanding the sensitivity of landscape metrics to error in the input data is an active topic of current research (e.g., Cardille et al., 1996; Wickham et al., 1997).

CAVEATS FOR LANDSCAPE PATTERN ANALYSIS, OR "READ THIS FIRST"

The widespread availability of spatial data in the past two decades—indeed, an almost exponential increase—has presented numerous opportunities for landscape patterns to be analyzed for many different purposes. Before embarking on the analysis of landscape pattern, however, several important issues should be determined. At present, it is easy to fall into the trap of generating a lot of numbers

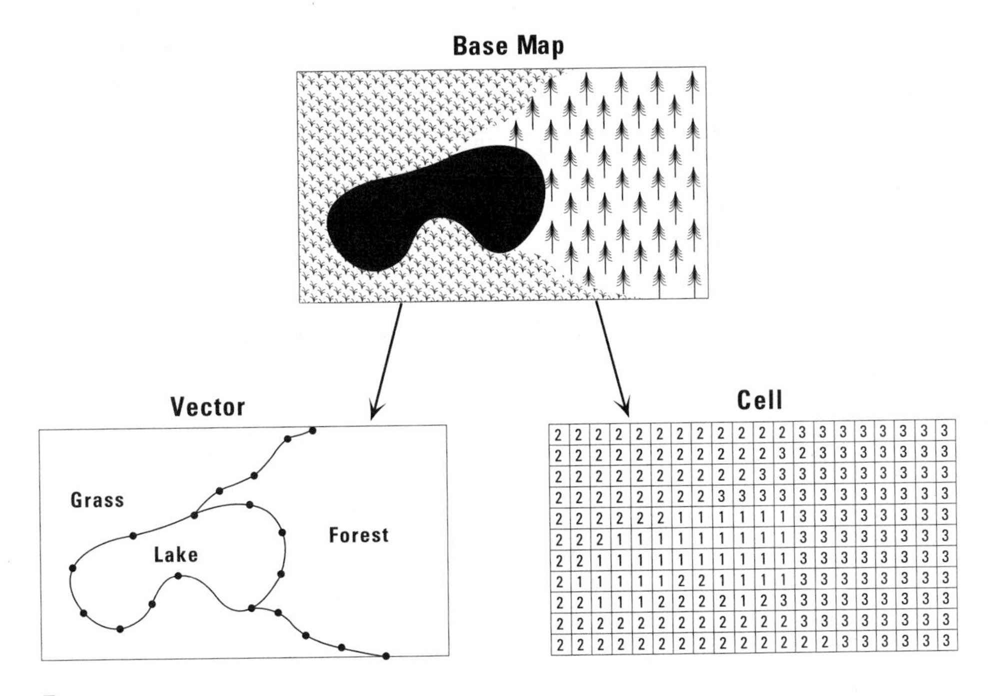

FIGURE 5.4.
Two principal methods for representing spatial data in a GIS: vector-based representation and raster, or cell-based, representation.

ADAPTED FROM COULSON ET AL., 1991.

TABLE 5.2.
SOURCES OF POTENTIAL ERROR IN GIS DATA.

Source	Explanation
Obvious sources	
Age of data	Some data change more rapidly than others (e.g., geological substrate versus land use). Old data may have been collected under different standards.
Aerial coverage	Coverage may be incomplete over a region of interest.
Map scale	Does the map scale match the resolution at which the data were originally collected? Is it appropriate for the question?
Political boundaries	Data characteristics may change across political (e.g., county, state) or administrative (e.g., agency, landowners) boundaries on maps formed as a composite.
Natural variation in original measurements	
Positional accuracy	Boundary lines or distinctions may not be precisely located due to field mapping or conversion between data formats (e.g., vector to raster) or spatial resolution.
Content accuracy	Are the cell attributes correct? In remote sensing interpretation, there are measurable errors associated with classification of the reflectance values.
Variation in data sources	Different interpreters may generate different maps; protocols may not be standardized; errors in data entry; natural variability.
Processing errors	
Numerical computation	Decimal precision and rounding errors in complex calculations.
Topological analyses	When combining map variables or coverages, use of logical operators multiplies errors in individual layers; conversion between vector and raster also may lead to error.
Classification	Errors in deriving the category assignments from aerial photos or satellite imagery; error typically estimated for entire map and for each category on the map.

ADAPTED FROM BURROUGH, 1986.

without having a clear idea of the purpose and limitations of the variety of available metrics. Without a clear a priori statement of the objectives of the analysis and/or hypothesized pattern changes (e.g., disturbances will cause a decline in the diversity of land-cover types), the comparisons among multiple metrics may be misleading. The problem of multiple comparisons is further exacerbated by the issue of pseudoreplication (Hurlbert, 1984; Hargrove and Pickering, 1992), which occurs when comparisons are made among samples that are not truly independent. The dangers of pseudoreplication are relevant to landscape ecology because the unique attributes of each landscape make statistical controls difficult and independent replicate samples nearly impossible.

The Classification Scheme Is Critical

The categories selected for use in the analysis have an extremely strong influence on the numerical results of any pattern analysis. For example, consider the pattern of vegetation mapped from remote imagery for a section of a landscape, but classified in different ways (Figure 5.5). It is clear that these two landscape representations look very different and that the quantitative descriptions of these two data sets would be quite dissimilar. Thus, *the choice of the categories to include in a pattern analysis is critical.* The classes must be selected for the particular question or objective and the classification rules rigorously applied across all landscapes being compared. For example, general categories (e.g., Level I in Anderson et al., 1976) would be appropriate to study landscape patterns in the eastern United States, but to study vegetation patterns within a forested landscape such as the Great Smoky Mountains National Park, descriptions of a variety of forest community types may be more desirable. Once classified, analyses of categorical maps generally assume homogeneity within each category. All categories may, in some sense, be considered an aggregate of a more detailed set of subcategories. As categories become more aggregated the relationship between pattern and process will become difficult to establish.

Even after the decision is made about the categories to be included in a data set, the actual interpretation of these categories from some primary data source may need further specification. Gustafson (1998) poses the example of studying the distribution of aspen in the Wisconsin Northwoods and the need to specify how much aspen must be present in order to be classified as an aspen stand. The threshold density for classification should be specified, and the use of different density levels among similar maps should be avoided. It is clear that each decision in the mapping process will affect the determination and analysis of spatial structure (Gustafson, 1998).

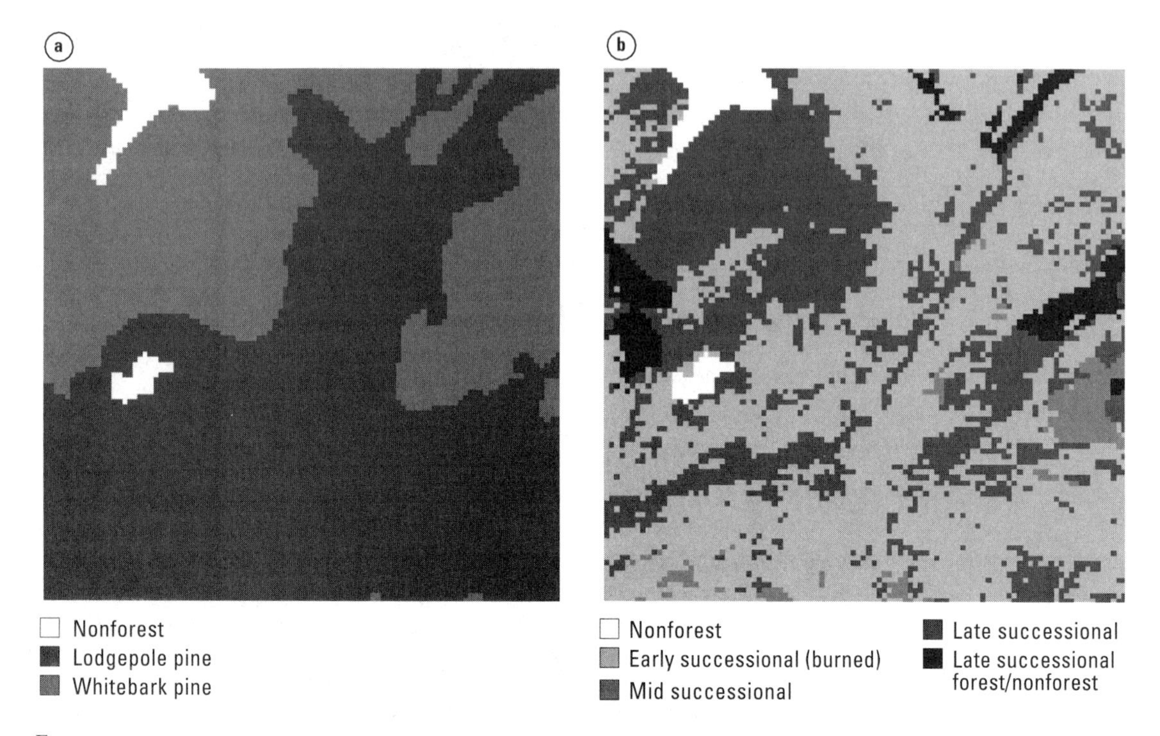

FIGURE 5.5.
Example of how the same landscape looks very different under different classification schemes. Both panels show a 5- by 5-km section (100 by 100 grid cells) of southwestern Yellowstone National Park. (a) The landscape is classified based on the forest community composition. (b) The landscape has been classified based on the successional stage of the forest stands. (Refer to the CD-ROM for a four-color reproduction of this figure.)

Scale Must Be Defined

The grain and extent of the data used in any analysis of landscape pattern will influence the numerical result obtained for a given metric (Turner et al., 1989b; Benson and MacKenzie, 1995; Moody and Woodcock, 1995; Wickham and Riitters, 1995; O'Neill et al., 1996). This sensitivity means that comparisons of landscape data represented at different scales may be invalid because results reflect scale-related errors, rather than differences in landscape patterns. As grain size increases (resolution decreases), cover types that are rare on the landscape typically become less well represented or may even disappear (see Box 5.1). The boundaries between different cover types also become underestimated with increasing grain size as the shapes depicted become less complex and details are lost (Figure 5.6). By changing grain in a Landsat image from 30- by 30-m to 1- by 1-km, Moody and Woodcock (1995) showed that scale-induced changes in the proportion of the landscape occupied by different cover types influenced landscape

BOX 5.1
ANALYTICAL EQUATIONS FOR AGGREGATION ERRORS

Significant errors in spatial data can occur as the result of aggregation of fine-grained information into coarser scales of resolution. The loss of information with successive levels of aggregation is a well-studied problem (e.g., Costanza and Maxwell, 1994; Henderson-Sellers et al., 1985; Meentemeyer and Box, 1987; Pierce and Running, 1995; Turner et al., 1989b; Mladenoff et al., 1997), and analytical equations have been developed to quantify aggregation effects (Turner et al., 1989b; Peitgen et al., 1992). The rule used to aggregate always results in error, but the nature of this error and the degree to which it affects landscape pattern can be quantified. Suppose that the resolution of a square grid composed of randomly placed cells of two types, *a* and *b*, will be changed by aggregation of 2 by 2 subsets of the grid into a single cell, thus reducing the number of cells in the aggregated grid to one-fourth of the original. If the proportion of cells of type *a* and *b* is equal to p and $1.0 - p$, respectively, then the proportion of cells of type *a* in the aggregated map, p, using the majority rule (at least three cells must be of type *a*) will be

$$p = p^4 + 4p^3(1.0 - p)$$

If a 50% rule for aggregation is invoked (at least two cells must be of type *a*) then

$$p = p^4 + 4p^3\,(1.0 - p) + 6p^2(1.0 - p)^2$$

Both rules result in substantial aggregation errors, shifting the proportion of cells of type *a* by as much as 22%. However, the values of p that produce the greatest error differ between rules; in both cases the error is zero when p is equal to 0.0 and 1.0 (an uninteresting result), but is also zero at $p = 0.76$ for the majority rule and at $p = 0.24$ for the 50% rule. A third rule, which produces zero error when $p = 0.50$, can be formulated as an average of the majority rule and the 50% rule. This averaging rule is equivalent to assigning the aggregated cell to type *a* or *b* at random when there are exactly two of each type in the 2 by 2 grid unit and using the majority rule otherwise. Consideration of the combinatorics for the random aggregation rule gives

$$p = p^4 + 3(p^3(1.0 - p)) + 3(p^2(1.0 - p) + p(1.0 - p)^3)$$

This rule is globally the most accurate rule because it preserves the value of p for all land-cover types, even over successive aggregations. Local errors (the accuracy in the representation of the amount and physical appearance of dominant land-cover types at fine scales) will increase. The effect of the random aggregation rule on local errors has not been thoroughly investigated. For additional details regarding these methods, see Milne and Johnson (1993) and Gardner (1999).

metrics, including patch size, patch density, and landscape diversity. Some metrics are more sensitive to changes in scale than others. In the Tennessee River and the Chesapeake Bay watersheds, Cain et al. (1997) found measures of landscape diversity, texture, and fractal dimension were more consistent across analyses in which grain size and the number of cover types varied, whereas measures of average patch shape or compaction were subject to change.

The spatial extent of the study area can affect landscape metrics independently of grain size. The smaller the extent of the map is the more serious the problem of artificial truncation of patches by the map boundary, resulting in biased measurement of patch size, shape, and complexity (Figure 5.7). However, it may not always be clear what the minimum map extent should be to prevent serious measurement errors. For example, imagine a square landscape of 400 by 400 cells (that is, the universe of interest) that is created by the random placement of a single cover type among the grid cells. Further imagine that the goal of the analysis is to estimate the

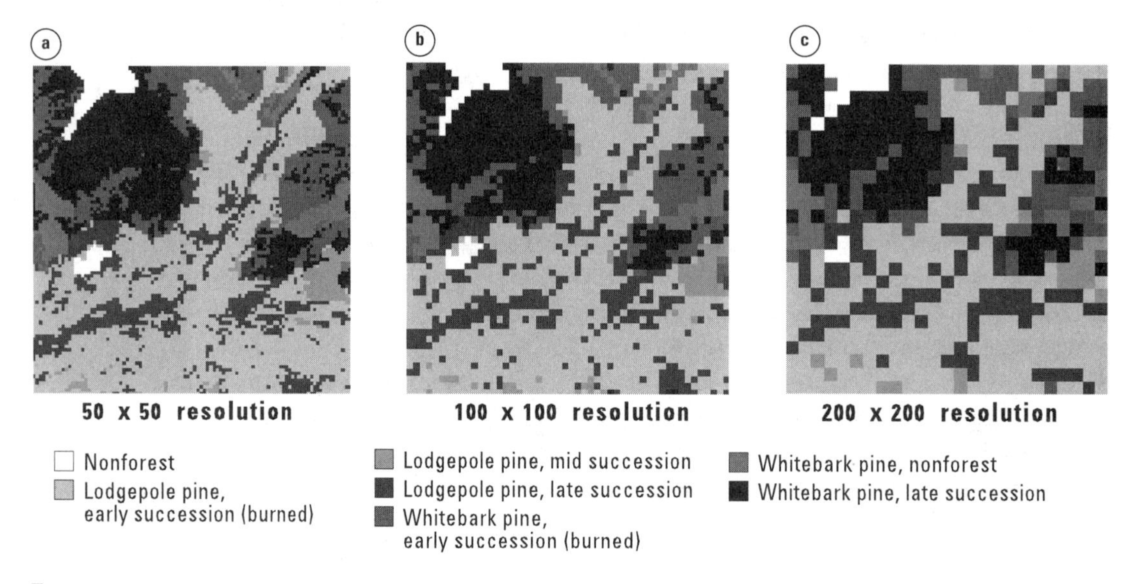

FIGURE 5.6.
Effects of changing grain size on a landscape map. Panels show a 5- by 5-km section (initially 100 by 100 grid cells) of southwestern Yellowstone National Park aggregated sequentially following a majority assignment rule. (a) The original landscape is shown with 50- by 50-m grid cells. (b) Grid cell size is 100 by 100 m. (c) Grid cell size is 200 by 200 m. (Refer to the CD-ROM for a four-color reproduction of this figure.)

number of patches in this universe by sampling with maps of either 50 by 50, 100 by 100, or 200 by 200 cells. The truncation effect of the smaller maps results in consistent overestimation of the total number of habitat patches. However, this bias is small (a maximum of 1.3% for the 50 by 50 map) for all map sizes when p, the fraction of the map occupied by the single habitat type, is also small. However, when p is greater than 0.6 (e.g., the 400 by 400 map has 60% of the cells randomly occupied by the single habitat type), sampling bias climbs to 23%, 9%, and 3% for the 50 by 50, 100 by 100, and 200 by 200 maps, respectively. The lessons from this example are that (1) it is difficult to specify *for all* situations what the biases will be for maps of different extent; (2) no single map extent is optimal for all analyses; (3) unless experiments can be performed to evaluate optimal sample sizes, we should sample with maps of the greatest possible extent. When landscapes contain multiple cover types, increases in map extent will usually increase the representation of rarer cover types, similar to the increases in species with increases in area sampled.

Is there a rule of thumb for selecting the appropriate grain and extent for an analysis? Using remotely sensed data for the southeastern United States and ex-

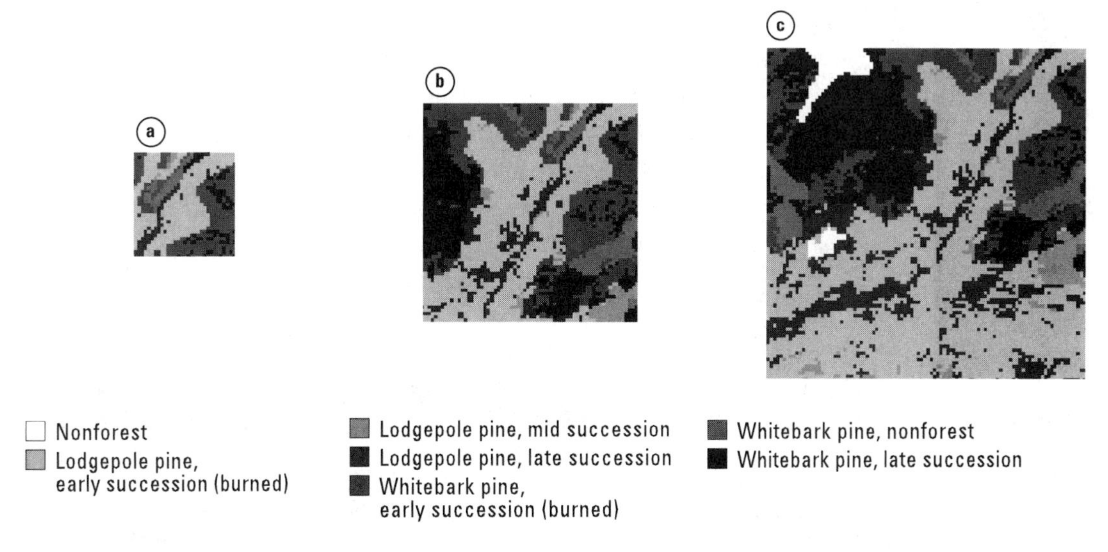

FIGURE 5.7.
Effects of changing extent on a landscape map of southwestern Yellowstone National Park. Note that the presence and relative proportions of the different land-cover types change as the extent of the map varies. (Refer to the CD-ROM for a four-color reproduction of this figure.)

amining the effects of changing the grain and extent of the maps on landscape metrics, O'Neill et al. (1996) proposed the following to avoid bias in calculating landscape metrics: The grain size of the map should be two to five times smaller than the spatial features being analyzed, and map extent should be two to five times larger than the largest patches.

Identifying a Patch

The concept of a patch is intuitive; we all seem to understand what constitutes a patch of grassland or forest. However, a clear definition is required before systematic (and repeatable) analysis of landscape pattern may begin. Forman and Godron (1986) defined a patch as "a nonlinear surface area differing in appearance from its surroundings." Converting this definition into a computer algorithm to identify patches on a gridded landscape, we have "a contiguous group of cells of the same mapped category." But what does contiguous (or touching) mean? Different rules have been established to define this simple concept. The most common method assumes that the four nearest neighboring cells all touch the cell of interest. The four-neighbor rule means that a string of cells of the same category arranged along diagonal lines do not touch and will be analyzed as a string of individual patches. The four-neighbor rule may be altered to consider next-nearest neighbors as touching sites (an eight-cell neighborhood); then diagonal neighbors sites are members of the same patch. Because different rules produce different mapped patterns (Figure 5.8), analysis programs have been specifically written to allow variable neighborhoods (Gardner, 1999).

The definition of a patch is also strongly affected by the grain of the data and by the classification scheme. Indeed, because classification of cover types or habi-

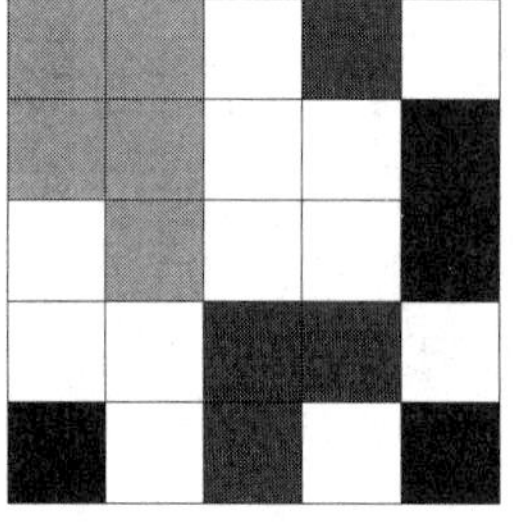

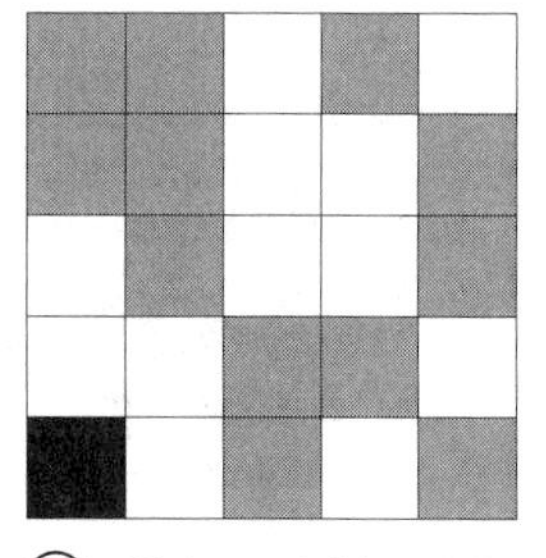

FIGURE 5.8.
Identification of patches (shaded) on the same map using either (a) a four-neighbor rule, in which the horizontal and vertical neighbors are considered, but the diagonal neighbors are not, and (b) an eight-neighbor rule, in which the horizontal, vertical, and diagonal neighbors are all considered.

tats can be dramatically different among species or for different ecological processes, a single map may need to be reclassified to perform patch-based analyses for various species or processes. Any subsequent quantitative analyses based on patches (e.g., average patch size, distribution of patches by size, perimeter-to-area relationships) are a function of all the factors inherent in defining patches, and this definition process should be well thought out. Patches are not fixed elements of the landscape, but rather are useful spatial constructs that vary with the objectives of a given study.

Most Metrics Are Correlated

As with the classification scheme and spatial scale of the data, the metrics chosen for analysis must also be selected based on the objectives of the study. In addition, many metrics are strongly correlated with one another, containing redundant information. Riitters et al. (1995) examined the correlations among 55 different landscape metrics by factor analysis and identified only five independent factors. Thus, many typical landscape metrics are *not* measuring different qualities of spatial pattern. In the Riitters et al. (1995) analysis, the five independent factors represented (1) the number of classes or cover types on the map, (2) whether the texture of the landscape pattern was fine or coarse, (3) the degree to which patches were compact or dissected, (4) whether patches were linear or planar, and (5) whether patch perimeters were complicated or simple in shape. Thus, the analyst should select metrics that are relatively independent of one another, with each metric (or grouping of metrics) able to detect ecologically meaningful landscape properties. It is often necessary to have more than one metric to characterize a landscape because there simply is not one number that "says it all." However, reporting ten highly correlated metrics does not yield new information, but only makes interpretations more difficult. Just because something *can* be computed does not mean that it *should* be computed!

What Constitutes a Significant Change?

One challenge still facing landscape ecologists is that the statistical properties and behavior of some landscape metrics are not well known. In cases where a single number is reported for a landscape, we may have little understanding of the degree to which landscape pattern must change to be able to detect an ecologically important or statistically significant change in the numerical value of the metric (Wickham et al., 1997). While this remains an important area of current research and development, the user should definitely consider the criteria that will be applied to determine whether an observed change is or is not meaningful. Presently, there is a need to build a collective library of empirical studies in which ecological responses are related to particular landscape configurations. Unfortunately, we have the power to measure and report more about landscape pattern than we can interpret in terms of effects on ecological processes. Therefore, serious attention must be paid to the rigor with which the metrics are applied and interpreted. Neutral landscape models (see Chapter 6) may be of particular use in elucidating the behavior of individual metrics and providing some measure of their statistical reliability.

METRICS FOR QUANTIFYING LANDSCAPE PATTERN

Caveats thus stated, numerous metrics can be computed for a landscape data set, too many for us to review in this text. The recommended readings listed at the end of this chapter provide excellent and very comprehensive treatments of the calculation and interpretation of nearly 100 metrics in total. Box 5.2 also contains practical information on some readily available software packages that can be used to analyze landscape pattern. In this section, we review some of the commonly used metrics within three broad categories: metrics of landscape composition; measures of spatial configuration, including contagion and patch-based metrics; and fractals. We then present examples of studies in which landscape metrics were used, and we address the challenge of the interpretation of multiple metrics.

Metrics of Landscape Composition

Metrics that quantify the composition of the landscape are not usually spatially explicit. That is, they measure what is present and in their relative amounts, or proportions, without reference to where on the landscape they may be located. However, nonspatial metrics may be important descriptors measuring landscape constraints of values of spatially explicit metrics. For example, the proportion of the landscape occupied by a given cover type limits the range of patch number and sizes (Gardner et al., 1987; Gustafson and Parker, 1992).

Fraction or Proportion (p_i) Occupied

A simple but useful number to calculate is the proportion, p_i, of the landscape that is occupied by each cover type i, where $i = 1$ to s, and s is the total number of cover types on the map. The p_i values are estimated by counting the number

BOX 5.2
THE PRACTICAL SIDE: SOFTWARE FOR SPATIAL PATTERN ANALYSIS

Most GIS systems have only a limited capability to perform the types of landscape pattern analyses described in the text, although this is changing rapidly. However, a variety of researchers has developed computer programs for spatial analysis that can be readily obtained. A widely used package is FRAGSTATS (McGarigal and Marks, 1995), developed at Oregon State University and available by anonymous ftp or on the CD provided with Gergel and Turner (2001). FRAGSTATS is comprehensive and available for both raster and vector maps, and McGarigal and Marks provided excellent documentation. Users interested in running landscape analyses might begin with FRAGSTATS. Another analysis program designed for both raster and ASCII input data is APACK, developed by Mladenoff and colleagues at the University of Wisconsin–Madison; interested users should visit the following web site for further information: http://flel.forest.wisc.edu/projects/apack/. For users of the GIS GRASS, Baker and Cai (1992) developed the **r.le** programs that interface with the GRASS system and are also available by anonymous ftp. A program called RULE (Gardner, 1999) can be used to explore the implications of different patch-definition rules on landscape maps, as well as to generate neutral landscape models (see Chapter 6); RULE is also available on the CD provided by Gergel and Turner (2001) or by ftp.

---

of grid cells of each cover type and then dividing by the total number of grid cells present on the entire landscape. The expected value of the p_i's is equal to $1/s$, making this statistic marginally useful as a landscape metric, although deviations of the p_i's from the expected value may be a useful indicator of landscape heterogeneity. The p_i's have a strong influence on other aspects of pattern, such as patch size or length of edge in the landscape (Gardner et al., 1987); therefore, these relationships will be further explored in the section on neutral models. The p_i's have been used to calculate a variety of landscape metrics, resulting in correlations among the metrics sharing this information.

Relative Richness

This simple metric calculates the number of cover types present, without regard to how they are spatially arranged, as a percent of the total number of possible cover types. Thus,

$$R = \frac{s}{s_{\max}} \times 100$$

where s = the number of cover types present, s_{max} is the maximum number of cover types possible, and the multiplication by 100 converts the fraction to a percent. The value of s_{max} is arbitrary (see prior discussion of classification scheme), but should be based on the maximum number of cover types observed for similar landscapes. Relative richness is especially useful for comparisons of landscapes before and after natural disturbances and the regeneration of vegetation cover types through time.

Diversity and Dominance

Metrics derived from information theory were first applied to landscape analyses by Romme (1982) to describe changes in the area occupied by forests of varying successional stage through time in a watershed in Yellowstone National Park, Wyoming. Two related indexes are dominance and diversity (O'Neill et al., 1988a), which provide the same information about the landscape. Diversity, or relative evenness, refers to how evenly the proportions of cover types are distributed. For example, if three cover types are present, does each occupy 33% of the landscape, or does one occupy 90% and the others only 5% each? The equation is given by

$$H = \frac{-\sum_{i=1}^{s} (p_i) \ln (p_i)}{\ln (s)}$$

where H = diversity, p_i = the proportion of the landscape occupied by cover type i, and s = the number of cover types present. Dividing through by $\ln (s)$ normalizes the index to range between zero and one. A high value of H indicates greater evenness, and a low value indicates less evenness.

Closely related to diversity is dominance, which is simply the deviation from the maximum possible diversity:

$$D = \frac{H_{max} + \sum_{i=1}^{s} p_i \ln (p_i)}{H_{max}}$$

where D = dominance, p_i = the proportion of the landscape occupied by cover type i, and $H_{max} = \ln (s)$, which is the maximum possible diversity for a landscape having s cover types. Again, this index ranges between zero and one, with a high value indicating dominance by one or a few cover types and a low value indicating that the cover types are present in similar proportions. Note that the summation in the numerator actually represents a deviation from H_{max} because the logarithms of values <1.0 are negative.

Two important points should be noted here. First, dominance and diversity are redundant, so any analysis need not report both. Second, and reflecting a more general issue for metrics of this type, dominance and diversity may return very similar numerical values for landscapes that are qualitatively different. For example, a landscape occupied by 80% agriculture, 10% forest, and 10% wetland would have the same diversity value as one occupied by 80% forest, 10% agriculture, and 10% wetland. Thus, the usefulness of these metrics is rather limited.

Connectivity

If a pattern across a landscape can be represented as a series of nodes and linkages, the gamma index described in Forman and Godron (1986) is a useful overall measure:

$$\gamma = \frac{L}{L_{max}} = \frac{L}{3(V-2)}$$

where L = the number of links in the network and V = the number of nodes in the network. This index can range between zero to one, with low values indicating less connectivity and high values indicated higher connectivity.

Measures of Spatial Configuration I: Contagion

Probabilities of Adjacency

The metrics described so far do not account for the spatial arrangement of habitat types. Probabilities of adjacency, that is, the probability that a grid cell of cover type i is adjacent to a grid cover type j, are sensitive to the fine-scale spatial distribution of cover types. These probabilities can be computed simply as

$$q_{i,j} = \frac{n_{i,j}}{n_i}$$

where n_i = the number of grid cells of cover type i and $n_{i,j}$ = the number of instances when cover type i is adjacent to cover type j. Note that this initial calculation assumes a single one-directional pass through the data matrix, e.g., horizontal or vertical. These probabilities can be computed directionally to detect directionality in the pattern (anisotropy), and average values can also be determined. If you calculate the probabilities simultaneously in four directions, the denominator must be modified to reflect the correct potential number of neighbors.

The $q_{i,i}$ values, which are the diagonals of the Q matrix, give a useful measure of the degree of clumping found in each cover type. The $q_{i,i}$'s give the likelihood that cells of the same cover type are found adjacent to each other. High $q_{i,i}$ values indicate a highly aggregated cover type, and low $q_{i,i}$ values indicate that the cover type tends to occur in isolated grid cells or small patches. Thus, this metric can be used to characterize relatively fine scale detail of the spatial pattern and is useful in providing data on each cover type.

Contagion

The contagion metric (O'Neill et al., 1988a; Li and Reynolds, 1993) distinguishes between overall landscape patterns that are clumped or rather dissected. The equation is given by

$$C = \frac{1 + \sum_{i=1}^{s} \sum_{j=1}^{s} (P_{ij}) \ln (P_{ij})}{2 \ln (s)}$$

where $P_{i,j}$ = the probability that two randomly chosen adjacent pixels belong to cover types i and j, respectively, that is, $P_{ij} = P_i P_{j/i}$; and s is the number of cover types on the landscape. Note that the equation for contagion can be formulated differently if the probabilities of adjacency are computed by another algorithm (see Li and Reynolds, 1993, for details). The metric ranges from zero to one, with a high value indicating generally clumped patterns of cover across the landscape and a low value indicating a landscape with a dispersed pattern of cover types. Contagion is useful in capturing relatively fine scale differences in pattern that relate to the texture or graininess of the map.

Measures of Spatial Configuration II: Patch-Based Metrics

Given a landscape data set in which the grid cells are assigned to discrete categories, patches, which are contiguous areas of the same cover type, can be identified and their distributions described. Again, the importance of the classification scheme used cannot be overstated. Patches are commonly identified in the data set by either of two rules. First, a patch may consist of adjacent cells in only the north–south and east–west directions on the grid, that is, the four nearest neighbors, excluding the diagonals. Alternatively, a patch may consist of adjacent cells from among the eight nearest neighbors, including both the adjacent and diagonal neighbors. Choice of this rule will influence the results of any patch-based measurements (see Figure 5.7). Patch-based measures of pattern include patch size, perimeter, number, shape, and density.

Patch Area and Perimeter

Once patches are located, some simple summaries can be computed easily. The area and perimeter of each patch can be determined. These results can then be summarized for the landscape or, more meaningfully for each cover type, by displaying a frequency distribution of numbers of patches by patch size, a cumulative frequency distribution of patch sizes, the simple mean and standard deviation of patch size, or the area-weighted mean patch size. The simple mean patch size is sensitive to the number of small or single-celled patches in the landscape; use of an area-weighted mean patch size avoids this problem.

Perimeter to area ratios (P/A) also serve as useful indexes of shape complexity. For a given area, a high P/A ratio indicates a complex or elongated boundary shape, and a low P/A ratio indicates a more compact and simple shape. However, P/A is sensitive to patch size, decreasing as patch size increases for a given shape. A variety of shape indexes are based on perimeter and area measurements, some of which correct for the size problem (e.g., see Baker and Cai, 1992). If computed for each patch, means, standard deviations, and frequency distributions can again summarize these ratios.

Connectivity

Once patches have been demarcated in the data set, some measures of connectivity for individual cover types can be computed. One very simple metric is simply the relative size of the largest patch of habitat k:

$$RS_i = \frac{LC_i}{p_i \times m \times n}$$

where LC_i = the size of the largest patch of cover type, p_i = the proportion of the landscape in cover type i, and $m \times n$ gives the size of the landscape that contains m rows and n columns. If all of cover type k occurs as a single patch, the value of the index equals 1.0, indicating complete connectivity. When the cover type is dispersed into very small patches, the index approaches zero.

Another measure of connectivity (or its inverse, fragmentation) of a habitat type is to calculate the average distance between patches. This is often done in one of two ways. First, the distance can be computed from the center of one patch to the center of the next nearest patch. This method requires an algorithm for determining patch centroids. Second, the distance can be computed from the grid cells on each of two patches that are closest to one another, thereby providing a minimum interpatch distance.

Proximity Index

The degree to which patches in the landscape are isolated from other patches of the same cover type may be of importance, especially when species habitat-use patterns are of interest. Gustafson and Parker (1992) developed an index that can be computed for each patch on a landscape to determine the relative isolation of the patches. This index is given by

$$PX_i = \sum \frac{s_k}{n_k}$$

where PX_i is the proximity index for focal patch i, and then within a specified search distance (which must be set by the user), s_k is the area of patch k within the search buffer and n_k is the nearest-neighbor distance between a grid cell of the focal patch and the nearest grid cell of patch k. This index is not normalized but returns an absolute number. Low values indicate patches that are relatively isolated from other patches within the specified buffer distance, and high values indicate patches that are relatively connected to other patches.

Area-Weighted Average Patch Size

The frequency distribution of patch sizes on many landscapes is frequently skewed; a few large patches will be found surrounded by many smaller patches. Under these conditions, the simple arithmetic average does not reflect the expected patch size that would be encountered by a simple random placement of points on the map. A more useful method of averaging is to weight patch sizes by area (Stauffer and Aharony, 1992). If there are n patches on the landscape and S_k is the size of the kth patch, then the area-weighted average patch size is

$$S_a = \frac{\sum (S_k{}^2)}{\sum (S_k)}$$

Area weighting of other indexes has been employed when skewed frequency distributions result in a disproportionate effect of small patches on the metric of interest (McGarigal and Marks, 1995).

Fractals

Fractals are considered separately here because, although their calculation is generally based on the spatial distribution of cover types or patches in the landscape, they may be reported in various ways: overall for the whole map or landscape, overall for each cover type, or for individual patches. In this section, we explain fractals briefly, focusing here on their use as a descriptor of spatial pattern. We begin by explaining very basically the development of fractals and some important concepts in their meaning. However, the use and application of fractals is very rich, and we refer interested readers especially to Milne (1991b) and Sugihara and May (1990) for more detailed treatments.

Many years ago, the scientist Lewis Fry Richardson studied the relation between the measured length of a coastline or lake perimeter and the scale at which it was mapped. Fry found that the length of the coastline as estimated by the map increased logarithmically with increasing map resolution. Why did this occur? As the resolution of the map was increased, more and more previously unresolved features could be delineated; in the limit, the length of the coastline is infinite (see Chapter 2). For shapes like coastlines, the curves are never actually as smooth as the drawn lines would have us believe. These findings were incorporated into the theory of fractals proposed by the mathematician Mandelbrot (1983).

The essence of fractals is the recognition that, for many phenomena, the amount of resolvable detail is a function of scale. An important corollary is that increasing the resolution does *not* result in an absolute increase in precision, but rather it reveals variation that passed unnoticed before. Consider, for example, two ideal fractal curves (Figure 5.9). If we measure the distance from A to B and measure by units of length x, we observe that in Figure 5.9a the distance = 4. If we decrease the resolution by a factor of 3 so that units are $x/3 = y$, more detail is seen; in Figure 5.9b, the distance from A to C is now 4 units of y, but the total distance between A and B will be longer in units of y than in units of x. Because the curves in Figure 5.9 behave similarly at all scales, once the properties are known at one scale they can be deduced from another merely by applying a scaling parameter. The level of variation present at all scales can be described by a single parameter, the fractal dimension, defined by Mandelbrot (1983) as

$$D = \frac{\log N}{\log r}$$

where N = number of steps used to measure a pattern unit length, and r is the scale ratio. In Figure 5.9a and b, $N = 4$ and $r = 3$, so $D = 1.2618$.

Figure 5.9.

Ideal fractal curves: (a) and (b) $D = \ln 4/\ln 3 = 1.2618$; (c) and (d) $D = \ln 5/\ln 3 = 1.4650$.

Redrawn from Burroughs, 1986.

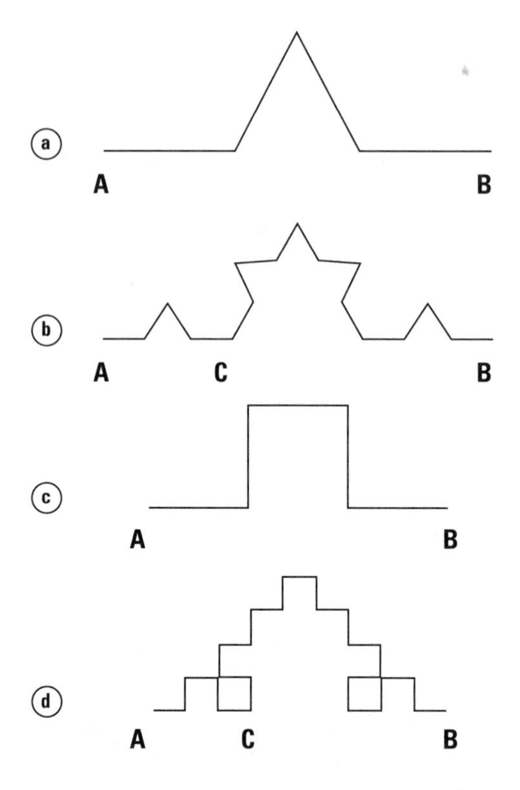

Figure 5.10.

Nested maps of soil patterns in northwest Europe at scales ranging from 100 km to 100 m. Note how natural variation may look similar across scales.

Redrawn from Burroughs, 1986.

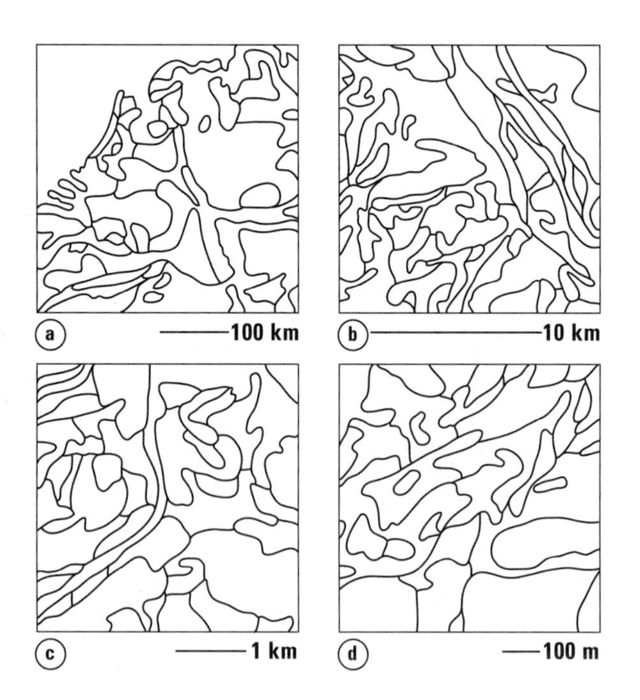

Mandelbrot (1985) defined a fractal as "a shape made of parts similar to the whole in some way." A fractal will look the same whatever the level of resolution used to observe the object. Fractals have two important characteristics: (1) they embody the idea of self-similarity, the manner in which variations at one scale are repeated at another; and (2) their dimension is not an integer, but rather a fraction, hence the *fractional dimension*, from which these objects acquired the name. Euclidean dimensions are familiar (e.g., a line has a dimension of 1; an area has a dimension of 2; and a volume, such as a sphere or cube, has a dimension of 3). The value of D for a linear fractal curve can vary between $D = 1$ and $D = 2$. When $D = 1$, it implies that the curve is in fact Euclidean, or a line. As D gets greater than 1, it implies that the line has an associated band of fuzziness or uncertainty that eats up a little of the second spatial dimension. When $D = 2$, the line has in fact become an area. The concept is easily extended from surfaces to volumes when the value of D ranges between 2 (a completely smooth two-dimensional surface) and 3 (infinitely crumpled three-dimensional object).

The idea of self-similarity embodied in the fractal concept implies that if geographical objects such as mountains or rivers are truly fractals their variations should be scalable. That is, we should be able to predict the patterns at different scales from knowing the pattern at one scale and the fractal dimension. The variation seen in landforms over a few meters, for example, should be statistically similar to that seen over hundreds or thousands of meters when transformed by a simple scaling parameter. The mapped patterns of soils at multiple spatial scales (Burrough, 1986) illustrates the way in which this concept might be applied (Figure 5.10). Applications of this idea are still in their infancy.

Fractal dimensions have been used as a metric of the complexity of landscape patterns in comparing different landscapes, changes through time in particular landscapes, and patches of different size. For example, Krummel et al. (1987) found that forest patches showed a distinct change in fractal dimension, with smaller patches having a simpler shape than larger patches (Figure 5.11). The reason appears to be that small patches were woodlots whose boundary was affected by human management; the large patches were more complex because they tended to follow natural boundaries, such as topography. A number of other studies have also found lower fractal dimensions in human-dominated landscapes or cover types (e.g., Turner, 1990; Mladenoff, et al., 1993). More recent studies have investigated the effects of fractal geometry on the coexistence and persistence of species (Johnson et al., 1992a; Palmer, 1992; Wiens et al., 1993; With and Crist, 1995; With et al., 1997).

The Fractal as a Measure of Patch Shape

The fractal dimension of a patch is easy to calculate and has proved to be a useful indicator of shape complexity (Krummel et al., 1987; Sugihara and May, 1990). However, unless care is taken in the estimation of this metric, serious errors can result. Avoiding these errors requires (1) a clear understanding of the algebraic relationships among area, perimeter, and the fractal dimension, (2) elimination of biases in the fractal dimension resulting from inclusion of small patches in the analysis, and (3) specification of the level of change of the fractal dimension that corresponds to a significant change in patch complexity.

Mandelbrot (1983) defined the relationship between perimeter and area for a two-dimensional object as $A = (kP)^d$, where A is the area and P is *some* measure of the perimeter of the patch, that is, the radius, diameter, length of the patch, or the actual measured perimeter; k is a constant that takes on different values de-

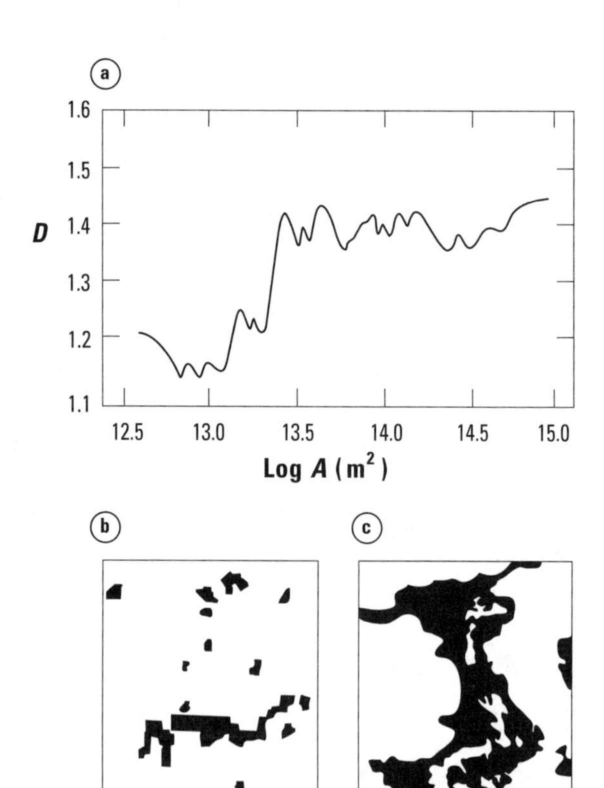

FIGURE 5.11.
(a) Fractal dimension (D) of forest patches in the vicinity of Natchez, Mississippi, as a function of patch size. (b) Section of the original map illustrating how small patches tend to be simple in shape. (c) Section of the original map illustrating the more complex shapes associated with the larger patches.

REDRAWN FROM KRUMMEL ET AL., 1987.

pending on the way that P is estimated; and d is the fractal dimension. To solve for d, we take the logs of both sides of the equation

$$d = \frac{\ln (A)}{\ln (P) + \ln (k)}$$

The fractal dimension, d, will have the value of 2.0 for two-dimensional objects such as circles and squares and will equal 1.0 for a straight line *if the length of one side of the object is used to estimate* P *and* k *is equal to 1.0*. If this is the case, then k can be dropped from the equation [$\ln (k) = 0.0$], and the method for estimating the fractal simplifies to

$$d = \frac{\ln (A)}{\ln (P)}$$

If the actual perimeter of the patch is used to estimate P, then k must be set equal to 0.25, and k cannot be dropped from the equation! Failure to properly relate k and P can produce errors much greater than the range of observed values of the fractal for patches of land cover.

Although the algebra used to estimate the fractal dimension is simple and straightforward, a variety of alternative formulations has been used to estimate the fractal dimension, resulting in considerable confusion when attempting to make comparisons between studies. One alternative formulation is the reversal of the relationship between perimeter and area (assuming that $k = 1.0$), which gives $kP = A^D$. It is clear from simple algebra that this *fractal, D,* is equal to $1/d$. Substituting $1/d$ and taking logarithms (assuming that $k = 1.0$), we obtain, as before, $d = \ln (A)/\ln (P)$. Other forms of the fractal dimension can be found, including the relationship, $P = (A^{1/2})^{d'}$ (Lovejoy, 1985; Burrough, 1986; Lovejoy et al., 1987; McGarigal and Marks, 1995). Obviously, there is no single correct (or preferred) form for estimating fractals, but authors should clearly state the algebra used to calculate this metric so that comparisons among studies can be reliably effected.

The fractal has often been estimated by regressing the logarithms of perimeter on area (or vice versa), with the slope of the line being a statistical estimation of the fractal. However, unless all patches are large (patches sizes greater than 100 cells), the grain size of the data set will bias the regression analysis. The value of k must also be considered by setting the regression intercept equal to $\ln (k)$. The extent of the map plays an important role in the estimation of the fractal, with smaller maps resulting in fewer patches that escape the truncation effects of the map boundary (Gardner et al., 1987). Presently, little effort has been made to link together these problems and establish a rigorous sampling theory for fractals (Loehle and Li, 1996). Only when samples are extremely large (e.g., Lovejoy, 1985; Krummel et al., 1987) can reliable estimates of the fractal be ensured.

Other Useful Metrics

There are numerous other quantities that can be assigned to a grid cell or polygon on a landscape map and that may have important ecological implications. For example, in regions that have road networks, which includes most landscapes, road length, road density, or distance to the nearest road may be important indexes of certain functional aspects of landscape structure. In a broad-scale analysis of habitat-use patterns of recolonizing gray wolves (*Canis lupus*) in northern Wisconsin, USA, Mladenoff et al. (1995) found road density to be an important predictor of where pack territories would be located.

The length of riparian vegetation (nonagricultural and nonurban) is another metric that may have important ecological implications because of the ability of riparian corridors to buffer nutrient transfers from land to water (e.g., Peterjohn and Correll, 1984). Population density, which is now readily available for recent years in digital format from the U.S. Bureau of the Census at the resolution of census tracts, may also serve as a useful metric of processes in a landscape (e.g., Turner et al., 1996; Wear et al., 1996).

Applications of Landscape Pattern Analyses

In numerous studies, landscape metrics have been used to describe changes in a landscape through time or to compare landscapes (e.g., Whitney and Somerlot, 1985; Iverson, 1988; Turner and Ruscher, 1988; LaGro and DeGloria, 1992; Kienast, 1993; and many others). We discuss just a few here to provide examples of the insights that have been produced by the application of landscape metrics; other examples of the use of landscape metrics in empirical studies and spatial models will be found in subsequent chapters.

Spies et al. (1994) studied the process of forest fragmentation in a 2589-km^2 managed forest landscape in northwest Oregon, USA, between 1972 and 1988 using Landsat data. Management for timber production in this region has converted extensive areas of old-growth forest to young conifer plantations. Spies et al. (1994) focused on changes in closed-canopy conifer forest (CF) and examined differences in landscape pattern among different classes of landowner (see Figure 5.3). Results revealed a decline in CF from 71% to 58% during the study period, but the rates of decline varied among landowners and were highest on private lands. The amount of edge habitat increased and the amount of interior habitat decreased on all ownerships. Large remaining patches (<5000 ha) of contiguous interior forest were restricted to public lands on which timber harvesting was restricted. In a study with similar objectives, Turner et al. (1996) compared the spatial pattern of forested landscapes between landowner categories in the southern Appalachian Mountains and the Olympic Peninsula, Washington. Differences in landscape pattern were observed between the two study regions and between different categories of land ownership. In both regions, private lands contained less forest cover, but a greater number of small forest patches than did public lands. Differences in landscape patterns were interpreted in terms of the management objectives of different types of landowner. This study demonstrated that the way in which human endeavors are organized through the institutions and scale of land ownership produces qualitatively different landscape patterns and dynamics (Turner et al., 1996).

In a study of forested landscapes of the upper Midwest, USA, Mladenoff et al. (1993) compared landscape patterns in an undisturbed old-growth area and a nearby area subjected to forest management and harvest (Figure 5.12). Managed forests were distinctly different from the natural old-growth landscape, which was dominated by large areas of eastern hemlock (*Tsuga canadensis*), sugar maple (*Acer saccharum*), and yellow birch (*Betula allegheniensis*). The disturbed landscape had smaller, remnant patches of old-growth ecosystems scattered among early successional forests. Forest patches were also significantly simpler in shape (lower fractal dimension) in the fragmented landscape. Important ecosystem juxtapositions of the old-growth landscape, such as hemlock associated with lowland conifers, were lost, and characteristic patterns of heterogeneity due to landform patterns in this glaciated region were altered by human disturbances.

Making Sense Out of Multiple Metrics

The description of landscape pattern requires more than one metric, so the question now becomes how to select a relevant subset. Determining how many metrics to use and how to combine the metrics so that the results are meaningful and interpretable remains challenging. The subset should explain pattern variability across the landscape, but redundancy should be minimized, particularly with indexes that may be highly correlated with each other. O'Neill et al. (1988a) observed that landscape scenes from across the eastern United States could be differentiated from one another by using three metrics, dominance, contagion, and

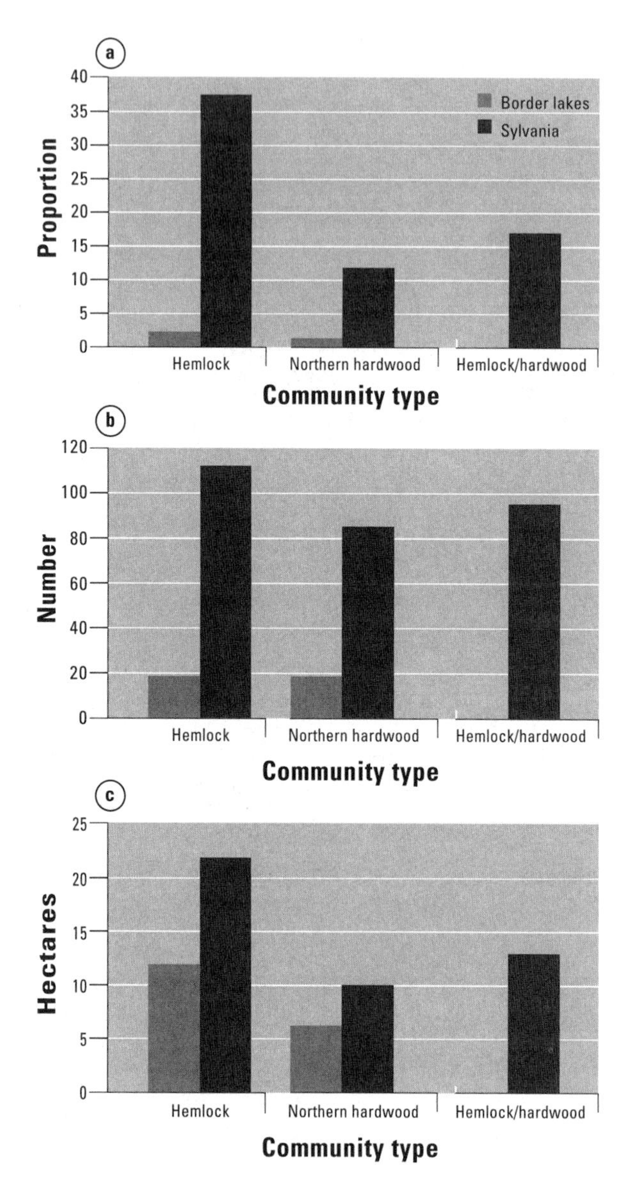

FIGURE 5.12.

Selected landscape metrics for three forest community types in the Border Lakes and Sylvania landscapes in the Great Lakes Region: (a) proportion of landscape occupied, (b) number of patches, and (c) mean patch size. Border Lakes is a disturbed landscape of second-growth forest, whereas Sylvania is landscape of primary old-growth forest.

DATA FROM MLADENOFF ET AL., 1993.

the fractal dimension, that were relatively independent, well distributed over their potential range, and collectively able to discriminate the geographic distribution of landscape pattern types. A variety of other studies (e.g., Iverson, 1988; Turner, 1990; Dunn et al., 1991; Mladenoff et al., 1993; Medley et al., 1995; Skinner, 1995) has found that patch-based measures applied to individual cover types seemed to perform well in documenting changes through time.

One method, and there are certainly others, of combining the information contained in several metrics is to identify a few independent indexes that capture aspects of the pattern that are relevant to the original question. For example, if we are quantifying a landscape and how it changed through time from the perspective of a particular animal, we might be interested in (1) the proportion of the landscape containing suitable habitat, (2) the number of patches of suitable habitat above the minimum size required by the animal, and (3) the connectivity of the suitable habitat as measured by interpatch distances. As the landscape changed through time or as alternative future scenarios were considered, the condition of the landscape could be plotted through a three-dimensional state space based on the coordinates of each of the three metrics considered. Furthermore, the volume within that state space that represented the zone of survival for the animal could be identified. We could then determine whether particular landscape changes result in a landscape trajectory that remains within or takes excursions beyond that zone of survival. Such an analysis could be done in n dimensions, although the visualization and interpretation will become increasingly complicated.

O'Neill et al. (1996) used a three-dimensional *pattern space* to show three subregions of the southeastern United States as points characterized by landscape indexes (Figure 5.13). Use of the pattern space effectively separated these landscapes based on dominance, contagion, and shape complexity. Simple geometry can be used to compute the distance between landscapes in the pattern space (O'Neill et al., 1996). What is particularly powerful about this approach is that the structure of a landscape can be plotted through time or compared to a desirable state and the Euclidean distance between points quantified. Both direction and magnitude of change through time can be plotted if repeated measurements are made for the same landscape. O'Neill et al. (1996) described two important constraints on this approach. First, the axes of the pattern space should be orthogonal, that is, as independent from each other as possible. Simple correlation analysis can be used to test for independence. Second, the sensitivity of the individual metrics used in the pattern space to landscape change must be established. The critical question is whether the indicator can detect small changes (changes that are not catastrophic or irreversible) such that it serves as a useful warning of undesirable landscape change (O'Neill et al., 1996). Simulated landscapes play important roles in evaluating the performance of landscape metrics with specified changes in spatial pattern (Li and Reynolds, 1994).

Multivariate statistics also offers a means of making sense out of multiple metrics. Riitters et al. (1995) compared 55 landscape metrics across 85 land-cover data sets. Pairwise comparisons revealed that many metrics have correlation coefficients greater than ±0.9. Eliminating the redundant measures reduced the candidates to 26. Factor analysis revealed five factors that all have eigenvalues greater than 1.0 and explain about 83% of the variance. Each factor contains several indicators. It is then necessary to choose a single indicator for each major factor. Based on the ease of calculation and interpretation, the following relatively independent indexes were recommended: (1) the total number of different land-cover types on the map; (2) contagion; (3) fractal dimension; (4) average patch perimeter–area ratio, given by

$$P = \frac{1}{m} \sum_{k=1}^{m} \frac{E_k}{A_k}$$

where there are a total of m patches and E_k is the perimeter of the kth patch and A_k is the area; and (5) relative patch area (average ratio of patch area to the area of an enclosing circle), which indicates how compact the patches are, is given by

$$R = \frac{1}{m} \sum_{k=1}^{m} \frac{A_k}{\pi L_k}$$

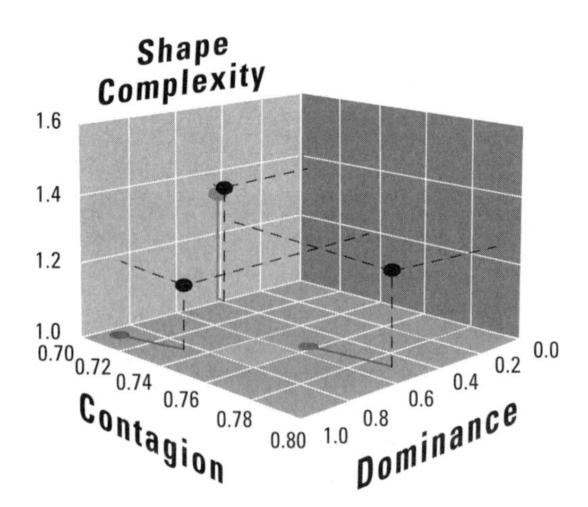

FIGURE 5.13.
Three-dimensional pattern space in which three subregions of the southeastern United States are characterized by three landscape metrics.

ADAPTED FROM O'NEILL ET AL., 1996.

where L_k is one-half of the longest straight line that can be drawn within the patch. Notice that the denominator is the area of a circle, with L_k as the radius, that will approximately enclose the patch. These five metrics provide a minimum set of independent measures of qualitatively different aspects of spatial pattern based on 85 landscapes scattered across the United States. Therefore, these five metrics should be useful in many applications, at least in temperate zones.

Other studies have also suggested a subset of measurements that might characterize landscape pattern. McGarigal and Marks (1995) also conducted a principal components analysis of 30 different metrics calculated for late-successional forests in the northwestern United States. Their analysis revealed three independent factors: (1) patch shape and edge contrast, (2) patch density, and (3) patch size. Li and Reynolds (1994, 1995) used theoretical considerations to propose five different aspects of spatial heterogeneity that could be computed: (1) number of land-cover types and (2) proportion of each type on the landscape; (3) spatial arrangement of patches; (4) patch shape; and (5) contrast between neighboring patches. We would then select a metric to quantify each of these fundamentally different aspects of pattern to avoid redundancy among metrics.

Methods for quantification of spatial pattern on categorical maps continue to develop rapidly. We have discussed a number of basic calculations here, but have not attempted to present all the metrics that have been proposed. Readers interested in additional metrics are referred especially to Baker and Cai (1992), Baskent and Jordan (1995), McGarigal and Marks (1995), Riitters et al. (1995), Haines-Young and Chopping (1996), and Gustafson (1998). In addition, readers interested in gaining practical experience in computing these metrics should consult Gergel and Turner (2000).

GEOSTATISTICS OR SPATIAL STATISTICS

In addition to the landscape metrics that we have discussed so far, methods based on spatial statistics (e.g., Ripley, 1981; Cressie, 1991) and geostatistical methods (e.g., Rossi et al. 1992; Deutsch and Journel, 1998) for point data are being increasingly applied in the analysis of landscape pattern. Spatial statistics are generally used (1) to identify the spatial scales over which patterns (or processes) remain constant (or, alternatively, the scales at which significant changes in pattern and process can be detected) and (2) to interpolate or extrapolate point data to infer the spatial distributions of variables of interest. Spatial statistics and geostatistics use point data for some property that is assumed to be spatially continuous across the landscape. Thus, they do not require categorization of the landscape nor do they assume a patchy structure or the presence of boundaries. In the example discussed earlier regarding mapping the distribution of aspen in the Wisconsin Northwoods, a spatial statistics approach would involve sampling aspen density at points across the landscape, rather than deciding a priori what density of aspen is sufficient to characterize the stand as aspen. As Gustafson (1998) notes, the categorical and point-data approaches to the description of spatial heterogeneity are seldom combined in most studies, leading to the appearance that they are not complementary. However, each approach offers advantages and disadvantages that depend both on the questions being asked and the nature of the system being measured. Two questions addressed by spatial statistics are of particular importance for landscape studies.

1. *What is the appropriate scale at which to conduct an analysis?* In this case, the objective is the identification of the spatial or temporal scale at which a sampled variable exhibits maximum variance. These applications are common in sampling design (Ball et al., 1993; Istok et al., 1993) to avoid problems in statistical analyses, because standard parametric statistics require that the data be spatially independent. Positive relationships at short distances distort statistical tests such as correlation, regression, and ANOVA, creating a greater possibility of Type II errors (statistical significance falsely detected). If the scale of autocorrelation can be identified, the spacing of field sampling or other measurements can be specified such that spatial independence of the dependent variable is assured, thereby allowing the appropriate use of traditional statistical methods. An example of this approach can be found in Pearson et al. (1996), where the semivariance was used to determine the spatial scale at which measurements were independent. A subsample of measurements was then selected for analysis as independent measures defined by the semivariance.

2. *What is the nature of the spatial structure of a particular variable?* In this case, quantification of the scale of variability exhibited by natural patterns of a variable of interest (e.g., soil type, pH, density, or biomass) is conducted to detect spatial structure. This approach has been used by ecologists seeking enhanced understanding of how patterns of environmental heterogeneity influence ecological processes (e.g., Legendre and Fortin, 1989; S. Turner et al., 1991; O'Neill et al., 1991b; Bell et al., 1993). The coincidence in the scales of variability of different ecological features, for example, plants and soil nutrients (Greig-Smith, 1979) or seabirds and their prey (Schneider and Piatt, 1986), can indicate the possibility of direct linkages. However, it is important to note that coincidence of the spatial structure does *not* mean causality, but rather suggests reason to test for causal

mechanisms. In practical terms, obtaining adequate data to characterize scale dependencies among multiple variables is difficult, because it requires simultaneous estimation of relevant physical and biological variables. Ocean sciences have had notable successes in this area, characterizing biotic changes as a result of frequency and extent of changes in the physical environment (Steele, 1989).

There is a growing variety of tools for the analysis of spatial scale, including correlograms (Legendre and Fortin, 1989; Legendre, 1993), semivariance analysis (Deutsch and Journel, 1992), lacunarity analysis (Plotnick et al., 1993; Plotnick and Prestegaard, 1995), spectral analysis (Platt and Denman, 1975), the paired-quadrat technique (Greig-Smith, 1983), and a variety of fractal-based methods (Sugihara and May, 1990). Discussion of all these methods is beyond the scope of this text, and the reader is referred to the review by Gardner (1998) and the recommended list of readings at the end of the chapter. Readers interested in greater depth are referred to Cressie (1991) and Upton and Fingleton (1985a, b). We briefly discuss the use of autocorrelation and semivariance because an understanding of general techniques may be frequently required.

Correlograms

The heart of spatial statistics is the concept of correlation of spatially distributed variables. A variable is said to be spatially autocorrelated if a significant association, as measured by the correlation coefficient, can be detected between points as a function of their spatial location. Estimating the autocorrelation requires the calculation of the covariance between points separated by distance h:

$$\hat{C}(h) = \frac{1}{N(h)} \sum_{i=1}^{N(h)} (x_i - m_i)(x_{i+h}) - m_{i+h}$$

where x_i and x_{i+h} are, respectively, the tail and head of the data pair, and m_i and m_{i+h} are the corrections for mean values of the tail and head, respectively. The estimated autocorrelation is then

$$\hat{D}(h) = \frac{\hat{C}(h)}{S_i S_{i+h}}$$

where S_i and S_{i+h} are the standard deviations of the data points comprising the tail and head, respectively. The correlogram provides a visualization of the change in spatial relationships by plotting the autocorrelation values on the ordinate and the lagged distances, h, on the abscissa. Indications of ecological scale can be verified by statistically testing the peak values of the correlogram (both positive and negative) for significant differences from zero (Carlile et al., 1989). The conditions for valid tests for the significance of these peaks are restrictive, requiring that (1) the data be equally spaced, (2) gradients of change or trends in the data be removed (see Legendre, 1993, for other restrictions in the analysis of gradients and autocorrelated data), and (3) residuals be normally distributed (Legendre and Legendre, 1983).

Correlograms are often analyzed by examining their shape, since characteristic shapes are associated with different spatial structures (Legendre and Fortin, 1989) (Figure 5.14). The alternating of positive and negative values is an indication of patchiness. However, correlograms cannot distinguish between real data representing a sharp step and a subtle gradient (Legendre and Fortin, 1989). In addition, correlograms cannot be used for interpolation between data points; this is where semivariograms and kriging techniques are needed. Correlogram analysis should not be computed using <50 sample points and should not be performed on data with many zeros or missing values. When data are lacking, the degree of autocorrelation will be overestimated. When Euclidean distances, d, are relative, the data plotted in the correlogram may be transformed by $1/d$ or $1/d^2$.

Semivariograms

One great difficulty in the study of landscapes is the development and use of methods for converting point data to spatial data, that is, the development of a spatial distribution for a variables (or set of variables) from a small sample of location-specific measurements. Although point data are valuable descriptors of landscape pattern and process (e.g., weather measurements, evapotranspiration rates, soil descriptions, and productivity estimates), the collection of such data is expensive and time consuming. Geostatistical techniques, such as variogram analysis and kriging, have been specifically developed to provide estimates of the spatial distributions from limited sets of point data. Rossi et al. (1992) provide an in-depth discussion of the assumptions, methods, and pitfalls of the application of these techniques to a variety of ecological problems. Semivariograms and related measures (e.g., covariogram and correlogram) used in spatial statistics may be applied to linear or two-dimensional data sets.

The semivariogram (Figure 5.15) defines the spatial scales over which patterns are dependent. The semivariogram

$$\gamma(h) = \frac{1}{2N(h)} \sum_{i=1}^{N(h)} (x_i - x_{i+h})^2$$

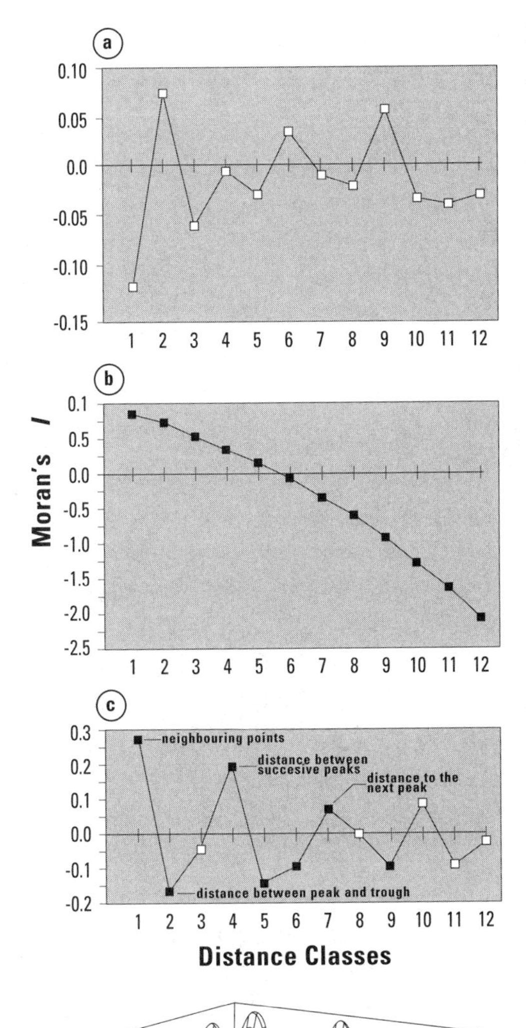

FIGURE 5.14.

All-directional spatial correlograms of artificial landscapes: (a) random landscape, (b) landscape with a gradient, and (c) landscape with a repeating pattern, the "nine fat bumps" shown below. Note that Moran's *I* behaves like a correlation coefficient.

ADAPTED FROM LEGENDRE AND FORTIN, 1989.

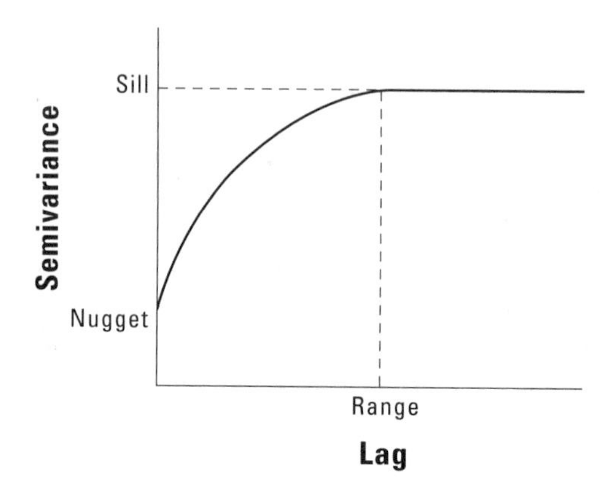

FIGURE 5.15.
Idealized semivariogram showing the nugget, sill, and range; see text for explanations.

is calculated by rearranging the data into N data pairs (x_i and x_{i+h}) separated by distance h, with $N(h)$ equal to the number of such pairs. The summation is made over all possible pairs and lagged distances, h. (In practice, at least 50 data pairs are also needed to adequately estimate the semivariance for each lagged distance, h.) Assuming that the two-dimensional data set is isotropic, data pairs can be constructed by taking transects in all directions. Alternatively, if anisotropic properties of the spatial data are of interest, then values of the semivariance for specific directions can be compared. A variogram is used to visualize the results by plotting the semivariance, $\gamma(h)$, on the ordinate against the lagged distances, h, on the abscissa. A relatively flat variogram indicates a pattern that lacks spatial dependencies (a random pattern). If a spatial data set is nonrandom and has been adequately sampled (that is, the spatial extent of the data provides an adequate representation of the pattern of interest), the variogram will ascend from an initial value at $h = 0$ to an asymptotic value. The geostatistical jargon for these two values are the nugget and the sill, respectively (Figure 5.15). The difference between the nugget and sill reflects the proportion of the total variance due to spatial dependencies within the data set. The value of h at which the semivariogram reaches an asymptote indicates the spatial extent, or range, over which spatial dependencies can be detected.

The semivariogram (or variogram) is related to spatial correlograms and is the basis for the kriging method of contouring, or interpolating between points. When prospecting, mining engineers needed a method for predicting the most likely place to strike a deposit of minerals or oil. In landscape ecology, if we cannot sample the entire landscape, these techniques can provide a means to extrapolate point observations over space. Variograms assume stationary data (that is, mean and variance in the data are the same in various parts of the data set) and may reveal a coarse-grained spatial structure while failing to detect finer-grained spatial structures. In addition, the variogram does not lend itself to any statistical test of hypothesis (in contrast to the correlogram). Like the correlogram, the variogram is often interpreted by visual inspection (Figure 5.16). A relatively flat variogram indicates a pattern that lacks spatial structure. If there are spatial dependencies, the value of h at which the semivariogram asymptotes (the range) indicates the extent over which spatial dependencies can be detected.

Spatial Statistics and Categorical Analyses: A Final Comparison

Spatial statistics use point data to analyze the scale of patchiness, identify hierarchies of scale in the data, and determine whether the spatial distribution is ran-

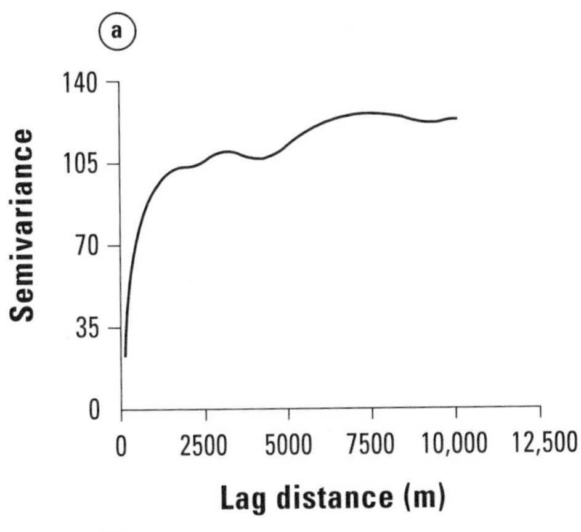

FIGURE 5.16.
Example semivariogram and correlogram computed for a landscape in northern Yellowstone National Park. Note that the shape of the correlogram is nearly identical, although inverted, to the shape of the semivariogram of the same data. These two approaches may be used interchangeably for exploratory analyses that do not require the use of statistical tests.

ADAPTED FROM MEISEL AND TURNER, 1998.

dom, aggregated, or uniform (Gardner, 1998; Gustafson, 1998). Spatial statistics also allow interpolation from point data to areas that have not been sampled. However, it remains difficult to collect sufficient data over broad spatial scales to detect multiple scales of variability in ecological data and relate these scales to the processes that generated the patterns (Carlile et al., 1989; Cullinan and Thomas, 1992; Meisel and Turner, 1998). Patch-based or categorical techniques, by definition, require an a priori assignment of the landscape into classes. This simplification affects the outcome of the analysis, but is often more amenable to interpretations with regard to ecological function. At the present time, it appears that the categorical analyses are more directly relevant to ecological theory than are point-based techniques. Thus, each method has its particular strengths, providing landscape ecologists with a diverse toolbox for tackling the problem of analysis of spatial pattern and its relationship to ecological processes.

SUMMARY

The quantification of landscape pattern is useful, indeed necessary, for understanding the effects of pattern on ecological processes and for documenting either temporal changes in a landscape or differences between two or more landscapes. A useful set of metrics to quantify landscape pattern should meet several criteria: (1) the metrics should be selected to answer a particular question or meet a particular objective; (2) the measured values of the metrics should be distributed over the full range of potential values and the behavior of the metrics should be known; and (3) the indexes should be relatively independent of each other. This can be easily tested by examining the correlation structure within a set of potential candidates. In addition, the analysis must recognize (and carefully choose) the classification scheme used to categorize the data and the spatial scale of the data. Collectively, this set of decisions places important constraints on the analysis and interpretation of landscape pattern. As a first approximation, the extent of the study landscape should be two to five times larger than landscape patches to avoid bias in calculating landscape metrics; grain size should be two to five times smaller than the spatial features of interest.

Many metrics are readily available for use in landscape pattern analysis. One metric is insufficient to characterize a landscape, yet there is no standard recipe for determining how many and which metrics are needed. One major limitation of overall metrics of pattern, that is, those reported for a whole landscape rather than by cover type, is that the same numerical value can be returned for a variety of qualitatively different landscapes. Based on ease of calculation and interpretation, we suggest a set of five metrics that are relatively independent of each other.

Spatial statistics reflect another large body of methods available for quantifying aspects of landscape pattern. These methods are typically used to detect the spatial scales of autocorrelation for landscape elements or to interpolate point data to infer spatial distributions of a variable of interest.

Although the development of landscape pattern analysis has been rapid, there are three major areas in which further understanding is sorely needed: (1) the statistical properties and behavior of metrics requires better knowledge, (2) the relative sensitivity of different metrics to detecting changes in the landscape is not known, and (3) the empirical relationships between landscape patterns and ecological processes of interest must be better documented. Collectively, progress in each of these three research areas will help ecologists to determine what is worth measuring and why and when a change in a metric is significant both statistically and ecologically.

DISCUSSION QUESTIONS

1. Landscape metrics are difficult to relate to landscape processes. Why does there continue to be a high level of interest in the development of new landscape metrics? Do you think the development of new metrics is a fruitful endeavor? Why or why not?
2. Imagine that two landscapes have been analyzed by a series of metrics and a number of differences have been detected. What assurances regarding the reliability and usefulness of the data should be examined before conclusions are drawn about differences between the two landscapes?
3. Classification of landscape data is required for most spatial metrics. How will alternative classifications affect the analysis of pattern? (*Hint*: You may use a landscape data set of your choice to explore this.) Design an experiment to test the effect of classification schemes on analysis results.
4. Why is the detection of spatial scales important in landscape studies? Given realistic constraints on data acquisition, is there a single most desirable method for defining spatial pattern and changes in pattern with scale?
5. Imagine that you are charged with designing the protocol for monitoring change through time in a large region (you should select a region on which to focus). Describe the steps that you would take to develop your monitoring scheme. How would you select the metrics to be included?

RECOMMENDED READINGS

Landscape Pattern Analysis

Gustafson, E. J. 1998. Quantifying landscape spatial pattern: what is the state of the art? *Ecosystems* 1:143–156.

Haines-Young, R., and M. Chopping. 1996. Quantifying landscape structure: A review of landscape indices and their application to forested landscapes. *Progress in Physical Geography* 20:418–445.

McGarigal, K., and B. J. Marks. 1995. *FRAGSTATS. Spatial Analysis Program for Quantifying Landscape Structure.* USDA Forest Service General Technical Report PNW-GTR-351.

Riitters, K. H., R. V. O'Neill, C. T. Hunsaker, J. D. Wickham, D. H. Yankee, S. P. Timmons, K. B. Jones, and B. L. Jackson. 1995. A factor analysis of landscape pattern and structure metrics. *Landscape Ecology* 10:23–40.

Wickham, J. D., and D. J. Norton. 1994. Mapping and analyzing landscape patterns. *Landscape Ecology* 9:7–23.

Spatial Statistics and Geostatistical Methods

Bell, G., M. J. Lechowica, A. Appenzeller, M. Chandler, E. DeBlois, L. Jackson, B. Mackenzie, R. Preziosi, M. Schallenberg, and N. Tinker. 1993. The spatial structure of the physical environment. *Oecologia* 96:114–121.

Cressie, N. A. C. 1991. *Statistics for Spatial Data.* John Wiley & Sons, New York. (*Note*: This is an in-depth reference book.)

Gardner, R. H. 1998. Pattern, process and the analysis of spatial scales. In D. L. Petersen and V. T. Parker, eds. *Ecological Scale: Theory and Applications,* pp. 17–34. Columbia University Press, New York.

Fortin, M.-J., P. Drapeau, and P. Legendre. 1989. Spatial autocorrelation and sampling design in ecology. *Vegetatio* 83:209–222.

Meisel, J. E., and M. G. Turner. 1998. Scale detection in real and artificial landscapes using semivariance analysis. *Landscape Ecology* 13:347–362.

Turner, S. J., R. V. O'Neill, W. Conley, M. R. Conley, and H. C. Humphries. 1991. Pattern and scale: statistics for landscape ecology. In M. G. Turner and R. H. Gardner, eds. *Quantitative Methods in Landscape Ecology,* pp. 17–50. Springer-Verlag, New York.

CHAPTER 6

Neutral Landscape Models

PROGRESS IN SCIENCE is ideally made by the sequential development of hypotheses and the execution of experiments designed to test these hypotheses (Platt, 1964; Quinn and Dunham, 1983). Indeed, hypothesis testing "is . . . characteristic of all experimentation" (Fisher, 1935). The simplest hypothesis that we may construct is the null hypothesis of no effect. A properly formed null hypothesis provides the required reference point against which alternatives may be contrasted. Strong (1980) noted that the physical (or "atomistic") sciences have been extremely successful with this approach, in part because they deal with systems that can be characterized by a limited number of variables, and examples of the success of this approach abound. Two examples cited by Strong (1980) are the following: (1) A null hypothesis in chemistry is that molecular properties of living systems are not unique. Therefore, any chemical synthesized by protoplasm can also be synthesized in the laboratory. Modern biochemistry has failed to disprove this null hypothesis. (2) In the post-Newtonian era of physics, it was stated that time is a variable that is independent of all other factors. Modern physics has rejected this hypothesis and replaced it with the alternative that time can be a function of space and relative velocities.

The ecological sciences, including landscape ecology, have not relied as strongly on the use of the null hypothesis. This difference is due in part to the large number of variables affecting ecological systems and the importance of stochastic effects that make ecological phenomena difficult to measure in a repeatable fashion. Thus, observed patterns for particular landscapes reflect complex histories of interactions between natural forces and events (e.g., climate, terrain, soils, water availability, biota, and natural disturbances) and those due to land-use alteration (e.g., urbanization, agriculture, and forestry management; see Chapter 4). The multitude of possible combinations of natural and anthropogenic forces and events produces landscape patterns that are recognizably unique. Under these circumstances, phenomenological and corroborative methods tend to dominate over the experimentation and hypothesis testing typical of sciences studying simpler systems (see Strong, 1980, for further discussion). The difficulty with corroborative studies is essentially the lack of design and replication. When treatment effects (e.g., different landscapes) are tested without true replication, the validity of the comparisons is often suspect (Hurlbert, 1984; Hargrove and Pickering, 1992).

How can null hypotheses be used in landscape studies to avoid these ills? One means of testing differences between landscapes when experimental manipulation and/or replication is not feasible is to develop a simple standard against which comparisons may be made. The simplest standard for landscape pattern is a random map (Figure 6.1), which lacks all factors that might organize or structure the pattern (Gardner et al., 1987). Tests of observed landscapes against replicate random maps can then reveal the magnitude and significance of differences due to the structure of actual landscapes. Random maps are neutral landscape models (NLM) against which effects of processes that structure actual landscapes may be tested. Studies of NLMs have shown that surprisingly rich patterns can be generated by random processes alone and also that actual landscapes may not always be measurably different from these random patterns (Gardner et al., 1993a).

The development of the first NLM is an interesting story of synthesis among multiple disciplines. In the summer of 1986, a Gordon Conference on fractals was held in New Hampshire. Our work at Oak Ridge National Laboratory on the use of fractals for landscape analysis was just beginning (Krummel et al., 1987). The Gordon Conference (attended by Gardner) provided an opportunity to hear and discuss with experts from diverse disciplines how to use fractal geometry to understand landscape pattern. One stimulating speaker at that conference was Dietrich Stauffer, who had recently written a text on percolation theory (Stauffer, 1985). Stauffer's presentation of randomly generated two-dimensional patterns and their analysis by percolation theory was remarkable. Percolation theory was designed to address questions regarding aggregation and clustering in material systems, but the examples, figures, and methods presented by Stauffer seemed to be adaptable to the needs of landscape ecologists to relate pattern and process in heterogeneous systems.

A second stimulus for the development of NLMs was the 1976 paper by Caswell, who compared alternative models of community formation against a model that was neutral to the biological interactions in question. Each model predicted that community structure and diversity were generated by biological interactions that made the community less susceptible to stochastic effects of a variable environment. The relative role of biological interactions in community formation was tested by Caswell by the use of a *neutral model.* Caswell's neutral model generated patterns of diversity under the assumption that species did not interact by either competition or predation nor did they differ in response to abiotic factors. Thus, departures of the neutral model's predictions from actual patterns of species diversity served as an estimate of the strength of the biological effect on the structure of communities. This concept of a complex model as a null hypothesis and Caswell's use of the term neutral model became the paradigm for our development of NLMs.

Once work began on NLMs, it became apparent that other early examples of neutral models could be found. For instance, Cole (1951, 1954) used random numbers to construct cycles similar to those observed in natural populations, Simberloff (1978) used island biogeographic theory to examine community patterns,

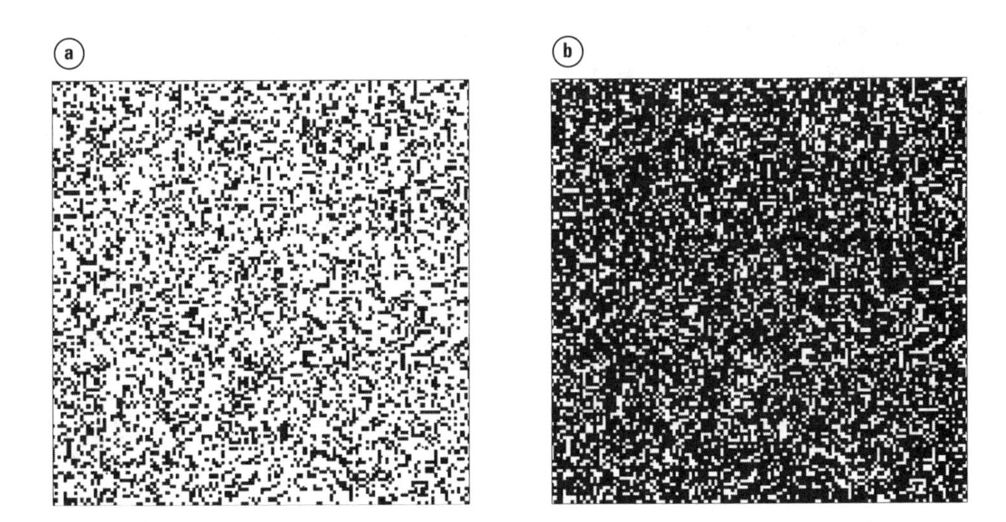

FIGURE 6.1.

Random maps (m, the number of rows and columns equals 128) generated with the probability, p, that grid cells contain the land-cover type of interest of (a) 0.3 and (b) 0.7. Occupied cells are shown in black.

Istock and Scheiner (1987) used random landscapes to test patterns of species diversity, and Nitecki and Hoffman (1987) produced an edited volume on the subject. More current examples include niche shifts in *Anolis* communities (Haefner, 1988a, b), community formation in fishes (Jackson et al., 1992), plant migration rates (Higgins and Richardson, 1999), hemlock regeneration and deer browsing (Mladenoff and Stearns, 1993), bird assemblages in fragmented landscapes (Sisk et al., 1997), tests of Holling's hypothesis of discontinuities in landscape pattern causing clumps and gaps in the distribution of body sizes within animal communities (Siemann and Brown, 1999), and the continued investigations of the formation of structure in natural communities (Wilson et al., 1995).

With and King (1997) have reviewed the use of NLMs and separated them into two categories: NLMs have been used (1) to determine the extent to which structural properties of landscapes (e.g., patch size and shape, amount of edge, connectivity, autocorrelation) deviate from some theoretical spatial distribution and (2) to predict how ecological processes, such as animal movement, seed dispersal, gene flow, or fire spread, are affected by landscape pattern. In this chapter, we describe the methods behind the generation of three varieties of random maps and review their application to landscape studies.

RANDOM MAPS: THE SIMPLEST NEUTRAL MODEL

The simplest possible method of generating a map is to randomly locate sites (or cells) of a single land-cover type within a two-dimensional grid. A simple matrix of 0's and 1's will suffice, with the 1's representing the cover type of interest and the 0's representing all other cover types. The only challenge is to locate the sites within the map randomly. The first step is to construct an array of m columns and m rows. The second step is to randomly assign values (0's and 1's) to the m^2 elements of the array. Each grid site is examined and a random number selected between 0.0 and 1.0. A probability value, p, is specified (say, $p = 0.4$) and the grid site is set to 1 if the random number is ≤ 0.4 or the site is set to 0 if the random number is $>p$. If the map is large, the proportion of sites set to 1 will be very close to the value of p, whereas the number of sites set to 0 will be approximately equal to $1 - p$ (e.g., $1.0 - 0.4 = 0.6$). The total number of matrix elements (grid sites or cells) occupied by the habitat (land cover) type of interest will be approximately equal to pm^2, while the number of sites of "nonhabitat" will equal $(1 - p)m^2$.

As the grid is filled with 0's and 1's, clusters, or patches, will be seen to form (Figure 6.1). Clusters are usually defined as groups of sites of the same land-cover type with at least one horizontal or vertical (but not diagonal) edge in common. This rule for patch identification is usually referred to as the four-neighbor or nearest-neighbor rule (see discussion in Chapter 5 and Figure 5.8). When a series of maps is generated with increasing values of p, the number of patches increases as p increases over the range $0.0 < p < 0.3$. Patch numbers decline as p continues to increase, because small patches begin to coalesce into larger ones (Gardner et al., 1987; see Figure 6.2a). The amount of edge on the map is also affected by p, with the maximum amount of edge occurring when $p = 0.5$ (Gardner et al., 1987; see Figure 6.2b).

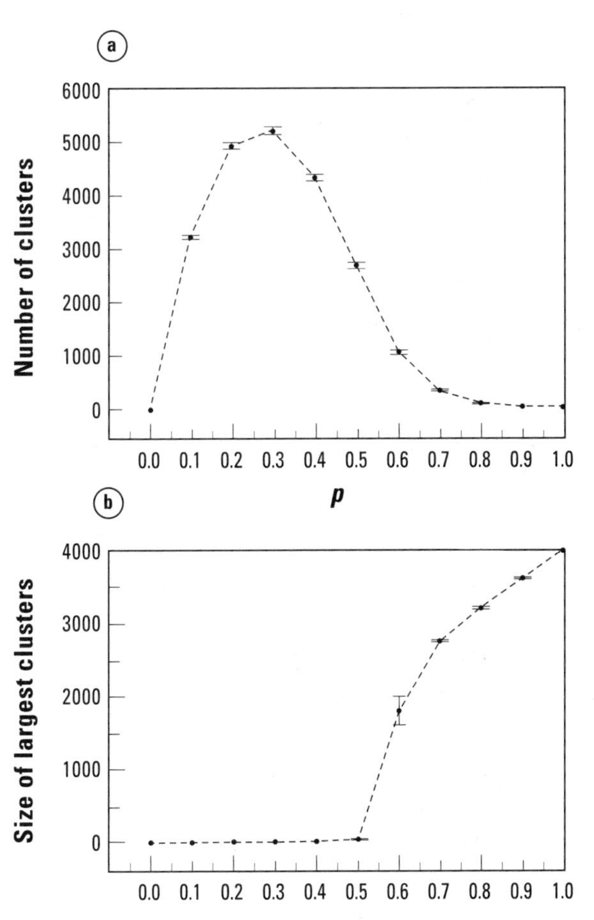

FIGURE 6.2.
(a) The number of clusters and (b) size of the largest cluster for random maps that vary in the probability, p, that a grid cell contains the habitat type of interest. Plotted from data reported in Tables 1 and 2 of Gardner et al. (1987) for maps with 200 rows and columns.

TABLE 6.1
AVERAGE NUMBER OF SITES COMPOSING A CLUSTER (PATCH SIZE IN GRID CELL UNITS) AS A FUNCTION OF MAP SIZE (NUMBER OF ROWS AND COLUMNS) AND p, THE FRACTION OF SITES OCCUPIED.

		P				
Map size	Map type[a]	0.1	0.3	0.5	0.7	0.9
64 by 64	Random	5.2	21.4	167	2,780	3,680
	$H = 0.2$	124	574	1,360	2,620	3,660
	$H = 0.8$	216	908	1,760	2,760	3,670
128 by 128	Random	6.3	27.4	255	11,200	14,700
	$H = 0.2$	482	2,140	5,600	10,500	14,600
	$H = 0.8$	1,110	3,720	7,160	11,100	14,700
256 by 256	Random	7.6	33.4	350	44,900	58,900
	$H = 0.2$	1,760	9,190	21,800	42,200	5,8600
	$H = 0.8$	4,450	15,000	28,400	44,300	58,800

AVERAGES BASED ON 100 INDEPENDENTLY GENERATED MAPS.

MAPS WERE GENERATED AND ANALYZED USING RULE (GARDNER, 1999). ALTHOUGH THE DOCUMENTATION OF RULE IS RECENT, THIS IS THE SAME PROGRAM ORIGINALLY USED TO GENERATE NEUTRAL MODELS (GARDNER ET AL., 1987).

[a]MAP TYPES: RANDOM = SIMPLE RANDOM MAP; $H = 0.2$, A MULTIFRACTAL MAP WITH THE VALUE OF H OF 0.2; $H = 0.8$, A MULTIFRACTAL MAP WITH THE VALUE OF $H = 0.8$. SEE TEXT FOR DISCUSSION OF MULTIFRACTAL MAPS.

The total extent of the map (the value of m representing the number of rows and columns) also affects measures of pattern. Smaller maps (lower values of m) will cause patches to be truncated by the map boundary. This effect is most noticeable when p is > 0.6 (Gardner et al., 1987). Table 6.1 illustrates the truncation effect for a variety of map types and sizes ($m = 64, 128, 256$). For random maps with $p < 0.5$, the size of clusters in smaller maps is approximately 80% of the size of clusters in the larger maps, indicating that truncation effects due to map size results in systematic underestimation of patch size. The truncation effect becomes more noticeable as the value of p increases. At $p = 0.5$, clusters are approximately 70% that of the next largest map size; and at $p = 0.7$ and 0.9 (Table 6.1), cluster sizes of the smaller maps are approximately 25% the size of the next largest map!

Why does the truncation effect depend on the value of p? Are there general rules of pattern formation in simple random maps that provide insight into the analysis of landscape patterns? These types of questions have been a primary focus of percolation theory (Stauffer and Aharony, 1992). A central concept to emerge from percolation theory to explain the relationship of pattern and p is the existence of a *critical threshold* value of p (symbolically defined as p_c). The concept of a critical threshold is a simple one. As maps are generated with successively greater values of p, the process of cluster formation is nonlinear. For maps formed by a square matrix with the nearest-neighbor rule used to identify clusters, the value of p_c will equal 0.59275 (Table 6.2). The reason for this threshold is that above p_c sites are so crowded that nearly all sites contact neighbors along one of their four edges, forming a single cluster that extends, or *percolates*, from one edge of the map to the other (Figure 6.3). In our previous example (Table 6.1), maps with values of $p > 0.59275$ *will always* result in truncation of the largest cluster on the map. If $p > p_c$, the percolating cluster will continue to increase in size as map dimensions increase. This phenomenon has lead to the concept of the *infinite cluster*. The growth of the cluster with increased map dimensions means that there is no finite dimension that will fully contain the cluster when $p > p_c$. Therefore, it is not surprising that a map one-half

TABLE 6.2
PERCOLATION THRESHOLDS FOR TWO-DIMENSIONAL MAPS WITH DIFFERENT NEIGHBORHOOD RULES.

Lattice geometry	Neighboring sites	p_c
Square	4	0.59275
	8	0.40725
	12	0.292
	24	0.168
	40	0.098
	60	0.066
Triangular	6	0.5
	12	0.295
	18	0.225
Honeycomb	3	0.6962
	12	0.3

ADAPTED FROM PLOTNICK AND GARDNER (1993).

the size of another will have an average cluster size that is only one-quarter as large (see Question 6.1 for further discussion).

The general dependence of cluster size on p and the existence of a critical threshold where small changes in p produce sudden changes in cluster sizes have important implications for both material systems (the original focus of percolation theory) and pattern and process within landscapes. The effect of a critical threshold may be easier to visualize by imagining the process of landscape fragmentation. If a landscape exists with $p = 1.0$, that is, a landscape entirely composed of a single land-cover type (say forest), then a reduction in p is equivalent to the random fragmentation of this landscape. As the value of p slowly declines (the forested lands are randomly converted to other land-cover types) from 1.0 to 0.90, isolated gaps in the continuous forested landscape occur with negligible effect on landscape pattern or process. As random clearing continues (values of p further decline from 0.9 to 0.6), forest gaps become more frequent and larger and the amount of edge increases, but nevertheless a single large cluster still dominates the landscape. It is still possible for organisms restricted to forests to move across the landscape; that is, the single large cluster still percolates. However, as the critical threshold is approached, the single large cluster becomes more and more dendritic. It is now possible to find numerous sites that if disturbed would disconnect the percolating cluster. The sudden disconnectance of the landscape resulting from the disturbance of a single site is most likely to occur when $p = p_c$.

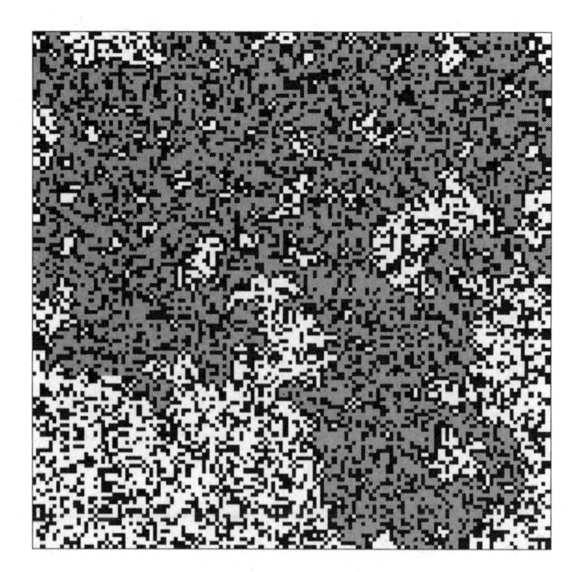

FIGURE 6.3.
Percolating cluster, shown in gray, of a random map (determined by a four-neighbor rule) with $p = 0.6$. The map has 128 rows and columns.

A physical analogue to the process of landscape fragmentation was presented by D. Stauffer at the 1986 Gordon Conference. The bombardment of delicate gold foil electrodes by micrometeorites in space has been a serious problem, because impacts with micrometeorites degrade the foil and ultimately result in failure of the electrode. The question is how much of the electrode can be destroyed before the electrical circuit fails. In laboratory experiments, bombarding electrodes with sand showed that a critical threshold existed as predicted by percolation theory ($p = 0.5982$). At the critical threshold, a single sand grain impacting the gold foil was likely to cause the circuit to fail.

The numerical value of a critical threshold depends on the neighborhood rule used to identify clusters. When an eight-neighbor rule is used to identify clusters, the value of p_c drops to 0.40725 (Table 6.2; see Figure 6.4). Because diagonal neighbors are now also counted as cluster members, potential neighbors are farther away from each other. With the inclusion of more distant neighbors within the cluster, large dendritic structures form and percolate across the grid at lower values of p. The ecological justification for the analysis of landscape pattern with different neighborhood rules is process dependent. For instance, if dispersal of a disturbance is slow and by immediate contact (e.g., some fungal diseases), then the nearest-neighbor rule might be applied and a critical threshold of the spread of the fungus would occur at $p = 0.59275$. However, short-distance dispersal of large seeds might cover a neighborhood of considerable area, resulting in a revised definition of connectance among neighboring sites. It may also be necessary to change the neighborhood rule if the resolution of the map were to change. For instance, a four-neighbor rule applied to maps with 90-m grid cells might be changed to a 21-neighbor rule if map resolution were increased to 30 m (other alternatives exist and may be explored in the lab exercises associated with this chapter). The value of the critical threshold has also been shown to vary with map geometry (Table 6.2), primarily because different map geometries have a different number of neighbors (e.g., a triangular grid has three neighbors associated with each site, whereas a honeycomb grid would have eight neighbors). Even though the value of the threshold may change, the general response of the system is similar no matter what rule is applied (Figure 6.4).

An initial misunderstanding in the use of random maps as NLMs was the idea that NLMs were intended to represent actual landscape patterns. When the original neutral model paper was presented at the U.S. IALE meeting in Charlottesville, Virginia, in the spring of 1987, one comment that was heard was "Why would anyone want to use random methods to represent actual landscapes? Landscapes aren't random—they are organized in complex ways." This question is a natural one, but the answer is simple: *NLMs do not represent actual landscapes, but provide the standard against which actual landscapes may be compared.* Interestingly, a comparison of NLMs with actual landscapes (Gardner et al., 1992b) shows many similar patterns. Aerial photographs for nine counties in Georgia taken at three different times (Turner and Ruscher, 1988) were used to develop 27 landscape maps. The nine counties included three from the piedmont and two each in the mountains, upper coastal plain, and lower coastal plain. The number of clusters and total edge of forested areas compared to NLMs are illustrated in Figure

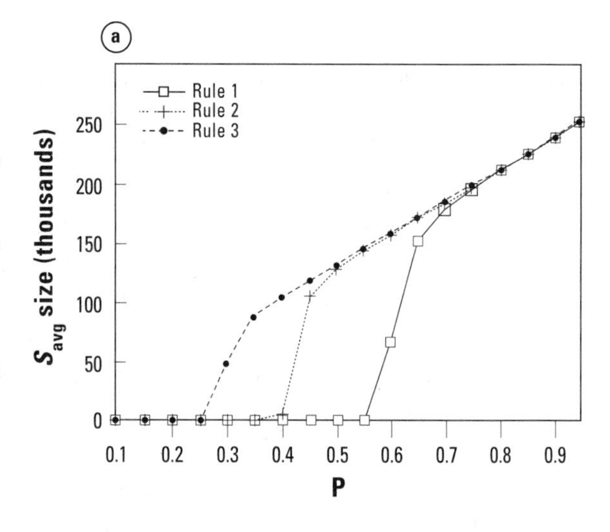

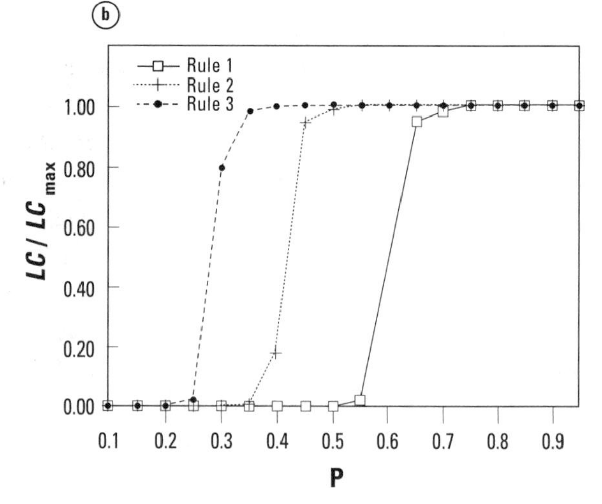

FIGURE 6.4.
Changes in the critical threshold as measured by (a) the area-weighted mean cluster size and (b) the relative size of the largest cluster with changes in the neighborhood rule. Rule 1 is the nearest-neighbor rule (four neighbors), rule 2 includes the next-nearest neighbor (eight neighbors), and rule 2 is the third nearest-neighbor (twelve neighbors). The relative size of the largest cluster is calculated as the ratio of the largest observed cluster (LC) divided by the largest cluster possible (LC_{max}), where LC_{max} is the potential size of the largest cluster if all available habitat were arranged into a single cluster.

ADAPTED FROM PEARSON ET AL., 1996.

6.5. A number of points illustrated by this comparison have been subsequently confirmed by analysis of data from other areas (Gardner et al., 1993a). The key points are the following:

1. It is trivially true that patterns of random and real landscapes are identical when p is equal to either 0.0 or 1.0. It is important to remember that the nearer the value of p is to these limits the more similar random and real landscapes become.
2. The total number of clusters of actual landscapes and NLMs is greatest when p is within the range of 0.1 to 0.3. Over the range of $p = 0.1$ to 0.5 the total number of clusters in actual landscapes is noticeably less than that of NLMs.

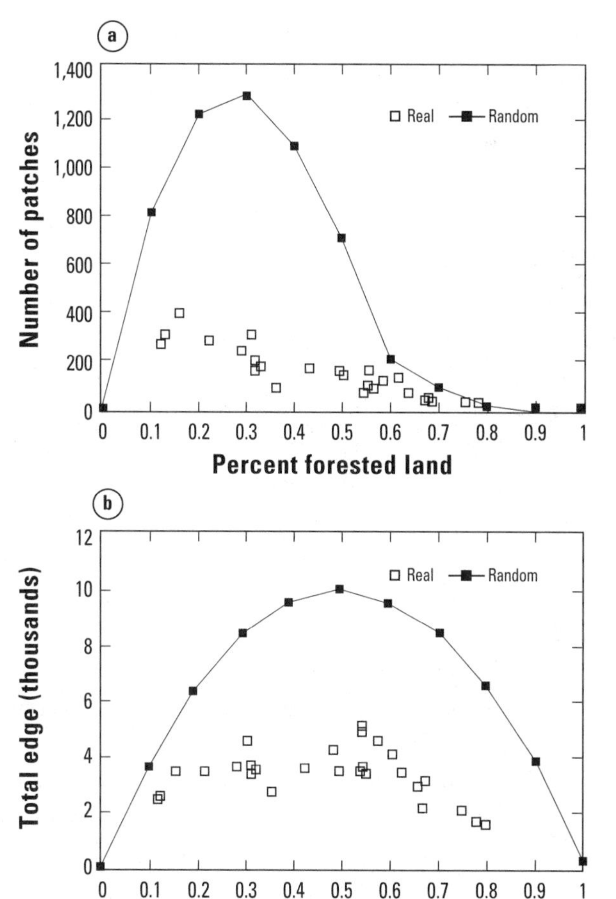

FIGURE 6.5.
Comparison of (a) the number of clusters and (b) total number of edges for random and actual landscapes.

ADAPTED FROM GARDNER ET AL., 1992B.

3. In actual landscapes and NLMs, the total amount of edge is at a maximum near $p = 0.5$, but, like the cluster numbers, the amount of edge in actual landscapes is much less than in random ones.
4. The degree of connectivity (as measured by the presence of a single cluster spanning the map) was equivalent to the NLMs in 25 of 27 actual landscapes. The two landscapes that differed from NLMs either percolated at $p = 0.43$ or failed to percolate at $p = 0.68$. The causes of this deviation were the interaction of topography (ridge and valleys) and the process of human land-use conversion.

The qualitative trends in pattern in actual landscapes and NLMs are similar although the magnitude (e.g., number of clusters, amount of edge) is less and the variability greater in actual landscapes. These differences are produced by a complex suite of factors that organize patterns on actual landscapes, causing noticeable differences from patterns produced by simple random processes.

The most fascinating aspect of percolation theory is the existence of critical thresholds—and the implications that these thresholds hold for relating landscape pattern to ecological processes. Analysis clearly shows that, although landscapes are not random, critical thresholds still exist where sudden changes in landscape pattern may occur with small shifts in disturbance regimes or changes in land use. The factors that organize actual landscapes make the prediction of the exact value of p_c for any given landscape uncertain (see Question 6.3). Nevertheless, critical thresholds for real landscapes do exist! Thus, above p_c we can expect landscape pattern to be dominated by a single very large cluster, whereas landscapes with values of p below p_c will be characterized by numerous, smaller, fragmented patches. If the value of p_c is estimated from a specific neighborhood rule (e.g., the neighborhood defined according to species or disturbance specific parameters), then ecological dynamics can be expected to shift at the critical threshold. Such shifts in pattern that result from small changes in land use have important implications for metapopulation dynamics and the conservation of species diversity (Dale et al., 1994b; Pearson et al., 1996; also see Chapter 8). Metapopulation dynamics exist in landscapes below p_c, whereas a single, large population dominates dynamics in landscapes where the amount of habitat is above p_c. Conservation efforts should be cognizant of the implications of critical thresholds and connectivity in actual landscapes. Because small changes in available habitat near the critical thresholds result in disproportionately large changes in the degree of landscape fragmentation, efforts to preserve continuous tracts of habitat are highly vulnerable to disturbance effects when the amount of habitat is near the critical threshold.

A practical application of NLMs has been their use to test the performance of different landscape metrics across a range of conditions (e.g., number of habitat types, map sizes, values of p). The evaluation of spatial indexes by NLMs before they are applied to actual landscapes and the systematic comparison among similar landscape indexes provide important information on the reliability of different metrics to identify unique patterns on actual landscapes (Gardner and O'Neill, 1991; Gustafson and Parker, 1992; Gardner et al., 1993a; Schumaker, 1996). Several essential messages from these studies bear repeating. The first is that the value of p is a dominant factor affecting the value of nearly all landscape metrics. Indeed, p often enters directly into the calculation of the metric itself (e.g., diversity, contagion) or indirectly as an indication of the amount of habitat found on the map. As p increases, the number of possible arrangements of land cover decreases. Obviously, differences in landscape pattern as a result of differences in p are not surprising. Relating these differences to processes that affect pattern must first account for processes that have altered the value of p! The second caveat is that the critical threshold causes a transition from many small to fewer large clusters on the map. Therefore, large differences in landscape metrics should be expected above and below this critical threshold. It is questionable whether metrics insensitive to this transition will provide useful insight into landscape pattern and process. Finally, it is very clear that the introduction of new indexes without prior testing by a series of neutral models should be regarded as a serious omission.

MAPS WITH HIERARCHICAL STRUCTURE

Analysis of actual landscapes frequently reveals shifts in pattern with changes in scale (see Chapter 1), and the factors that contribute to landscape pattern operate over a wide range of scales (see Chapter 4). Thus, many landscapes exhibit patterns that can be detected at multiple spatial scales. For instance, Anderson (1971) found multiple scales of pattern in three Australian dry-land communities; Krummel et al. (1987) used the fractal dimension to reveal two distinct scales of landscape pattern; changes in pattern with scale were found by O'Neill et al. (1991a) when the change in variance of habitat types was plotted against sample area; and an analysis of replicate transect data revealed three to five scales of distribution of vegetation (O'Neill et al., 1991b). The contagion index (see Chapter 5) can be used to control pattern at a single spatial scale by producing varying degrees of habitat clumping (Gardner and O'Neill, 1991; Fall and Fall, 1996).

Random patterns constrained by contagion result in shifts in patch number, size, and shape (Gardner and O'Neill, 1991).

A general and flexible method of generating NLMs with scale-dependent patterns is by recursive procedures used to construct maps with hierarchical patterns (O'Neill et al., 1992a; Lavorel et al., 1993). A recursive procedure is one that references itself. For instance, we can use recursion to calculate a factorial ($n!$) using the simple relationship that $n! = n \times (n - 1)!$ and calling this procedure repeatedly. This repeated calling of the same routine as n becomes successively smaller (assuming that n is ≥ 0) is a simple example of a recursive algorithm. Hierarchically structured random maps, also known as *curdled* systems in fractal geometry (Mandelbrot 1983), can be produced by recursion (for more examples of fractal geometry and recursion see Barnsley et al., 1988). Recursion in the curdling algorithm (Figure 6.6a) simply defines a series of scales and then successively generates random pattern within each scale. The sequence of steps requires several parameters to control the process, including the number of hierarchical levels L and the size, m_i, and probability, p_i, of sites being set to 1 at each level. The resulting patterns (Figure 6.6b) allow a variety of scale-dependent structures to be simulated, while maintaining the overall size (M) and the probability (P) of habitat within the map (see Lavorel et al., 1993, and Plotnick and Gardner, 1993, for additional details on map generation procedures).

Hierarchical structures result in fewer clusters and more aggregated landscapes than similar random maps, especially when the probabilities at any specific level (p_i) are greater than p_c. Hierarchical maps also produce multiple patterns (and hence greater variability) at the same level of p. Because these maps are structured, they also demonstrate a significant downward shift in the percolation threshold. For a large map with L levels, the percolation threshold (using the four-neighbor rule) will be equal to 0.59275^L (Chayes et al., 1988). Gardner et al. (1993a) found that hierarchically structured maps displayed remarkably similar patterns to maps of forest habitat measured from remotely sensed data, whereas the structure of these maps was different from simple random maps with the same fraction of suitable habitat. Because real landscapes exhibit scaled patterns that reflect a variety of causal factors (that is, real maps are structured), hierarchically structured neutral models provide a useful means of identifying changes in process (e.g., species dispersal and persistence or spread of disturbances and disease) that occur as a result of scale-dependent patterns (Lavorel et al., 1993; Lavorel et al., 1995). The bottom line is that random maps provide the simplest form of neutral model but lack structures typical of most landscapes. If we wish to study the effect of structure, then hierarchical maps, which generate random structures, might be the most useful neutral model.

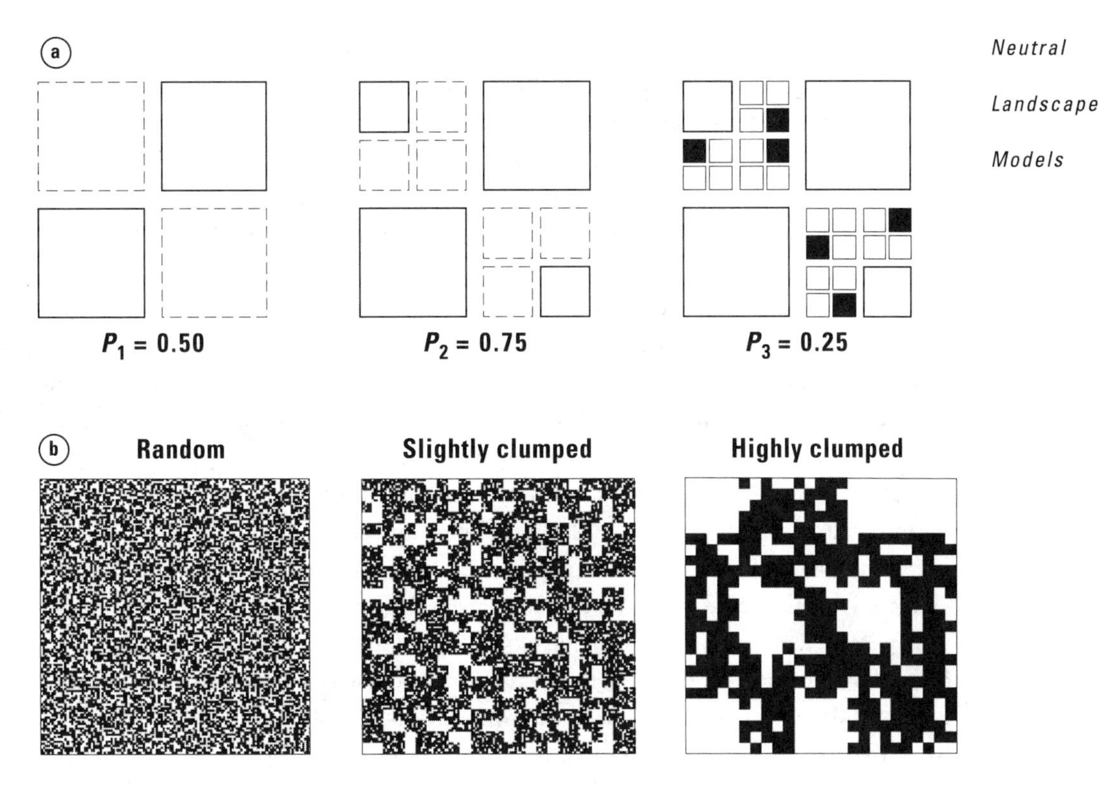

FIGURE 6.6.

(a) Recursive method for generating a hierarchical neutral map (see text) and (b) examples of three maps produced using this method. In (a) the dashed lines indicate successive division of the map with random assignment of habitat to successive levels (solid lines). Shaded cells at the fine scale (level 3) contain suitable habitat. Each map in (b) has $p = 0.52$.

ADAPTED FROM PEARSON AND GARDNER, 1997.

FRACTAL LANDSCAPES

Most landscapes are composed of multiple habitat or land-cover types, requiring a more complex neutral model to characterize these patterns. Often the arrangement of multiple land-cover types is directly linked to the topography of the region. For instance, wetlands and riparian forests are usually associated with rivers and floodplains and found at lower elevations, whereas drier conditions and habitats occur along ridge tops. Conditions between these extremes are intermediate in elevation and usually intermediate in soil moisture and temperature levels. Because habitat characteristics (land-cover types) vary with these elevational gradients, many landscapes with multiple cover types are characterized by a strong autocorrelation between habitat types. Methods that can generate patterns of continuous change would provide a useful neutral model for landscapes with multiple land-cover types.

One method for representing continuous, autocorrelated variation of patterns is the generation of maps by fractal Brownian motion. A fractal Brownian motion in one dimension is produced by creating a series of steps, X_t, whose distance from the previous step $(X_{t+1} - X_t)$ is randomly determined from a Gaussian distribution. A three-dimensional map may be produced by allowing steps to occur in both the X and Y directions, with the random displacements recorded as elevation (the Z direction). The midpoint displacement method (MPDM) for creating fractal surfaces has been extensively used to model three-dimensional patterns (Barnsley et al., 1988). The fractal of fractal Brownian motion is controlled by two parameters: the variance of displacement of points, σ^2 (usually set to 1.0), and H, which controls the correlation between successive steps (Saupe, 1988; Plotnick and Prestegaard, 1993). Because the successive displacement of points results in an expected difference between any two points equal to $(E[X_1 - (X_1 - d)] \propto d^H$ (Plotnick and Prestegaard, 1993), the difference between two points will be proportional to the square of the distance, d, and the correlation, $C(d)$, between the points $[C(d) = 2^{2H-1} - 1]$ (Mandelbrot, 1983; Feder, 1988). The fractal dimension, D, of maps generated by the MPDM is equal to $D = 3.0 - H$ (Saupe, 1988). When $H = 0.5$, successive displacements in the Brownian walk are not correlated; when $H < 0.5$, successive displacements are negatively correlated and maps appear to have a very rough surface; and when $H > 0.5$, steps are positively correlated and the maps have a smooth surface (Figure 6.7). Habitat maps may be generated from the continuous numbers produced by the MPDM by scaling the real numbers and assigning ordinal values to each grid square proportional to the fraction of the map, p_i, occupied by each habitat type (Gardner, 1999). This process of generating a neutral model with fractal maps is summarized in three steps: (1) generation of a topographic map with roughness controlled by H, (2) slicing the topography into contours with the area of each contour equal to the proportion of the map occupied by that habitat type, and (3) assigning ordinal habitat (land cover) values to sites within each contour.

The realistic nature of the fractal maps is the direct result of the autocorrelated

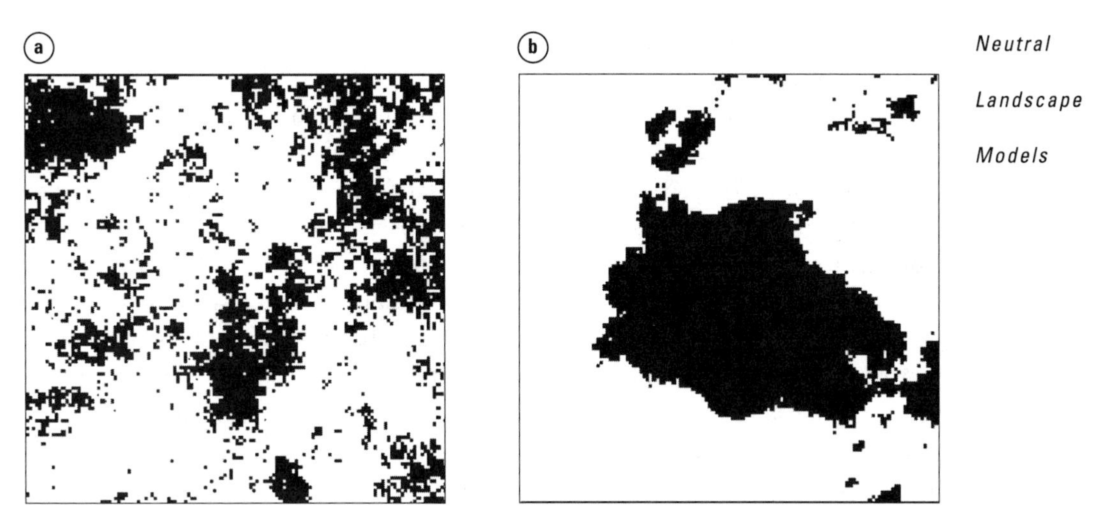

FIGURE 6.7.
Two examples of fractal maps with (a) $H = 0.2$ and (b) $H = 0.8$. Each map has 128 rows and columns, and the value of p (shaded cells) for each landscape type equals 0.33.

process of map generation (Figure 6.7), which results in realistic associations between habitat types (e.g., riparian forests will not be found along ridge tops). Although the patterns are constrained by this autocorrelation, map patterns are still random. The creation of multiple maps with the same set of parameters (map size, H, number of habitat types, and the value of p for each habitat) produces dramatically different patterns. However, successive habitat types will always be associated with each other. The positive autocorrelation of maps with high values of H creates larger average cluster sizes than maps with smaller values of H when $p < p_c$ (Table 6.1). Above the critical threshold, average cluster sizes are similar among all map types.

The generation of spatial patterns with fractal maps has a number of intriguing applications. Fractal landscapes have been used to represent the degree of spatial dependence of actual landscapes (Milne, 1991b, 1992; Palmer, 1992), the effect of landscape fragmentation on patterns of population distribution (With et al., 1997), the effect of spatial contagion on dispersal success (With and King, 1999a), the projection of potential distributions of cesium contamination in aquatic environments (Pearson and Gardner, 1997), and the development of experimental systems to measure the effect of habitat connectivity on insect movement patterns (With et al., 1999). In most of these examples, the effect of structure on habitat arrangement was quantified by comparison of fractal maps with simple (nonstructured) random maps. For instance, the objective of With et al. (1997) was to examine how landscape structure affected the patterns of population dispersion of mobile organisms. Variation in landscape structure was created by generating maps that differed in the number and proportion of habitat types and the methods used to generate the spatial patterns. Simple random maps created pattern without an underlying structure, whereas fractal maps with different values of H created differently structured maps. The results showed that landscape structure had a large effect on the distribution of simulated patterns of species distributions. Although population size remained fairly constant over all simulations, patterns of distribution shifted owing to the aggregation of individuals within specific habitat types. The control of landscape structure created by the comparison of neutral models allowed the effect of pattern and scale to be evaluated.

Pearson and Gardner (1997) used randomly generated fractal landscapes for an entirely different purpose: to determine the consequences of spatial variation in the patterns of ^{137}Cs contamination in a Tennessee reservoir. The spatial pattern of contamination was important because sites within the reservoir with high contaminant levels (hot spots) could affect cleanup strategies. It was believed that contaminant hot spots should be spatially correlated, but the degree of correlation was not known. Fractal maps with varying levels of H were produced to assess the effectiveness of various sediment sampling schemes on the detection of these hot spots. The results showed that spatial patterns could be detected accurately in maps with a large degree of spatial autocorrelation using relatively few samples. However, as autocorrelation declined, the number of samples required to achieve the same degree of accuracy increased dramatically. A comparison of fractal maps with ^{137}Cs distributions estimated by a sedimentation model showed that contaminant levels were positively correlated within deposition zones (areas with similar hydrodynamics), but uncorrelated across different deposition zones.

Other applications of fractal landscapes as neutral models, including the exploration of edge effects (With 1997a, b) and source–sink relationships (Milne, 1992; With, 1997a), have been suggested. Because multiple realizations of these neutral models can be generated, systematic application allows the effect of one component of landscape structure, the autocorrelation among habitat types, to be determined.

NEUTRAL MODELS RELATING PATTERN TO PROCESS

As the preceding examples demonstrate, neutral models have not been restricted solely to the generation of random maps and the analysis of landscape patterns. Coupling NLMs with dynamic ecological models allows insight into the relationships between pattern and process within heterogeneous landscapes. In these circumstances, the ability to generate replicate maps creates a control over the variation in spatial heterogeneity not possible with traditional sampling methods. An important use of this information has been the design of efficient empirical studies (Gardner et al., 1991; Pearson and Gardner, 1997; With et al., 1999). Other studies have used NLMs to define the scale-dependent relationships between landscape pattern and the dispersal of organisms (O'Neill et al., 1988b; Gardner et al., 1989; Gardner et al., 1991; see also Chapter 8). Comparisons of model results with natural landscapes have confirmed the importance of critical thresholds on patterns of habitat utilization (O'Neill et al., 1988b; Gardner et al., 1991). Disturbance dynamics have also been studied with NLMs showing that disturbance propagation depends on the structure of the landscape as well as the nature of the disturbance regime (Turner et al., 1989c; Lavorel et al., 1994a; see also Chapter 7). The application of NLMs to landscape issues seems limited only by imagination and is certainly an economic precursor to more expensive empirical studies.

GENERAL INSIGHTS FROM THE USE OF NLMs

The use of NLMs has illustrated a number of important lessons that should be kept in mind when analyzing landscape pattern and relating these patterns to ecological processes. Two of the most important lessons are (1) variation in the amount of habitat, p, on a map will almost always produce significant differences in map patterns, and (2) thresholds exist where small changes in p (and associated changes in pattern) can cause dramatic changes in connectivity. Other related lessons learned from NLMs include the following:

- *Map dimensions*: Map boundaries affect pattern by the truncation of map patches. The truncation effect becomes more serious as map dimensions decline and p increases. Patterns for maps that are smaller than 100 rows and columns may be seriously affected by these truncation effects.

- *Patch structure*: Simple random maps have the greatest number of patches, with the number of patches determined by p. When patterns with contagion are generated (positive or negative associations between sites on the map), the number of patches decreases. For instance, curdled maps generally have fewer patches than random maps, because the hierarchical structure of map generation affects the contagion between map sites.
- *Thresholds of connectivity*: Simple random maps are likely to have a single cluster that spans the map (percolates) when $p \sim 0.6$. Random maps with very high or very low contagion will percolate at $p > 0.6$. Random maps with moderate levels of contagion will percolate at $p < 0.6$. When a landscape is above the threshold of connectivity, patches tend to be large and contiguous, and there is less difference between different patterns. When a landscape is below the threshold of connectivity, patches tend to be small and fragmented, and there may be greater differences between different maps. For instance, curdled maps can percolate when the overall value of $p \sim 0.6$, but each level must also percolate. On all maps (random or real), the probability of percolation is directly related to the size of the largest patch.
- *Connectivity and scales*: Connectivity of sites across a map is defined by the relationships between map pattern and the process of interest that connects adjacent sites. Therefore, connectivity is directly related to habitat abundance (p), the spatial arrangement of suitable habitat, and the resource utilization rule of the process being considered. On random maps, thresholds in connectivity occur near 0.6, 0.35, and 0.25 for successively larger neighborhoods of 4, 8, and 12 neighbors, respectively. Connectivity may be expected to vary most at intermediate levels of habitat abundance (e.g., 0.3 to 0.6).

The use of NLMs for landscape studies has also generated a number of misunderstandings. In a review of neutral landscape models, With and King (1997) pointed out several of the misuses (or pitfalls in the use) of NLMs.

- Agreement of a NLM with a set of observations is not proof that the NLM is true (Caswell, 1976). Agreement may suggest hypotheses that can be experimentally tested to establish their validity.
- The lack of agreement between an NLM and a set of observations does not prove that the excluded processes are responsible for the observed pattern (Caswell, 1976).
- NLMs are theoretical constructs that may not be directly applicable to actual landscapes. For instance, it would be a misuse of NLMs to design a

conservation reserve with the proportion of habitat equal to 0.59275. "On the other hand, approaching the design of the reserve with an appreciation of the importance of connectivity . . . would be an appropriate application" (With and King, 1997).

- It is a misunderstanding of the role of NLMs to reject them as artificial and hence misleading simply because they fail to be good predictors of a particular ecological process (see Schumaker, 1996). No single NLM will be appropriate for all situations. Rather, the NLM should be designed to provide the appropriate null hypotheses against which actual patterns may be tested.

SUMMARY

A neutral landscape model (NLM) is any model used to generate pattern in the absence of the specific processes being studied. Predictions from NLMs are not intended to represent actual landscape patterns, but rather to define the expected pattern in the absence of a specific process. Comparison of the results of NLMs against actual landscapes provides a standard against which measured departures may be compared. If real landscapes do not depart from an NLM, there may be no need for a more complex model. The types of NLMs that may be generated are diverse (see Keitt, 2000, for a unified approach to the generation of NLMs). Random maps provide the simplest NLM, but more complex neutral methods, including hierarchical random maps and fractal maps, have been used to provide insight into the effect of structured patterns of land cover on ecological dynamics.

Studies utilizing NLMs have been important in the development of theory and the testing of methods for the analysis of landscape patterns. Results of these studies have been helpful for exploring the implications of landscape patterns for ecosystem processes, population dynamics, disturbances, management decisions, and conservation design. Neutral models are particularly useful for testing differences between landscapes when experimental manipulation and/or replication is not feasible and also serve as an economical means for designing expensive empirical studies. Because NLMs are useful in identifying the domain where landscape structure matters, they can often be used to find the domain where spatial pattern is inconsequential (With and King, 1997).

NLMs have played an important role in the development of theoretical landscape ecology (e.g., Gardner et al., 1987; Turner et al., 1989c; O'Neill et al., 1992a, b; With and King, 1997) and in understanding the behavior of a variety of measures of landscape pattern (e.g., Gardner et al., 1987; Gustafson and Parker, 1992; Meisel and Turner, 1998). The difficulty of implementing landscape studies will undoubtedly result in a continued growth in the variety of problems to which NLMs will be applied.

DISCUSSION QUESTIONS

1. It was stated in the text that it is not surprising that a (simple random) map one-half the size of another will have an average cluster size that is only one-quarter as large. Provide an algebraic proof that this will always be the case for simple random maps when $p > p_c$. Is this a scaling rule? Explain why this is not the case for simple random maps when $p < p_c$ (see Table 6.1).
2. Can theoretical or empirical rules relating pattern to map size be defined for fractal maps? How would you go about establishing an empirical scaling rule for fractal maps?
3. Percolation theory predicts a critical threshold when $p \geq 0.59275$. What are the assumptions behind the use and application of this value? Do these assumptions apply to actual landscapes?
4. Table 6.1 shows that average cluster sizes of random and fractal maps are nearly the same when $p = 0.7$ or 0.9. Why is this true? Will other measures of landscape pattern also be similar for these values of p? Will the effects of landscape change be undetectable unless p falls below the critical threshold?

RECOMMENDED READINGS

Gardner, R. H., and R. V. O'Neill. 1991. Pattern, process and predictability: the use of neutral models for landscape analysis. In M. G. Turner and R. H. Gardner, eds. *Quantitative Methods in Landscape Ecology,* pp. 289–308. Springer-Verlag, New York.

Gardner, R. H., B. T. Milne, M. G. Turner, and R. V. O'Neill. 1987. Neutral models for the analysis of broad-scale landscape pattern. *Landscape Ecology* 1:5–18.

Milne, B. T. 1992. Spatial aggregation and neutral models in fractal landscapes. *American Naturalist* 139:32–57.

With, K. A., and A. W. King. 1997. The use and misuse of neutral landscape models in ecology. *Oikos* 79:219–229.

CHAPTER 7

Landscape Disturbance Dynamics

Disturbance is an important and integral part of many ecosystems and landscapes. Disturbances create patterns in vegetation by producing a mosaic of seral stages (Figure 7.1) that ecologists have long recognized as important to landscape-level patch mosaics (e.g., Cooper, 1913; Leopold, 1933; Watt, 1947; Reiners and Lang, 1979; White, 1979). The causes, patterns, dynamics, and consequences of disturbances are major research topics in landscape ecology (Romme and Knight, 1982; Risser et al., 1984; Turner, 1987b, 1989; Baker, 1989a, 1989c; Turner and Dale, 1998). Indeed, disturbances may even be required for the maintenance of community structure and ecosystem function (White, 1979; Mooney and Godron, 1983; Sousa, 1984; Glenn and Collins, 1992; Collins et al., 1998). For example, hurricanes contribute to the maintenance of species diversity in many tropical forests, and regular fires may maintain some landscapes, like prairies. Disturbance has been increasingly recognized by ecologists as a natural process and source of heterogeneity within ecological communities, reflecting a real shift in perception during the latter half of the 20th century from an equilibrial to nonequilibrial view of the natural world (Wu and Loucks, 1995).

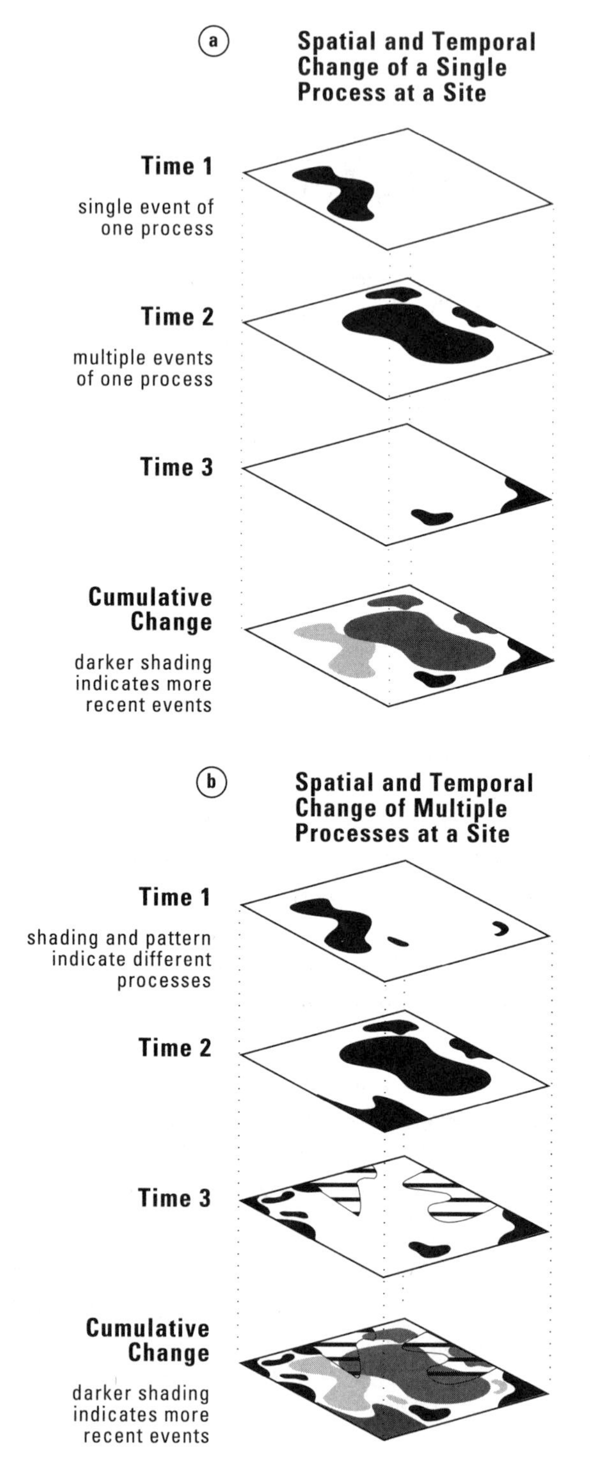

FIGURE 7.1.
Diagrams illustrating patch dynamics. (a) Representation of a process that varies in space and time in a hypothetical landscape. Layers represent the site at different points in time, with the patches representing a disturbance affecting different parts of the landscape at each time interval and the cumulative pattern of the disturbance. Each patch may differ in age, depending on the time it was last disturbed. (b) Representation of multiple processes acting on the same landscape through time and cumulatively. When viewed through time, the landscape looks like a changing patchwork in which patches result from disturbances that differ in frequency, intensity, size, and shape.

ADAPTED FROM PARKER AND PICKETT, 1998.

There are several reasons for considering landscape disturbance dynamics in a separate chapter. Disturbances are unusual in that they both create and respond to landscape pattern. Disturbances usually result in open space, such as gaps in otherwise continuous forest, and they often alter levels of resources such as light and nutrients. By creating these open spaces, disturbances create patchiness in a landscape. However, the effects of disturbance on the biota often depend on the state of the system before it was disturbed. For example, tree height influences the extent of uprooting and stem snapping that occurs in a forest affected by catastrophic wind. The successional stage of a community when it is disturbed may control the availability of propagules that determine, in part, the composition of the postdisturbance community. Disturbances are also of tremendous importance in land and resource management. Fire is a prominent example, but managers must consider a wide range of natural disturbances. Managing human disturbances so that they mimic the spatial and temporal patterns of natural disturbances and reduce undesirable deleterious effects has also been widely discussed (e.g., Hunter, 1993; Attiwill, 1994; Delong and Tanner, 1996). Of course, meeting such an objective requires understanding the dynamics of the natural disturbance regime in a given landscape. Finally, in landscapes subject to large, infrequent disturbances, the spatial pattern imposed by a disturbance event will structure the landscape until the next disturbance occurs. The eruption of Mount St. Helens in 1980 and the Yellowstone fires of 1988 are examples where the large disturbance established the template for species and ecosystem processes in those landscapes for decades or centuries to come.

In this chapter, we focus on how disturbances interact reciprocally with landscape pattern. We begin by providing common terminology and then consider how landscape pattern influences the susceptibility of sites to disturbance and how the patterns of disturbance spread. We next discuss the spatial mosaic created by disturbance, how this spatial heterogeneity influences succession, and the effects of changing disturbance regimes on landscape pattern. Finally, we consider concepts of equilibrium as applied to disturbance-prone landscapes.

DISTURBANCE AND DISTURBANCE REGIMES

A *disturbance* is defined as a relatively discrete event that disrupts the structure of an ecosystem, community, or population and changes resource availability or the physical environment (White and Pickett, 1985). Disturbances happen over relatively short intervals of time: hurricanes or windstorms occur over hours to days, fires occur over hours to months, and volcanoes erupt over periods of days or weeks. In origin, disturbances may be abiotic (e.g., hurricanes, tornadoes, or volcanic eruptions), biotic (e.g., the spread of an exotic pest or pathogen), or some combination of the two (e.g., fires require conditions suitable for ignition and burning, which are abiotic, as well as a source of adequate fuel, which is biotic). Ecologists distinguish between a particular disturbance event, like an individual storm or fire, and the *disturbance regime* that characterizes a landscape. The disturbance regime of a landscape refers to the spatial and temporal dynamics of disturbances over a longer time period. It includes characteristics such as spatial distribution of the disturbances; disturbance frequency, return interval, and rotation period; and disturbance size, intensity, and severity (Table 7.1).

The literature on *patch dynamics* (Watt, 1947; Levin and Paine, 1974b; Whittaker and Levin, 1977; Pickett and Thompson, 1978; Pickett and White, 1985; Levin et al., 1993), in which ecological systems are conceptualized as mosaics of patches generated by disturbance, was an important precursor to the explicit treatment of disturbance in landscape ecology and offered a useful conceptual framework; a concise overview of patch dynamics can be found in Wu and Levin (1994). Patch dynamics implies a relatively distinct spatial pattern (Figure 7.1), but it does not establish constraints on patch size, persistence, composition, or geographic location, which may shift through time (Levin and Paine, 1974a, b). In addition, it implies spatial relationships among patches and between disturbed patches and the surrounding matrix or interspersed undisturbed areas (Figure 7.1). Finally, patch dynamics emphasize change. Disturbance dynamics and succession are intertwined in their effects on landscape patterns and change, and the successional changes that follow disturbance are main components of our understanding of disturbance in a landscape context.

It is important to recognize that the definition of disturbance is scale dependent (Allen and Starr, 1982; Pickett et al., 1989; Turner et al., 1993b; Wu and Loucks, 1995). Allen and Starr (1982) laid this out nicely in their book, *Hierarchy*. They argued that a disturbance can only affect a holon, or particular level within a hierarchy, if the scales of both the disturbance and the holon are similar (see Chapter 2). If the disturbance is small and ephemeral relative to the entity of interest, it passes unnoticed. If the disturbance is large and long in duration or return time, two alternatives may occur. First, the holon may live out its entire life in the presence of the disturbance. Second, the holon may never experience the cycle of the disturbance at all. If, however, the holon and the disturbance are characterized by almost identical scales, then the holon will indeed be perturbed by the transition from the undisturbed to the disturbed state.

TABLE 7.1.
DEFINITIONS OF COMPONENTS OF A DISTURBANCE REGIME.

Term	Definition
Frequency	Mean or median number of events occurring at an average point per time period, or decimal fraction of events per year; often used for probability of disturbance when expressed as the decimal fraction of events per year.
Intensity	Physical energy of the event per area per time (e.g., heat released per area per time period for fire or wind speed for storms); characteristic of the disturbance, rather than the ecological effect.
Residuals	Organisms or propagules that survive a disturbance event; also referred to as biotic legacies. Residuals are measures of severity and thus (at least within one disturbance) an index of intensity.
Return interval	Mean or median time between disturbances; the inverse of frequency; variance may also be important, because this influences predictability.
Rotation period	Mean time needed to disturb an area equivalent to some study area, which must be explicitly defined.
Severity	Effect of the disturbance event on the organism, community, or ecosystem; closely related to intensity, because more intense disturbances generally are more severe.
Size	Area disturbed, which can be expressed as mean area per event, area per time period, or percent of some study area per time period.

ADAPTED FROM WHITE AND PICKETT, 1985, AND TURNER ET AL., 1998B.

What is especially interesting about this hierarchical framework is that a particular disturbance observed at one scale may be a disruptive force, whereas it may be a stabilizing force when observed at a different scale. Consider the phenomenon of forest fire. A tree burns to death once in its lifetime, and if the holon is the tree, then fire is a major perturbation, and the fact that other trees burn at other times is beside the point (Allen and Starr, 1982). If we move from the single-tree holon to the forest holon, however, the components might be species rather than individuals, and the forest integrates multiple fires. Species may be adapted to survive within the forest as a mosaic of patches of differing ages and compositions, such that fire may even be required for maintenance of forest diversity (Allen and Starr, 1982). Thus, as with other aspects of landscape ecology, we must be cognizant of the spatial and temporal scales of the phenomenon under consideration.

INFLUENCE OF THE LANDSCAPE ON DISTURBANCE PATTERN

Landscape Position and Susceptibility of Sites to Disturbance

The term *landscape position* typically refers to the topographic position of a site or group of sites, including relative elevation, landform, slope, and aspect (see Chapter 4). Are various spatial locations in the landscape differentially susceptible to disturbance? If so, can we predict which areas are more or less susceptible to particular types of disturbance? Susceptibility to disturbance of sites located at particular landscape positions can be evaluated by comparing the probability or frequency of occurrence of a particular disturbance at many places in a landscape. A variety of field studies has addressed these questions in different types of ecosystems.

Runkle (1985) studied the disturbance regime in cove forests of the Southern Appalachian Mountains and found it to be determined by regional and local topographic position. Cove forests occur in sheltered areas at middle elevations and are dominated by mesophytic species (e.g., sugar maple, *Acer saccharum*; yellow buckeye, *Aesculus sylvatica*; yellow birch, *Betula lutea*; American beech, *Fagus grandifolia*; silverbell, *Halesia carolina*; white basswood, *Tillia heterophylla*; and eastern hemlock, *Tsuga canadensis*). Regionally, Runkle (1985) found that the occurrence of fire was lower in the mountains of eastern Tennessee and western North Carolina compared to the piedmont farther to the east. Within the mountains, fires occurred primarily on south-facing slopes near ridgetops, especially on ridges at lower elevations. North-facing slopes and sheltered ravines had the lowest incidence of fire. Wind-related disturbances were found to be dominated by small disturbance events, resulting in the deaths of one or a few canopy trees (a *forest gap*) at any given location; interestingly, the rate of repeat disturbance was high. Having initially sampled vegetation in 273 gaps, Runkle revisited these gaps 6 to 7 years later. He found that new gaps often were forming close to the old gaps such that the changed environmental conditions (e.g., greater sunlight) were maintained, and the process of gap closure was slowed. Thus, landscape position influenced some aspects of the disturbance regime, and new gap disturbances were more likely to occur in the vicinity of old gaps.

A series of studies in old-growth forests of New England also demonstrated that disturbance acts selectively within a landscape and that sites can be arranged along exposure gradients. Foster (1988a, b) examined a natural disturbance regime characterized by frequent local events, such as windstorms, pathogens, and lightning strikes, and occasional broad-scale damage by hurricanes and winds. Foster found that slope position and aspect controlled the susceptibility of a site to disturbance. For example, the hurricane winds that affect the region typically come from the southwest toward the east, and site susceptibility to hurricane damage was controlled by the degree of exposure to these directions (Figure 7.2). Following a major hurricane in 1938, exposed southeastern slopes and northwest lakeshores had the greatest damage, and exposed hilltops were also strongly affected. Further work combined analysis of remotely sensed historical and field data with a meteorological model and a topographic exposure model (Foster and Boose, 1992; Boose et al., 1994). Results of these integrated studies demonstrated that forest damage due to hurricanes resulted from characteristics of the storm (e.g., wind directions and maximum gusts), exposure, and the height and com-

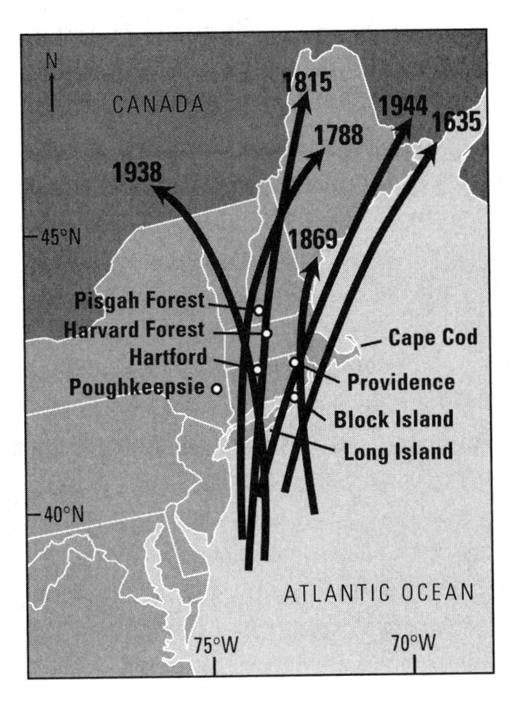

FIGURE 7.2.
Tracks of six severe hurricanes that caused significant forest damage in New England.

ADAPTED FROM BOOSE ET AL., 1994.

position of the vegetation. Other studies have observed similar effects. For example, levels of hurricane damage in Hawaii varied substantially along an elevational gradient (Harrington et al., 1997), and forest stands of similar composition in Minnesota suffered different amounts of windstorm damage based on their location (Dyer and Baird, 1997). After the eruption of Mount St. Helens, topographic positions where snow accumulated and protected the meristems of plants revegetated more quickly than other areas (Dale, 1991; del Moral and Bliss, 1993).

In contrast to the observations of Runkle, Foster, and others, a study addressing these questions in Upper Michigan found no evidence for significant effects of landscape position on disturbance susceptibility. Frelich and Lorimer (1991) studied three tracts of old-growth forest in which the disturbance regime included effects of fire, wind, drought, insects, disease, ice storms, and lightning strikes. They tested for differences in the disturbance regime between upland and lowland sites, aspect classes, slope classes, and the hemlock zone along Lake Superior and the interior forests dominated by sugar maple. Extensive field sampling was conducted, but no significant effects of these factors were observed. The upper Midwest has relatively little topographic relief, and Frelich and Lorimer (1991) suggested that the lack of topographical influence on the disturbance regime in this landscape may reflect the types of windstorms that occur in the Upper Great Lakes Region as compared to those that occur in the eastern United States. The windstorms that do the greatest damage in the Upper Great Lakes have strong downward components, compared to the horizontal winds of hurricanes; even the Southern Appalachians, where Runkle conducted his work, are influenced by hurricanes, albeit infrequently. Studies of the spatial pattern of tornadoes have also demonstrated an absence of physiographic control (Peterson and Pickett, 1995). Downbursts (strong downdrafts of air from a thunderstorm or large cumulus cloud that generate an outflow of wind that travels away from the point where it hits the ground) and tornadoes can hit any position in the landscape, resulting in little differentiation among topographic positions.

The extensive 1988 fires in Yellowstone National Park provide another example in which topographic position exerted little influence on susceptibility of sites to disturbance. Topography can influence fire spread: flames burn more readily uphill than downhill because of the tendency for hot air to move upward. Therefore, leeward slopes often burn with less intensity than windward slopes (Heinselman, 1996). In addition, areas with less fuel or less flammable fuel can function as firebreaks. However, the burning conditions that occurred in Yellowstone during 1988 were so extreme that slope position, fuel, and natural firebreaks (e.g., streams, rivers, and even the Grand Canyon of the Yellowstone River) did not impede the fire. Thus, landscape position had little influence on susceptibility to fire under the extreme drought and wind that occurred in Yellowstone during the summer of 1988 (Turner and Romme, 1994).

In landscapes subject to periodic burning, the probability of ignition may vary spatially (e.g., Burgan and Hartford, 1988; Chou et al., 1993). In Glacier National Park, there is a high frequency of lightning ignitions on ridgelines and south-facing slopes (Habeck and Mutch, 1973). Topographic position does indeed influence fire ignitions, but human influences in the landscape may also produce spatial variability in fires. In the upper midwestern United States, Cardille (1998; Cardille et al., 2001) investigated the relationship between wildfire origin locations and environmental and social factors for >18,000 fires between 1985 and 1995. Results revealed that fires were more likely to occur in locations having higher human population density and higher road density and less likely to occur at greater distances from interior forests. Studies in Alaska, USA, have also demonstrated a strong relationship between fire occurrence and nearness to roads (F. S. Chapin III, personal communication).

Landscape position appears to influence susceptibility to disturbance when the disturbance itself has a distinct directionality (e.g., hurricane tracks) such that some locations are usually more exposed than others. In addition, landscape position may influence susceptibility if the disturbance is of moderate intensity, such that its spread is influenced by subtle differences in the landscape. However, if the disturbance itself has no spatial directionality (e.g., downbursts) or is sufficiently intense that its spread is unaffected by differences in the landscape (e.g., high-intensity crown fire), then landscape position does not influence susceptibility to the disturbance.

Effect of Landscape Pattern on the Spread of Disturbance

Understanding the effect of landscape heterogeneity on the spread of disturbance was identified by Risser et al. (1984) as a fundamental question in landscape ecology, and this was the theme of the first U.S. Landscape Ecology Symposium, held at the University of Georgia in January 1986 (Turner, 1987b). Risser et al. (1984) noted that spatial homogeneity often enhances the spread of a disturbance; consider the spread of pests through agroecosystems, the perpetuation of wildfire, or epidemics such as Dutch elm disease. They also noted that other disturbances may be enhanced by landscape heterogeneity; for example, fragmented forests harbor larger populations of deer that disturb surrounding crops or overbrowse native forest species. But landscape heterogeneity enhances the rate of recovery by providing refuges for organisms that recolonize disturbed areas. A variety of studies suggests an interaction between landscape heterogeneity and the spread of disturbance, although its direction (e.g., whether it enhances or retards spread) differs among disturbance types (Turner and Bratton, 1987; Castello et al., 1995). We will consider several examples that illustrate different aspects of this important interaction.

Theoretical Development

An influential conceptual study by Franklin and Forman (1987) examined the probability of disturbance (e.g., wildfire, windthrow, and pests) as a function of the spatial heterogeneity imposed on a forested landscape by clear-cutting. Indeed, numerous subsequent studies have focused on the mosaic that results from alternative spatial arrangements of clear-cuts in landscapes (e.g., Li et al., 1993; Wallin et al., 1994; Gustafson and Crow, 1996). Franklin and Forman (1987) explored the consequences of forest cutting along a gradient of forest conditions from primeval to a completely clear-cut landscape. They used simple geometric models (Figure 7.3) to evaluate how patch size, number of patches, and lengths of edge changed under different cutting patterns; temporal dynamics (e.g., succession) were not considered. Next, the implications of these alternative arrangements of forest and cutover lands for disturbances were evaluated. Windthrow susceptibility was assumed to increase with the amount of edge, the isolation of forest in small patches, and increasing wind fetch. Results of the model demonstrated that windthrow potential would increase initially with forest cutting and continue to increase as forest patches became isolated on all sides (Figure 7.4a). Wind fetches then progressively increase, especially after half the landscape is cut over; after 80% of the original forest is cleared, windthrow risk to all remaining patches was considered to be maximal. Predictions were also made for susceptibility to both fire ignition and spread, along with risk of particular pest and pathogen outbreaks (Figure 7.4b and c). Results suggested a striking influence of landscape heterogeneity on a variety of disturbances, although the specific effects varied by disturbance type.

Another conceptual framework for studying the effects of landscape heterogeneity on disturbance was developed by Turner et al. (1989c) based on the neutral model approach (see Chapter 6). In this model, the landscape was represented as a grid of 10,000 cells containing habitat that either was or was not susceptible to a given disturbance. Susceptible habitat was distributed at random and occupied different proportions, p, of the landscape ranging from 0.1 to 0.9. Disturbance was then simulated by two simple parameters: f, the probability of initiation of a new disturbance in a susceptible site, and i the probability that the distur-

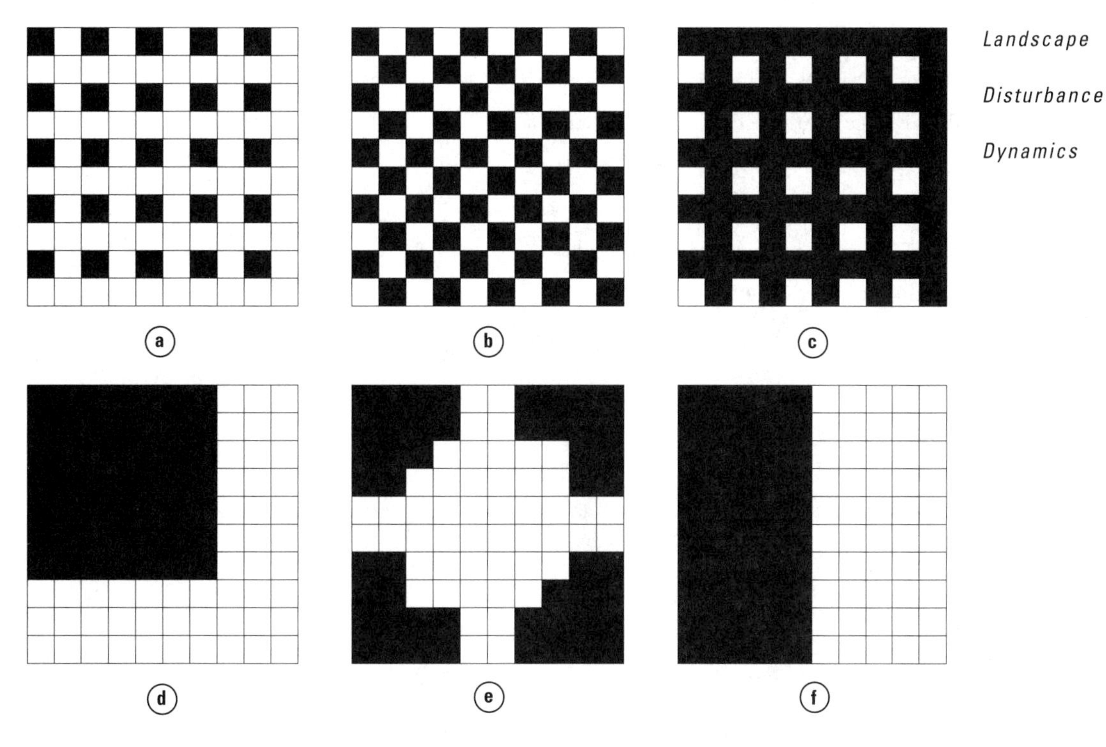

FIGURE 7.3.
Patterns of clear-cutting developed under various models by Franklin and Forman (1987). In (a)–(c), a dispersed cut pattern is used in which the amount of cutover area (black) varies, but there is a regular distribution across the landscape. In (d)–(f), the cutover area is 50%, but it is arranged as a single-nucleus, four-nucleus, or progressive parallel cutting system.

ADAPTED FROM FRANKLIN AND FORMAN, 1987.

bance, once initiated, would spread to adjacent sites of the same habitat. Numerous simulations were conducted in which p, i, and f were varied, and the disturbance was continued until it was extinguished or could not spread any farther. The extent and spatial arrangement of the postdisturbance landscape was then analyzed. Results of these simulations demonstrated a qualitative shift in the influence of the landscape on disturbance spread with changes in p (Figure 7.5). When p for susceptible habitat was less than the critical threshold of connectivity, the percent of available habitat that was disturbed was affected most by f, the probability of new disturbances being initiated, and i had little effect. That is, the frag-

FIGURE 7.4.

Predicted susceptibility of forests in the Douglas fir region to various types of disturbance as a function of the percent of the landscape that is clear-cut, as shown by the checkerboard model (see Figure 2A–C). (a) Potential for windthrow in residual forest patches. (b) Potential for wildfire ignition and spread. (c) Susceptibility to insect and fungus pests.

ADAPTED FROM FRANKLIN AND FORMAN, 1987.

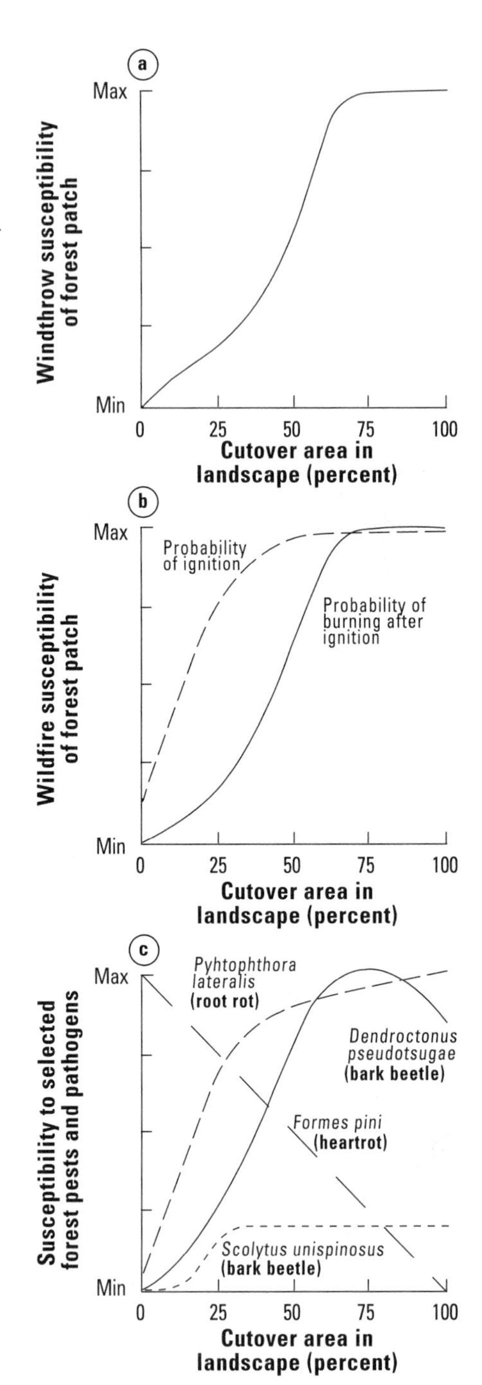

mentation of susceptible habitat into small, isolated patches prevented disturbance spread across the landscape; the only means for affecting a large proportion of the habitat was to initiate disturbance in many patches. In contrast, once p for susceptible habitat was greater than the critical threshold of connectivity, the probability of spread, i, controlled the percent of habitat that was disturbed. Under these landscape conditions, susceptible habitat formed large, continuous patches, and even a single disturbance could potentially spread across the entire landscape.

A class of models also used to evaluate the spatial spread of a disturbance is derived from the theory of epidemiology (Bailey, 1975; Gardner and O'Neill,

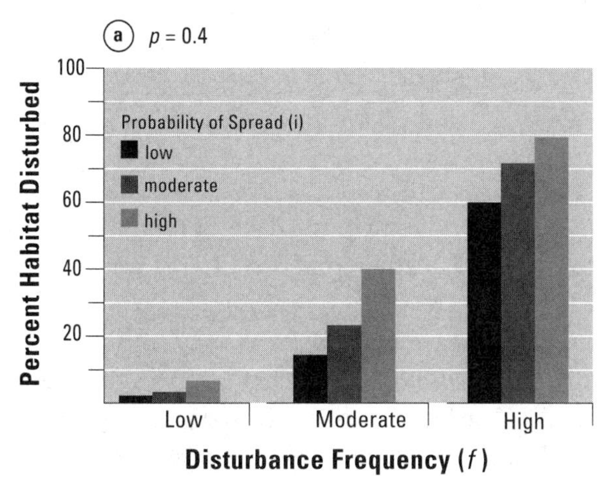

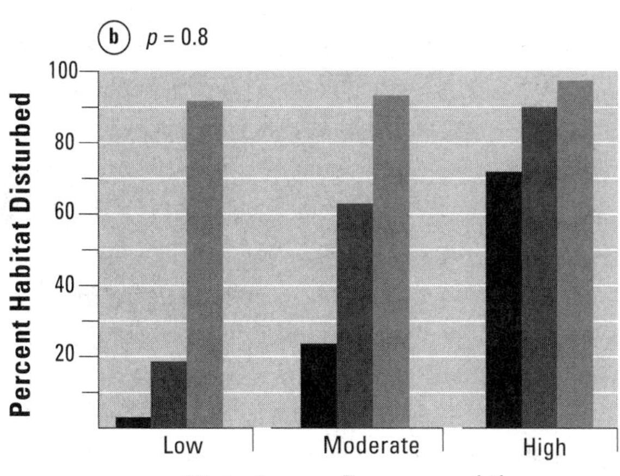

FIGURE 7.5.
Simulated percent of susceptible habitat disturbed as a function of f, the probability of disturbance initiation, and i, the probability of the disturbance spreading to adjacent susceptible sites. (a) Initial proportion, p, of the landscape occupied by susceptible habitat is 0.4, below the threshold of connectivity. (b) Initial proportion, p, of the landscape occupied by susceptible habitat is 0.8, above the threshold of connectivity.

ADAPTED FROM TURNER ET AL., 1989C.

1991). Epidemiology, which deals with the spread of a disease from individual to individual through a population, predicts factors such as the rate of spread and the proportion of the population that is affected by the disease. Models are available that trace the spread of an epidemic through space (e.g., Radcliffe, 1973), but model sites are considered to be equally susceptible to the disease. Thus, in its general form, the *general epidemic theory* can be considered to be a neutral model with respect to the possible effects of spatial heterogeneity in the landscape on the spread of a disturbance. O'Neill et al. (1992b) developed a model derived from epidemiology theory and applied it to the spread of disturbance in a landscape. Results of this model also demonstrated that the spatial pattern of susceptible sites, particularly as related to their connectivity, would be expected to determine the total extent of a single disturbance event. Spatial models of disease spread are proving to be useful tools in understanding and predicting the spread of pests, pathogens, and disease (e.g., Hohn et al., 1993; Liebold et al., 1993; Castello et al., 1995; Nicholson and Mather, 1996).

Empirical Studies

Many studies have focused on the spatial spread of natural disturbances, with pest or pathogen dynamics and fire receiving much attention. There is a rich and varied literature on the subject, and we cannot review it in its entirety.

Landscape heterogeneity due to forest fragmentation was found to enhance outbreaks of the forest tent caterpillar (*Malacosoma disstria*) in northern Ontario, Canada (Roland, 1993). The outbreaks in boreal mixed-wood forests were of longer duration in areas that had higher landscape heterogeneity resulting from forest clearing and fragmentation. This caterpillar exhibits cyclic population outbreaks and declines with a period of about 10 years, and outbreaks occur in forests that have at least some aspen (*Populus tremuloides*). Following an outbreak, the decline from peak density is associated with high mortality caused by a virus and a parasitic fly. However, the duration of the high-density outbreak phase can vary between 2 and 9 years among regions. Using aerial survey data on the spatial extent of three complete caterpillar outbreak cycles over an area of 26,623 km^2, Roland (1993) calculated mean outbreak duration in 261 townships in eight forest districts and compared this to measures of forest and landscape structure. Results demonstrated a strong effect of forest fragmentation as measured by edge density (km forest edge/km^2). Within townships, each 1 km/km^2 increase in edge density increased duration of the outbreak by 0.92 year. Outbreaks in townships with continuous forest cover lasted only 1 to 2 years, whereas townships with 2.0 to 2.5 km/km^2 of edge lasted 4 to 6 years (Figure 7.6). Among districts also, the outbreaks were longer

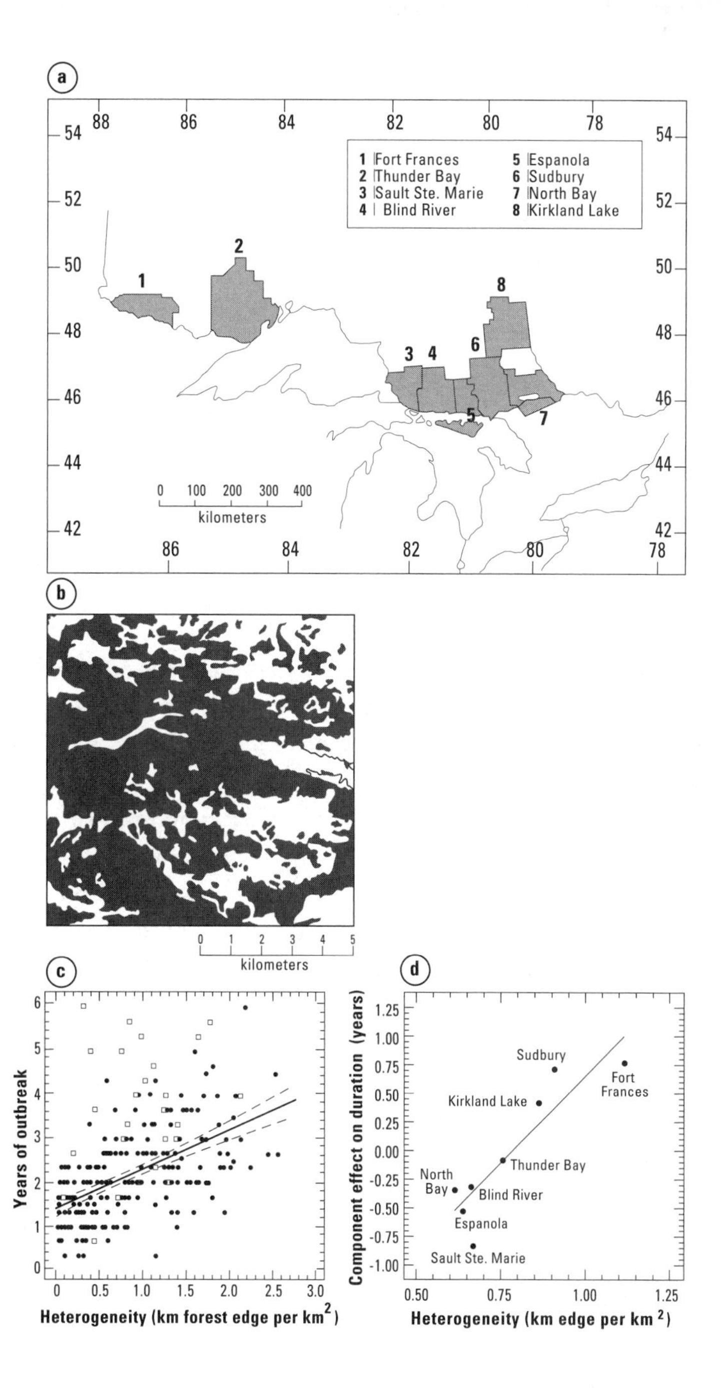

FIGURE 7.6.

(a) Areas of Ontario, Canada, in which duration of forest tent caterpillar outbreaks was related to forest and landscape structure. (b) Example from one township (MacPherson Township, North Bay District) showing the distribution of forest (green) and nonforested (white) land. (c) Mean duration of forest tent caterpillar outbreak for 261 townships as a function of edge density for all townships. (d) Effect of mean habitat heterogeneity on outbreak duration within the eight forest districts.

REDRAWN FROM ROLAND, 1993.

in those districts that had high average fragmentation (Figure 7.6). At both levels, the amount or proportion of aspen was not significant. Roland (1993) suggested several explanations for these results. Broad-scale fragmentation of the forest may affect the interaction between the natural predators on the caterpillar, in that dispersal of the pathogens may be limited by forest fragmentation. In addition, many species of Lepidoptera lay more eggs along the edges of host-plant patches than within the interior, so the forest tent caterpillar abundances may also be greater initially along the forest edges. Warmer microclimatic conditions along the patch edges may also lead to more rapid development of the insect.

Reconstructions of regional outbreaks of western spruce budworm (*Choristoneura occidentalis*) in the western United States during the past three centuries suggest that landscape heterogeneity decreases the spread of this pest. Swetnam and Lynch (1993) found that the 20th century had the longest intervals of reduced budworm activity, and the most recent outbreak that occurred through the 1970s and 1980s was unusually severe. Also, budworm infestations and epidemic periods appeared to be most synchronous during the 20th century, meaning that they were likely to occur simultaneously in many different geographic locations. Budworm infestations develop and spread under conditions of high tree density and connectivity among forest stands, and this was pointed out by interesting variation in the chronologies. For example, a widespread outbreak that occurred between 1900 and 1920 was missing from the Colorado Front Range and the Sangre de Cristo Mountains. Swetnam and Lynch (1993) attributed this absence to the rapid changes that had occurred in the Southern Rocky Mountain mixed conifer forests during this period. Extensive logging and previous fires had reduced conifer densities substantially, and the resulting landscape heterogeneity, in which forests were sparser and less connected, retarded spread of the budworm. Subsequent fire suppression, reduced sheep grazing, and favorable climatic conditions allowed host-tree seedlings to become established, setting the stage for a dramatic future increase in tree density and forest connectivity. During the first part of the century (including the period of the widespread outbreak), these developing forests were less susceptible to budworm outbreaks because they contained few mature trees, and the open stand structure limited dispersal of the budworm. By the 1940s, however, these mixed conifer forests had greater canopy closure, the mature host trees became an important component of stand composition, and the forests were more spatially continuous across the landscape than they had been in presettlement times. Budworm outbreaks subsequently became more widespread and more severe than in earlier periods. Thus, decreased landscape heterogeneity, induced by human activities, resulted in increased spread and synchrony of spruce budworm outbreaks. Swetnam and Lynch (1993) suggested that regional patterns of budworm outbreak, observed as synchrony among widely dispersed stands, were related to climate control on budworm dynamics, primarily through spring rainfall. However, differences in local patterns resulted especially from land-use history in which stand density, stand age, and landscape pattern may override the effects of broad-scale climatic influences.

The role of landscape heterogeneity in controlling patterns of fire spread has been explored in a variety of systems, both through modeling and empirical study. In some coniferous forests, heterogeneity in the pattern of different forest age classes tends to retard the spread of fires (e.g., Givnish, 1981; Foster, 1983; Foster and King, 1986). If flammability is related to stand age (e.g., through stand density and the dead woody fuel mass), the spatial distribution of old and young stands may constrain or enhance fire spread. In California, for example, fires in chaparral were observed to burn well in old stands and become diminished as they spread toward patches of younger vegetation (Minnich, 1983). There may be critical thresholds in environmental constraints that determine whether landscape heterogeneity will influence the spread of crown fire (Renkin and Despain, 1992; Turner and Romme, 1994). Landscape pattern may have little influence on crown fire behavior when burning conditions are extreme (Turner and Romme, 1994) (Figure 7.7). Under conditions of extreme drought and high winds, all fuels across the landscape become highly susceptible to burning and may render the occurrence of large stand-replacing fires inevitable (Fryer and Johnson, 1988; Johnson, 1992; Bessie and Johnson, 1995).

Synthesis

Given the wide range of studies on the subject, can we generalize about whether landscape heterogeneity does or does not enhance the spread of disturbance? Turner et al. (1989c) suggested that the answer depends on whether the disturbance spreads *within* the same cover type, such as the spread of a species-specific parasite through a forest, or whether it crosses boundaries and spreads *between* different cover types. If the disturbance spreads within the same cover type, then greater landscape heterogeneity should retard the spread of disturbance. This is observed in the spruce budworm example (Swetnam and Lynch, 1993) and the spread of fires under moderate burning conditions (Turner and Romme, 1994). If the disturbance spreads between cover types or is otherwise enhanced by edge effects, then increased landscape heterogeneity should enhance the spread of the disturbance. This was observed in the forest tent caterpillar example (Roland, 1993) and windthrow (Franklin and Forman, 1987). Recent results also suggest that there are circumstances in which land-

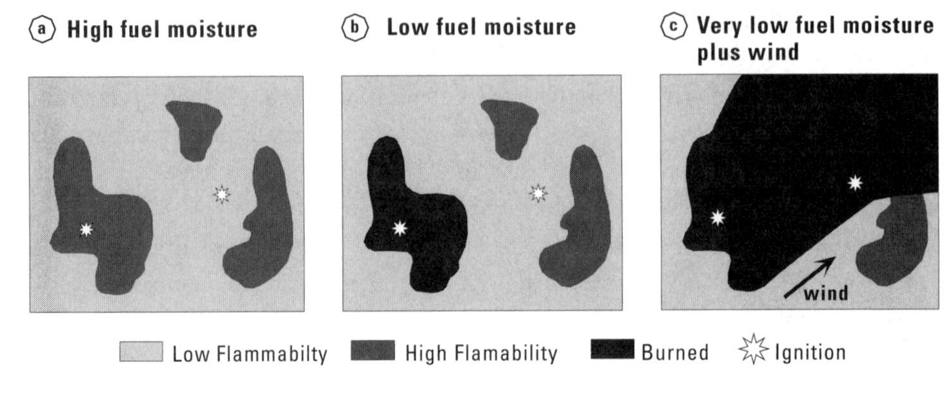

FIGURE 7.7.
Interaction between hypothesized thresholds in both meteorological conditions and landscape pattern that interact to produce large crown fires. (a) If fuel moisture is high, lightning strikes (stars) are unlikely to initiate a fire, even if the strike occurs in highly flammable forest (dark green). Landscape pattern does not control fire spread. (b) If fuel moisture is low, but burning conditions are not extreme, then crown fires (black) are likely to be constrained by the spatial distribution of highly flammable patches in the landscape. (c) If fuel moisture is extremely low and there are strong winds, crown fires (black) are likely to burn through a variety of fuel types. Under these conditions, the patterning of more flammable stands does not constrain fire spread.

ADAPTED FROM TURNER AND ROMME, 1994.

scape heterogeneity does not influence disturbance spread, in that the magnitude of the abiotic control on disturbance overrides the local landscape controls; large infrequent crown fires (Turner and Romme, 1994) and tornadoes (Frelich and Lorimer, 1991; Peterson and Pickett, 1995) provide examples.

INFLUENCE OF DISTURBANCE ON LANDSCAPE PATTERN

The Disturbance-Generated Mosaic

What do we mean by the disturbance-generated mosaic? When a disturbance occurs in a landscape, it does not act uniformly throughout. Rather, disturbances create very complex heterogeneous patterns across the landscape in which the disturbance may affect some areas but not others, and the severity of the disturbance often varies considerably within the affected area. When we talk about the *disturbance mosaic*, or the heterogeneity created by disturbance, we refer to the spatial distribution of disturbance severities across the landscape. For example, the 1988 fires in Yellowstone National Park created a complex pattern of burned and unburned areas across the landscape (Figure 7.8), and the burned areas themselves had widely variable severities within them (Christensen et al., 1989; Turner et al., 1994b). Even very large crown fires rarely consume an entire forest, because variations in wind, topography, vegetation, and time of burning result in a mosaic of burn severities (effects of fire on the ecosystem) and islands of unburned vegetation across the landscape (Rowe and Scotter, 1973; Wright and Heinselman, 1973; Van Wagner, 1983).

Of tremendous importance for the ecological effects of the disturbance and the subsequent patterns of succession are the legacies and residuals that remain after the disturbance (Turner and Dale, 1998). Ecological legacies of disturbance have

FIGURE 7.8.

The landscape mosaic created by the 1988 Yellowstone fires as observed from the air in October 1988, shortly after the fires had been naturally extinguished. (Refer to the CD-ROM for a four-color reproduction of this figure.)

PHOTO BY M. G. TURNER.

both biological and physical components. *Biotic legacies*, or *residuals*, refer to the types, quantities, and patterns of organisms and biotic structures that persist from the predisturbance ecosystem. For example, residuals may include surviving individuals, standing dead trees, vegetative tissue that can regenerate, seed banks, litter, carcasses, and microbial and fungal soil organisms. *Abiotic legacies* are physical modifications of the environment that may result from the disturbance, such as mudslides or slope failures, lava flows, or movements of rocks or boulders in streams. Understanding the nature of the disturbance mosaic and the factors controlling these landscape patterns is essential for predicting ecosystem dynamics and vegetation development in disturbance-prone landscapes.

Foster et al. (1998) examined the landscape patterns and legacies that develop from several types of large, infrequent disturbances that influence forests (Figure 7.9). There has been considerable recent interest in the role of large, infrequent disturbances (Turner et al., 1997a; Turner and Dale, 1998) following a number of natural disturbances that received considerable attention from both the ecological research community and the general public (e.g., the eruption of Mount St. Helens in 1980, the 1988 Yellowstone fires, the 1993 floods in the midwestern United States, and hurricanes Hugo in 1989 and Andrew in 1992). Here, we draw on the synthesis by Foster et al. (1998) to illustrate the variety of landscape patterns generated by catastrophic wind, fires, and floods.

Hurricanes produce a patchwork of forest age and height structure, uproot mounds and downed boles, standing broken snags, and leaning and damaged trees (Figure 7.9) (Foster, 1988a; Foster et al., 1998). The survival, releafing and sprouting of windthrown and damaged trees may influence subsequent ecosystem development. In addition, increased accumulations of fine woody debris and leaves may increase the likelihood of fire occurring in the same area (Patterson and Foster, 1990; Paine et al., 1998). In contrast to hurricanes, tornadoes are relatively small and short-lived, although they are violent and unpredictable. A grouping of tornadoes that affected Pennsylvania, Ohio, New York, and Ontario in 1985 illustrates the landscape pattern of severe tornadoes (Peterson and Pickett, 1995). Tracks of the tornadoes were oriented eastward and northeastward, resulting in more than 800 km of tornado damage. Path widths averaged 500 m and ranged from <200 to >2750 m. The damage patterns of tornadoes are remarkable for the sharpness of the edges between intact forest and completely windthrown areas (Peterson and Pickett, 1995). Hurricanes, tornadoes, and downbursts are extremes in the gradient of size and severity of wind damage; however, all storm types have a gradient of intensities and severities and vary in the spatial extent of damage (Foster et al., 1998).

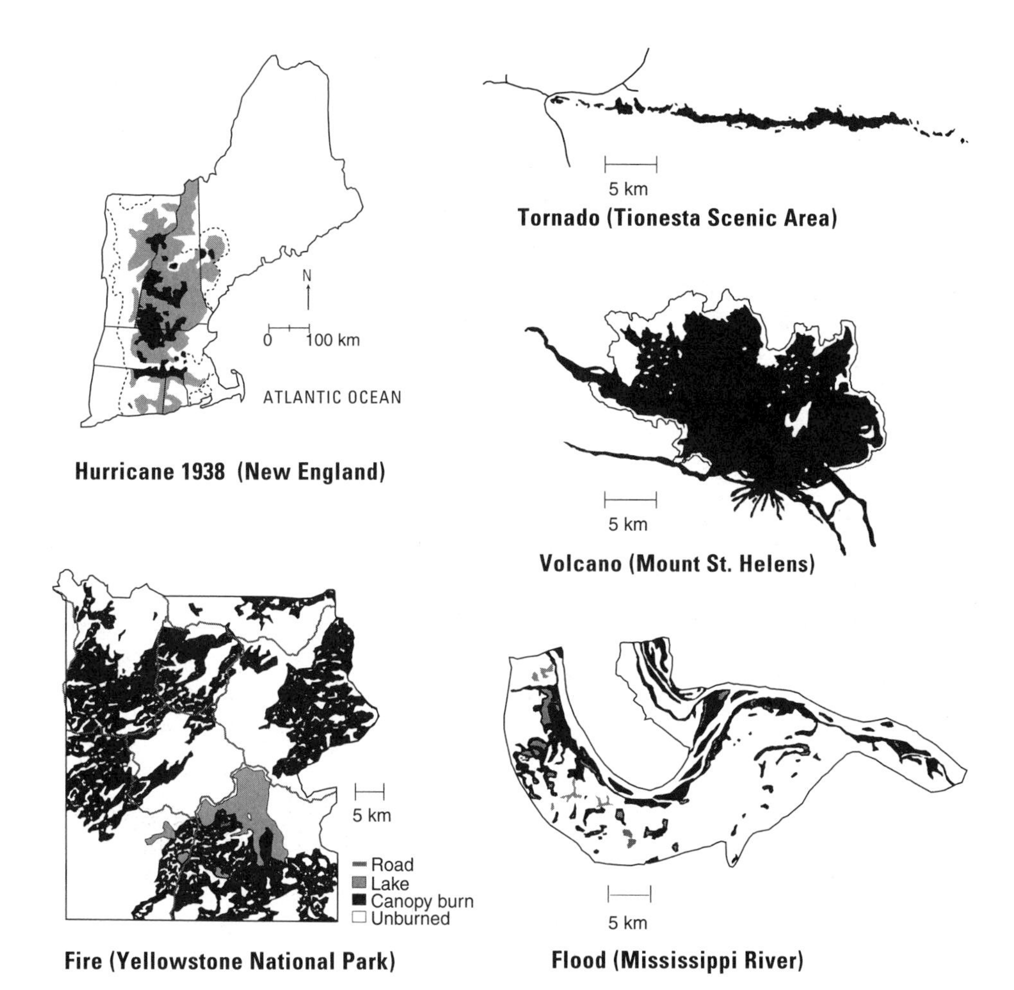

FIGURE 7.9.

Landscape and regional scale patterns of forest disturbance resulting from five contrasting, large, infrequent disturbances: the 1938 hurricane in New England, the Yellowstone fires of 1988, the eruption of Mount St. Helens in 1980, the tornado at the Tionesta Scenic Area in Pennsylvania in 1985, and floods in the Mississippi River in 1993. The areas of greatest disturbance are shown in black. Lesser disturbance severity is shown in gray.

FROM FOSTER ET AL., 1998.

Seasonal flooding is a natural process in many river systems, and the suppression of floods is actually a major disturbance to most river–floodplain ecosystems. However, exceptional floods may create extensive and heterogeneous patterns of damage in the landscape. In large river–floodplain landscapes, for example, flood duration varies spatially across the floodplain with land elevation and is a critical influence on survival of biotic populations (Sparks et al., 1998). In mountain landscapes, areas affected by large movements of soil, sediment, and wood following a large flood are interspersed with refuge sites that experience minor flood effects (Swanson et al., 1998). The severity of flood effects is related to flood duration, geomorphology, species composition, and species physiology. Not surprisingly, the disturbance mosaic created by floods is often correlated with topography (Sparks, 1995).

Like hurricanes, tornadoes, and floods, fires are a climatically driven disturbance. In northern and montane forest landscapes, large infrequent fires account for <3% of all fires, but more than 95% of the land area burned (Johnson, 1992). These fires leave an irregular long-lasting mosaic of burned and unburned vegetation, as well illustrated by the heterogeneity created by the 1988 Yellowstone fires (Christensen et al., 1989; Turner et al., 1994b). As the intensity of a fire lessens, for example, during the evening when humidity is higher, or on leeward slopes, or near the edge of a burn, the proportion of trees and other plants that are killed is reduced. The complex of dead trees, organic matter, and surviving organisms and the pattern of variable-sized patches of unburned vegetation and areas of different burn severities comprise the complex disturbance mosaic that is important in many forested landscapes (Heinselman, 1973; Turner et al., 1994b; Foster et al., 1998; Spies and Turner, 1999).

The landscape patterns resulting from forest harvest strategies have been evaluated by a number of researchers (e.g., Franklin and Forman, 1987; Li et al., 1993; Gustafson and Crow, 1996), and a number of comparative studies have examined the differences in the landscape mosaic resulting from wildfire and forest harvesting. Delong and Tanner (1996) compared the spatial characteristics of landscapes in British Columbia subjected to regularly dispersed 60–100-ha clear-cuts with the historic patterns generated by wildfire. They found that wildfires created a more complex landscape mosaic that included a greater range of patch sizes and more complex disturbance boundaries. In addition, individual wildfires were often >500 ha in size, but unburned forest patches remained within the perimeters of the fire (Delong and Tanner, 1996). Contrasting results were observed for forests in northwestern Ontario, Canada, where Gluck and Rempel (1996) observed that patches in clearcut landscapes were larger in size and more irregular in shape than patches in a wildfire landscape. There has been considerable discussion in the literature about developing management strategies that mimic natural disturbances in a particular landscape (Hunter, 1993; Attiwill, 1994), with the implicit assumption that ecological processes will be better maintained in this way. For example, Runkle (1991) suggested that temperate deciduous forest should be harvested in a pattern that mimics small treefall gaps, whereas Hunter (1993) recognized that boreal forests would require very large clear-cuts if they were to imitate the size and arrangement of boreal fires. Improved understanding is needed of the nature and dynamics of disturbance mosaics in a wide variety of landscapes and how these differ from those generated by human disturbances.

Disturbance and Spatial Patterns of Succession

Effects of the Disturbance Mosaic on Succession

Disturbance and succession are inextricably linked when we consider landscape dynamics. Ecologists have been trying to develop a framework for understanding and predicting vegetation change since the very beginning of the discipline, and an excellent treatment of the development of successional concepts and our contemporary understanding can be found in Glenn-Lewin et al. (1992). Recovery following disturbance can be very sensitive to spatial pattern created by disturbance and is strongly influenced by the spatial pattern of biotic residuals left behind. Investigations into mechanisms of plant succession following fire and other disturbances have often emphasized the autecology and life-history attributes of individual plants and species (e.g., Connell and Slatyer, 1977; Noble and Slatyer, 1980; Peet and Christensen, 1980; Pickett et al., 1987a, b; Halpern 1988, 1989; Peterson and Pickett, 1995). These studies also demonstrated that species responses may vary with different kinds and severities of disturbance and with the larger spatial and temporal context of the disturbance (also see Pickett, 1976; Finegan, 1984; Glenn and Collins, 1992). Patch size, heterogeneity, and distance from undisturbed sites may affect species in a manner dependent on their life-history characteristics (Denslow, 1980a, b; Hartshorn, 1980; Miller, 1982; Malanson, 1984; Green, 1989; Peterson and Carson, 1996). A thorough understanding of succession must include understanding of how successional processes vary with respect to disturbance intensity, size, and frequency (van der Maarl, 1993).

Life-history traits related to the ability of the predisturbance populations to resist or tolerate a particular type of disturbance interact with disturbance intensity to influence the species composition of residuals. For example, mobility and degree of adaptation to flooding were critical in determining the effects of the 1993 floods on taxonomic and functional groups of organisms in the midwestern United States (Sparks et al., 1998). Virtually all individuals of tree species that could not tolerate the anoxic conditions that developed under extended soil saturation during the growing season died (Sparks et al., 1998). In contrast, some species of aquatic plants survived by growing upward into the lighted zone as the flood rose, and a rare species of false aster (*Boltonia decurrens*) that requires fresh mudflats for seed germination increased dramatically (Smith et al., 1998).

Residual plants that reestablish vegetatively following disturbance often achieve large sizes more quickly than those that start from seed, and species with abundant or larger residual seeds have a head start on those that must disperse into the disturbed area from the surroundings. One example comes from the regeneration of a large forest windthrow. In 1985, a powerful tornado created a 400-ha area of windthrow in the old-growth hemlock–northern hardwoods forest of the Tionesta Scenic Area in northwest Pennsylvania. During initial revegetation following the windthrow event, thickets of surviving advance regeneration of *Fagus grandifolia* and *Acer pennsylvanicum* had a substantial size advantage over individuals that germinated after the disturbance, and this size advantage has been maintained (Peterson and Pickett, 1995). These thickets of advance regeneration have severely inhibited local colonization by *Betula alleghaniensis,* which is abundant in other areas of the blowdown having a lower density of residuals (Peterson and Pickett, 1995).

An interesting question regarding the influence of the disturbance mosaic on succession is whether or under what conditions the size of the disturbance patch influences succession. Following the 1988 fires in Yellowstone, measurement of the vegetation during the first 5 years after the fires demonstrated significant effects of patch size and burn severity on early succession (Turner et al., 1997b). More severely burned areas had higher cover and density of lodgepole pine seedlings (*Pinus contorta*, the dominant tree species), greater abundance of opportunistic species, and lower species richness of vascular plants than less severely burned areas. Larger burned patches had higher cover of tree seedlings and shrubs, greater densities of lodgepole pine seedlings and opportunistic species, and lower species richness than smaller patches. The most enduring legacy of the mosaic of burn severities may be spatial variability in lodgepole pine density across the landscape. Areas of severe surface burn may develop persistent high-density stands of lodgepole pine that grade into areas of lower density (Turner et al., 1997b).

Why might disturbance size be important for succession? Disturbance-induced changes in the biophysical environment are subject to edge effects related to disturbance size (Turner et al., 1998b). The centers of large, disturbed patches are likely to experience very different physical conditions than small patches or disturbed areas near intact vegetation. In tropical moist forest, for example, relatively larger gaps experienced higher air temperatures, lower humidity, higher wind speeds, and reduced soil moisture (Denslow, 1987). However, it is on the availability of propagules that disturbance size may exert its strongest effect (Turner et al., 1998b).

The availability of propagules is a fundamental determinant of successional patterns (Clements, 1915; Pickett et al., 1987a, b) and one that can be especially sensitive to the combination of high intensity and large size (Turner et al., 1998b). In small disturbed areas, the surrounding intact community is likely to provide sufficient propagules for succession, even if biotic residuals are few. However, the density of propagule inputs from the surrounding undisturbed area into a disturbed area decreases with distance (Johnson, 1992; Nepstad et al., 1990; Aide and Cavelier, 1994; da Silva et al., 1996), so the proportion of disturbed area beyond the zone of high propagule input decreases as disturbance size increases. If dispersal from outside the disturbed area is important, then the size, shape, and configuration of disturbed patches will influence propagule availability and thus vegetation composition. Distance from the edge of the disturbed patch, which is controlled in part by patch size, has a particularly strong effect (McClanahan, 1986; Bergeron and Dansereau, 1993; Galipeau et al., 1997). For example, the abundance of conifer recruitment following fire in some boreal forests may be strongly influenced by distance to seed sources (Bergeron and Dansereau, 1993).

Turner et al. (1998b) suggested that effects of disturbance size and frequency must first be considered within the context of disturbance intensity as it influences the abundance of residuals. They suggested that succession will be relatively predictable following disturbances of any size when residuals are abundant and the effects of local environmental attributes (e.g., nutrient availability, soil texture, and soil moisture) are considered (Figure 7.10). Spatial effects of disturbance (disturbance size, shape, and arrangement) become increasingly important when residuals are few or sparse and the disturbance is large; under these conditions, colonization and hence succession become slower and more difficult to predict (Figure 7.10). Furthermore, if the frequency of large, high-intensity disturbances increases such that residuals decrease in abundance or change in composition with successive disturbance events, successional pathways may shift qualitatively (Figure 7.10). Landscape context may interact with species life-history traits to initiate different successional trajectories within similar abiotic environments because of local variation in disturbance intensity or availability of plant propagules (Glenn-Lewin and van der Maarel, 1992; Baker and Walford, 1995; Fastie, 1995). Predictability of successional trajectories has received relatively little study (Peet, 1992;

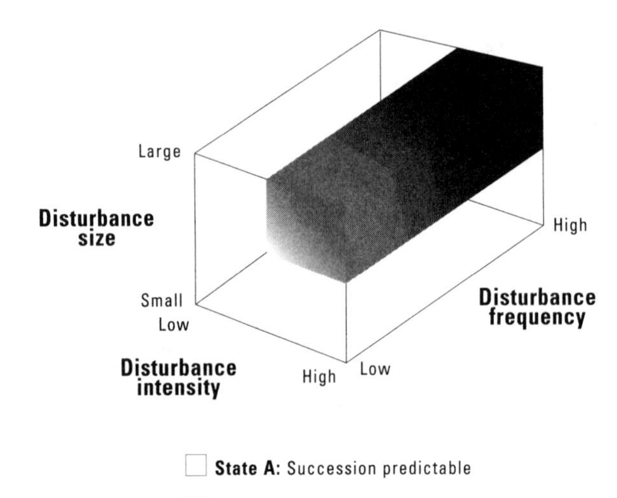

FIGURE 7.10.
Conceptual state-space diagram for succession following disturbances varying in size, intensity, and frequency. Succession is more predictable and spatial attributes of the disturbance are less important whenever disturbance intensity is low (such that residuals are abundant) or disturbances are small (state A). Succession is initially less predictable and determined by disturbance size, shape, and configuration when disturbance intensity is high (such that residuals are scarce) and disturbances are large (states B and C). Successional pathways may be qualitatively altered if high-intensity, large disturbance increase in frequency (state C).

FROM TURNER ET AL., 1998B.

Wood and del Moral, 1993), and it remains challenging to identify the factors controlling vegetation dynamics at multiple spatial scales.

INTEGRATED MODELS OF DISTURBANCE AND SUCCESSION

Models exist in vegetation science to produce reasonably good predictions of successional dynamics at the canopy gap scale (0.1 to 10 ha) (Glenn-Lewin et al., 1992). In temperate and tropical forests, a substantial body of observation and theory has developed around the role of small gaps (e.g., Brokaw, 1982, 1985, 1987; Runkle,

1982, 1985; Lertzman, 1992). In the small-gap paradigm, differences in vegetation dynamics are primarily attributed to gap size, whereas shade tolerance differentiates tree species into early-successional and late-successional species (Swaine and Whitmore, 1988). The forest gap models (Botkin et al., 1972; Shugart and West, 1981) have been applied to a broad spectrum of community types, extrapolating landscape-scale effects from multiple simulations of small (0.1 ha) forest plots. Although simulation of disturbance effects has been included in the gap models (Shugart and Noble, 1981; Bonan and Shugart, 1989), inclusion of spatial effects of dispersal and fire has been accomplished by linking multiple plots (replicate models) within a spatial network (Urban et al., 1991; Acevedo et al., 1996).

Applying the theoretical and empirical advances in understanding of how disturbance and succession interact with large, heterogeneous landscapes is challenging because of the spatial interactions and the long time scales involved (He and Mladenoff, 1999; Mladenoff and Baker, 1999). Integrated models of disturbance and succession can reveal trends and dynamics in landscapes that cannot be easily observed empirically, and such models may be particularly useful for addressing questions of broad-scale disturbances or global change. Extrapolating models across spatial scales, however, remains difficult (McKenzie et al., 1996). An example of an integrated model of disturbance and succession is LANDIS, developed for a 500,000-ha landscape in northern Wisconsin, USA (Mladenoff et al., 1996; He and Mladenoff, 1999; Mladenoff and He, 1999). The study landscape is located in the transitional zone between boreal forest to the north and temperate forest to the south; the region was glaciated, has little topographic relief, and is largely forested. The model represents landscape heterogeneity caused by spatial variation in environmental conditions and disturbance rates, as well as the effects of past human uses in the landscape. Multiple disturbance types are simulated, including fire, windthrow, and forest harvest, and succession is represented at the species level. Successional dynamics are based on life-history attributes (e.g., shade tolerance, disturbance susceptibility, vegetative reproduction, time to sexual maturity, longevity) of up to 30 species, and spatial processes such as seed dispersal are included.

LANDIS was initialized with the current landscape pattern and used to simulate landscape dynamics for 500 yr, with fire as the only disturbance (He and Mladenoff, 1999). Results demonstrated that, even when the presettlement disturbance regime was reestablished, some species, such as hemlock, yellow birch, oak, and pine, did not recover their presettlement proportions of the forest community for 100 to 500 yr. They also found that landscape recovery could be detected at broad spatial scales as the composition of the forest communities differentiated on different land types. Interestingly, the landscape showed greater alteration on the more mesic landforms in which disturbances were infrequent but severe, as compared to the more xeric landforms in which disturbance was more frequent but less severe. Similar results have been observed in other systems (Gardner et al., 1996; Turner et al., 1998b). It is important to remember that these modeling approaches are not used deterministically to predict what will happen with specific individual events. Models such as LANDIS and Fire-BGC (Keane et al., 1996a, b) can incorporate feedbacks among species, disturbance, and environmental variability and are valuable tools for examining complex interactions of species and disturbance over large areas and long time periods.

Effects of Changing Disturbance Regimes on Landscape Pattern

Because disturbances can be such a strong source of landscape structure, intentional or unintentional shifts in the disturbance regime may dramatically alter the landscape. For example, past climatic changes of small magnitude have caused significant changes in fire regimes in forested landscapes (Green, 1982; Hemstrom and Franklin, 1982; Clark, 1988, 1990; Bergeron, 1991; Campbell and McAndrews, 1993), and fire suppression during the past century has lengthened the fire return interval and altered successional pathways in many regions (e.g., Glitzenstein et al., 1995; Linder et al., 1997). Changes in the frequency and extent of flooding have resulted in ecological responses in many river–floodplain systems. However, enhancing our quantitative understanding and ability to predict the effects of changing disturbance regimes on landscape structure remains a current topic of active research (Graham et al., 1990; Miller and Urban, 1999). In particular, little is known regarding which components of landscape structure are most sensitive to change or how directional changes in landscape structure can be detected.

Spatial models have proved valuable in exploring the sensitivity of landscape patterns to changes in disturbance regime. Using a GIS-based spatial model and data on historical changes in fire sizes, Baker (1992) simulated the effects of settlement and fire suppression on landscape structure in the Boundary Waters Canoe Area of northern Minnesota, an area in which fire dynamics have been well studied (Heinselman, 1996). Prior to European settlement, fires were relatively large and infrequent. As the upper Midwest was settled by Europeans, fire frequency increased substantially because of indiscriminate burning by early settlers, land speculators, and prospectors. Subsequently, there was an extensive period of active human fire suppression. The periods of settlement and fire suppression, which represented substantial shifts from the presettlement disturbance regime, produced significant effects on landscape structure (Figure 7.11). Settlement produced immediate effects on some metrics of landscape pattern, including patch age, patch shape, and Shannon diversity, but no effect on other measures, including patch size or fractal dimension. Fire suppression resulted in immediate responses in a few metrics, but a delay of several decades in patch age and fractal dimension and a delay for hundreds of years in others, including patch size. Effects on landscape structure were more likely to be immediate when fire size declined but frequency increased (as with the change from presettlement to settlement regimes) (Figure 7.11). Results of this study suggested that alteration of the fire regime of large, infrequent fires to smaller, more frequent fires would be un-

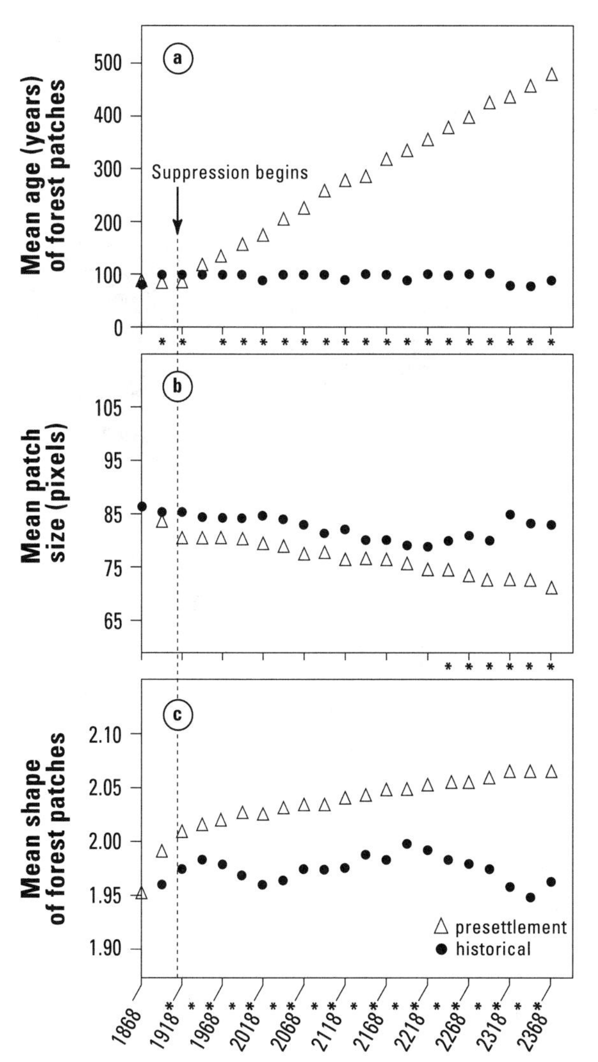

FIGURE 7.11.

Changes in landscape structure in the Boundary Waters Canoe Area, northern Minnesota, USA, under changing disturbance regimes. (a) Mean age of forest patches, (b) mean patch size, and (c) mean shape of forest patches.

ADAPTED FROM BAKER, 1992.

likely to restore the landscape to presettlement conditions. For the Boundary Waters Canoe Area, unusually large fires would actually hasten the return of the landscape to its presettlement pattern of change (Baker, 1992).

In another study in which spatial simulation modeling was used, Wallin et al. (1994) evaluated changes in landscape structure in response to alternative disturbance regimes. In this case, the disturbance regimes were alternative forest cutting plans in which the size, spatial location, and timing of clear-cuts were varied (Figure 7.12). This study was motivated by the question of how long a pattern imposed by a particular forest cutting pattern would persist once the disturbance regime was changed, for example, if forest harvest was to cease or if return intervals were to be lengthened considerably. Results demonstrated that the landscape pattern created by dispersed disturbances was difficult to erase without a substantial reduction in disturbance rate or a reduction in the minimum stand age eligible for disturbance (Wallin et al., 1994). Even after only 20 yr of dispersed

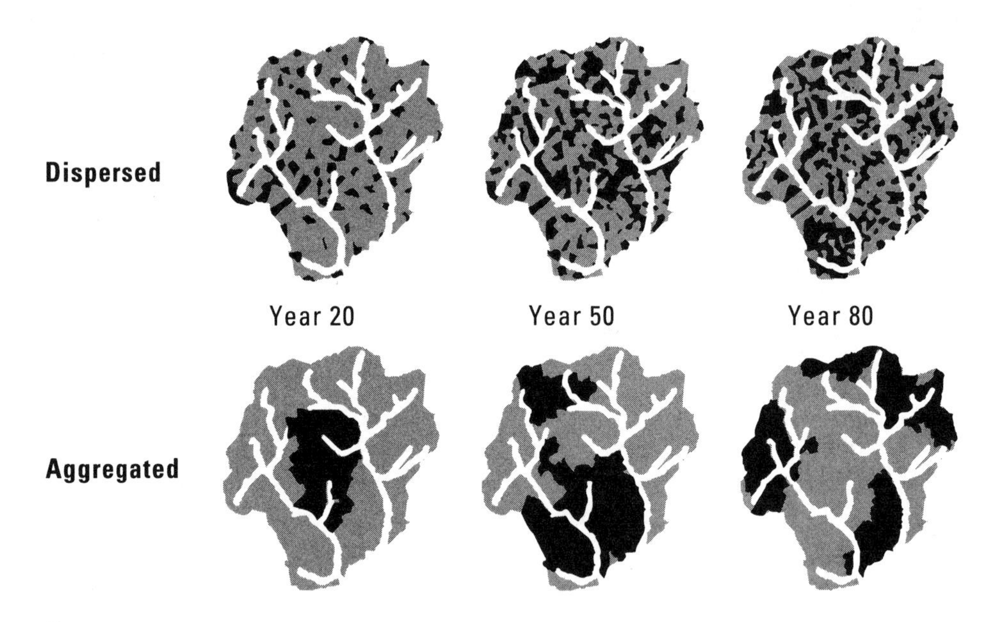

FIGURE 7.12.

Landscape pattern development for one pair of simulations comparing the dispersed and aggregated cutting plans. Green shading represents recently cut units; gray represents closed-canopy forest, and white areas are riparian zones.

REDRAWN FROM WALLIN ET AL., 1994.

cutting, the switch to aggregated cutting produced only small changes in landscape pattern as reflected by edge density and mean size of interior forest patches (Figure 7.13); after 40 or 60 years of dispersed cutting, the change in disturbance regime produced an even smaller effect. This study demonstrated that the response of landscape pattern can lag substantially behind a change in the disturbance regime.

In a changing climate, the plant communities that become established following disturbance may also differ from those present at the time of the disturbance. On Mt. Rainier, for example, Dunwiddie (1986) demonstrated that fires that occurred during the mid-1800s burned through an *Abies amabilis* and *Tsuga mertensiana* forest that had persisted for centuries. The mid-1800s were characterized by warming temperatures that caused earlier seasonal snowmelt and longer growing

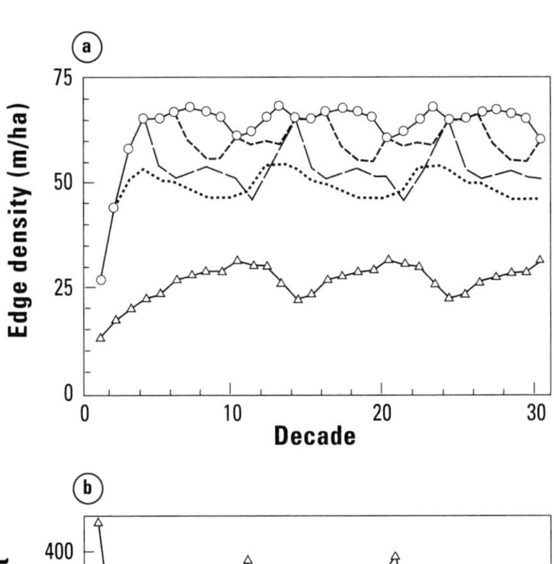

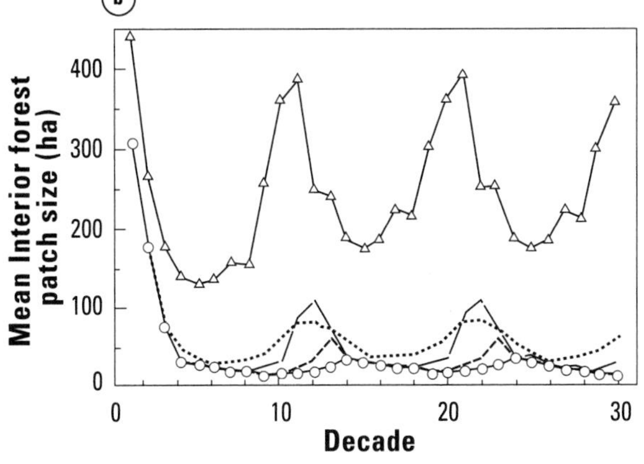

FIGURE 7.13.

Three-hundred-year simulations using a 100-yr rotation length (cutting rate of 10% per decade) and a canopy-closure age of 30 yr. All runs were initiated using the dispersed-cutting rule set and then switched to the aggregated-cutting rule set after 20 (····), 40 (- - -), or 60 (—) yr of cutting. Each line represents the mean of five replicates. Mean response curves for 300-yr of dispersed (- - -triangle- - -) and aggregated (—triangle—) cutting are included for comparison. (a) Edge density; (b) mean interior closed-canopy forest patch size.

ADAPTED FROM WALLIN ET AL., 1994.

seasons. These climatic conditions allowed *Tsuga heterophylla* to become abundant briefly after the fires, and *T. heterophylla* was then replaced by *Abies lasiocarpa*. Because long-lived mature trees may survive short-term climatic fluctuations, species that are best adapted to the current climate may only be able to enter the forest in open habitats following severe fires, and forest composition may respond to climatic changes primarily after disturbance (Dunwiddie, 1986). A study by Cwynar (1987) also suggests that, although the ultimate cause of postglacial vegetation change in the Pacific Northwest was climate change, the proximate cause of some postglacial vegetation changes was an altered fire regime. A small change to a drier climate probably triggered a relatively large change in the disturbance regime by increasing fire frequency (Cwynar, 1987). Potential climatically induced alterations in crown fire regimes may therefore lead to substantial landscape changes, in terms of both the characteristic vegetation mosaic and the species composition of particular regions. Understanding the sensitivity of disturbance regimes to climate change remains an important topic of current research.

One general result to emerge from these studies is the importance of using a variety of metrics of landscape pattern to detect changes resulting from alterations of disturbance regimes. Some aspects of pattern are likely to be affected more than others or to respond more quickly, yet we presently do not have sufficient data from which to recommend a narrow set of metrics sufficient to characterize or predict disturbance effects. The paucity of data and the impossibility of experimental manipulation at landscape scales make the development and testing of hypotheses problematic.

CONCEPTS OF LANDSCAPE EQUILIBRIUM

The notion of equilibrium in ecological systems has inspired a long history of interest and controversy in ecology (e.g., Egerton, 1973; Bormann and Likens, 1979; Connell and Sousa, 1983; Wiens, 1984; DeAngelis and Waterhouse, 1987). Pickett et al. (1994:159) identified equilibrium as one of few overarching paradigms in ecology, and one of the oldest and most pervasive, that affect the dialog between observable phenomena and conceptual constructs in all the more specialized areas of ecology. Six tenets of this paradigm were identified (Pickett et al. 1994), in which ecological systems were considered: (1) to be essentially closed, (2) to be self-regulating, (3) to possess stable point or stable cycle equilibria, (4) to have deterministic dynamics, (5) to be essentially free of disturbance, and (6) to be independent of human influences. The shift in ecology from an equilibrium view of nature to the nonequilibrium paradigm occurred gradually throughout the 20th century. In the nonequilibrium paradigm, ecological systems are thought to be open, to be regulated by both intrinsic and extrinsic factors, to lack a stable point equilibrium, to be nondeterministic, to incorporate disturbance, and to admit human influence. If equilibrium is observed, it may only appear at certain spatial and temporal scales (Pickett et al., 1994). Because these ideas represent such a fundamental shift in ecological thinking and because they are so relevant for both basic and applied ecology, we review the development of the concepts as they apply to landscape ecology in some depth.

Questions of whether equilibrium could be detected on landscapes subject to disturbance and how large a landscape must be to incorporate a given disturbance regime have been important themes in landscape ecology (Shugart and West, 1981; Romme, 1982; Baker, 1989a). The controversy regarding equilibrium has stemmed in part from inconsistent definitions and criteria used by investigators and in part from disagreement about whether it is valid to define the existence of an equilibrium state at all (Turner et al., 1993b). DeAngelis and Waterhouse (1987) provide an excellent review of the treatment of these concepts in ecological models. The critical issue is that concepts of equilibrium are confounded by problems of scale, and landscapes can exhibit a suite of dynamics of which equilibrium is but one (Turner et al., 1993b).

Equilibrium points can be precisely defined mathematically, but equilibrium and stability are not well defined when applied to real ecological systems (DeAngelis and Waterhouse, 1987). Properties used to evaluate equilibrium fall into two general categories: *persistence* (nonextinction or presence) and *constancy* (no change or minimal fluctuation in numbers, densities, or relative proportions). Persistence can be applied to all species, as is often done in many population-oriented models (e.g., DeAngelis and Waterhouse, 1987), or to the presence of all stand age classes or successional stages in a landscape (e.g., Romme, 1982). Constancy may refer to the number of species (e.g., MacArthur and Wilson, 1967), the density of individual species (e.g., May, 1973), the standing crop of biomass (e.g., Bormann and Likens, 1979; Sprugel, 1985), or the relative proportions of seral stages on a landscape (e.g., Romme, 1982; Baker, 1989a, c). There are fundamental differences in considering species composition versus structural attributes such as biomass, age classes, and seral stages. Seral stages or age classes do not become extinct because they can be regenerated by disturbances, provided the species comprising each stage do not become extinct.

Equilibrium in the sense of *absolute constancy*, where there are no changes through time, is the simplest concept that might be applied to a landscape. However, disturbance and change are ubiquitous in ecological systems as disturbances reset succession back to earlier stages, and any concept of landscape equilibrium therefore must incorporate disturbance. Even in the absence of absolute constancy, there may be a particular aspect of a landscape that is invariant. In the *shifting mosaic steady-state* concept (Bormann and Likens, 1979), the vegetation present at individual points on the landscape changes, but the proportion of the landscape in each seral stage is relatively constant, that is, is in equilibrium when considered over a large area or long time period (Figure 7.14). Bormann and Likens (1979) suggested that, prior to settlement in northern hardwood forests of New England, the standing crop biomass of a watershed or other landscape unit varied slightly around a mean, although the biomass present at any small plot within the watershed fluctuated through time due to treefalls and subsequent regrowth. The

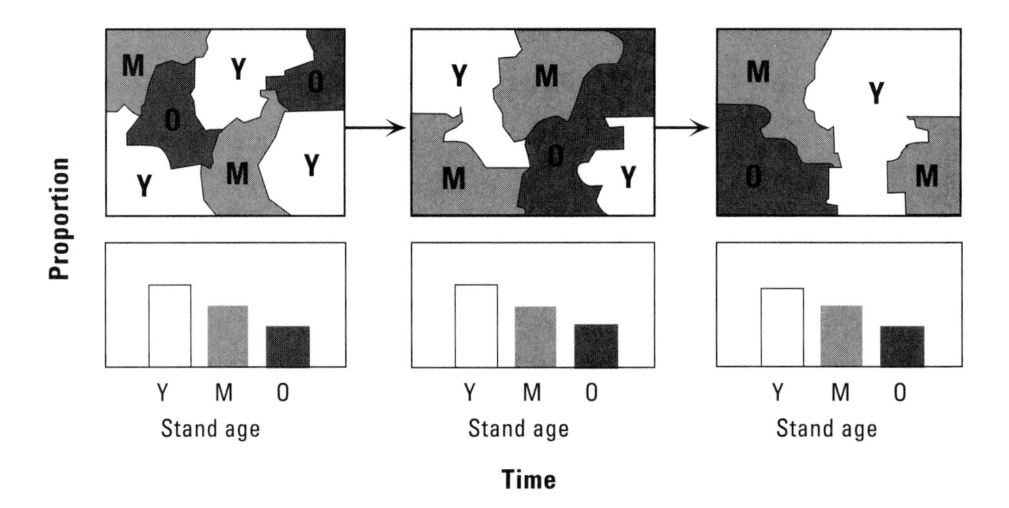

FIGURE 7.14.
The shifting mosaic steady-state concept. Upper panels show a landscape at different times in which the shadings indicate different stand ages (Y = young, M = mature, O = old) and their locations through time. The lower panels depict the proportion of the landscape occupied by each age class, which remains constant through time. The shifts occur in response to disturbance and succession.

ADAPTED FROM LERTZMAN AND FALL, 1998.

shifting mosaic steady-state concept has been difficult to test empirically, but it has been suggested to apply to other systems. Studies of wave-generated fir forests in the northeastern United States have suggested a steady-state condition over the entire system despite widespread local patterns of community degradation and regeneration (Sprugel, 1976; Sprugel and Bormann, 1981). Zackrisson (1977) suggested that the forest mosaic in a boreal forest in northern Sweden remains unchanged, even though the spatial distribution of postfire succession is always changing. The concept applies best when disturbances are small and frequent in a large area of homogeneous habitat (Pickett and White, 1985). Large areas may be more likely than small areas to exhibit a stable mosaic (Zedler and Goff, 1973; Connell and Sousa, 1983; DeAngelis and Waterhouse, 1987). Shugart and West (1981) suggested that a quasi-steady-state landscape was likely only where the landscape was at least 50 times the average size of a disturbance, although Baker (1989a) failed to find equilibrium in the Boundary Waters Canoe Area even at a scale 87 times the mean disturbance patch size.

Another concept considers landscape equilibrium to be a *stationary process* (a stochastic process that does not change in distribution over time or space) with episodic perturbation (Loucks, 1970). Loucks (1970) suggested that communities may appear unstable at any particular time because community composition is changing, but that the entire long-term sequence of changes constitutes a stable system, because the same sequence recurs after every disturbance. In fire-dominated landscapes, for example, the statistical distribution of seral stages, time intervals between successive fires, or similar parameters can be determined (e.g., Van Wagner, 1978; Johnson, 1979; Yarie, 1981; Johnson and Van Wagner, 1985; Johnson and Gutsell, 1994). This concept explicitly acknowledges the stochastic nature of disturbance, but assumes that the distribution of disturbance intervals and the proportion of the landscape occupied by different seral stages remains more or less constant through time. However, the distribution of intervals between disturbances may not be the same, and the probability of disturbance may change with time since last disturbance (Clark, 1989).

A concept related to the stationary process is that of *stochastic* or *relative constancy* through time. Botkin and Sobel (1975) suggested that a system that changes but remains within bounds is a stochastic analog of equilibrium that is applicable to ecological systems. Harrison (1979) also suggested that the concept of a system remaining within acceptable ranges in spite of environmental uncertainty was most relevant to ecology. However, even this concept is scale dependent. For example, long-term directional changes in climate due to global warming or glacial cycles would eventually move a landscape out of preset bounds.

The shifting mosaic steady-state concept, which has been found to describe some systems well, provided considerable impetus for studies of landscape dynamics in disturbance-prone landscapes. Romme (1982) studied a 7300-ha watershed on the subalpine plateau in Yellowstone National Park affected by a natural crown-fire regime. He used fire history methods to date fires, age forest patches, and reconstruct this landscape over a 200-yr period. He found wide fluctuations in landscape composition and diversity during that time and failed to find any evidence for the existence of a steady-state mosaic (Figure 7.15). The cycle of extensive fires occurring at intervals of approximately 300 yr suggested a landscape characterized by continual change (Romme, 1982; Romme and Knight, 1982). Romme concluded that this landscape is more appropriately viewed as a nonsteady-state system characterized by cyclic, long-term changes in structure and function. Romme and

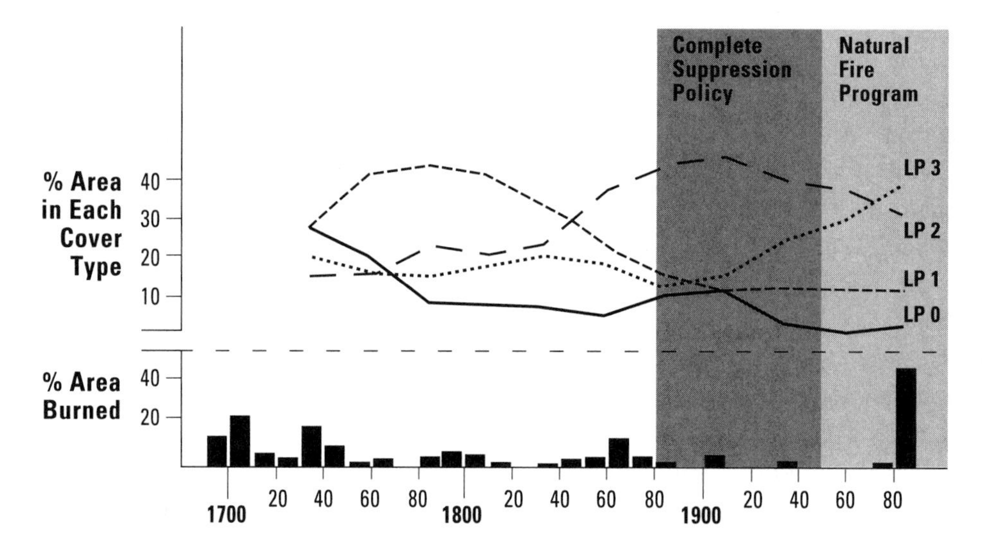

FIGURE 7.15.

Bottom: Percent of a 129,600-ha study area within Yellowstone National Park burned by stand-replacing fires in each decade from 1690 to 1988. *Top:* Percent of the study area covered by each successional stage from 1735 to 1985. Note the wide fluctuations through time. LP 0 is earliest successional stage in the lodgepole pine forests, extending to about 40 yr postfire; LP 1 extends from approximately 40 to 150 yr, LP 2 from ~150 to 250 yr, and LP 3 >250 yr.

ADAPTED FROM ROMME AND DESPAIN, 1989.

Despain (1989) expanded this study to an area of 129,600 ha, but still found constant fluctuation in the patch mosaic during the past 250 yr.

Similarly, Baker (1989a, c) tested for a stable patch mosaic in the 404,000-ha fire-influenced Boundary Waters Canoe Area (BWCA), but did not find a stable patch mosaic at any of five spatial scales. Baker (1989a) used a Markov chain approach to examine potential long-term trends in the vegetation mosaic in the Boundary Waters Canoe Area, Minnesota, based on fire-year maps published by Heinselman (1973). The model was developed for heuristic purposes and assumed that the fire regime that occurred during the 141-yr study period (1727–1868) remained constant into the future. Transitions occurred between forest age classes. Baker used the model to compare the final distribution of age classes at five spatial scales and to evaluate the presence of a stable mosaic and the minimum area needed for a stable mosaic to be observed. Similarity between the simulated final distributions of age classes and the steady-state final distribution was generally low across all spatial scales. He suggested that the lack of a steady-state mosaic was due to (1) spatial heterogeneity in the fire regime, whereby ignition sources, drought severity, fuel load, and fire spread probability would vary across the landscape and (2) a difference in the scales of fire patches and environmental heterogeneity. Baker (1989a) concluded that the Boundary Waters Canoe Area landscape was a "mosaic of different non-steady-state mosaics." Indeed, crown-fire-dominated systems may generally be considered as nonequilibrium landscapes (Turner and Romme, 1994; Boychuk et al., 1997).

Turner et al. (1993b) used a simple spatial model to develop a view of landscape dynamics that considered the spatial–temporal scales of disturbance and the resultant landscape dynamics and that could be applied across a range of scales. A simple spatial model was developed that incorporated four major factors characterizing landscape dynamics: disturbance frequency, recovery time, spatial extent of disturbance events, and size of the landscape of interest. These four factors were reduced to two key parameters representing ratios of time and space. The use of ratios in both parameters permits comparison of landscapes across a range of spatial and temporal scales.

The temporal parameter (T) was defined by the ratio of the disturbance interval (the time between successive disturbances) to the recovery time (the time required for a disturbed site to achieve recovery to a mature stage). Defining the temporal parameter as a ratio permitted three qualitatively different states to be considered, regardless of the type or time scale of the disturbance: (1) the disturbance interval is longer than the recovery time ($T > 1$), so the system can recover before being disturbed again; (2) the disturbance interval and recovery time are equal ($T = 1$); and (3) the disturbance interval is shorter than the recovery time ($T < 1$), so the system is disturbed again before it fully recovers.

The spatial parameter (S) was defined similarly by a ratio of the size of the disturbance to the size of the landscape of interest. Two qualitatively different states were of importance, again regardless of the type of disturbance: (1) disturbances that are large relative to the size of the landscape and (2) disturbances that are small relative to the extent of the landscape. As defined by Turner et al. (1993b), the parameter S could range from 0 to 1; that is, disturbance events larger than the size of the landscape were not considered. A state space was then constructed, with T on the ordinate and S on the abscissa.

A wide range of simulations of disturbance and recovery was conducted, and results were tabulated by tracking the proportion (p) of the simulated landscape occupied by different successional changes through time (Turner et al., 1993b). Landscape equilibrium was observed under conditions of small disturbance size and relatively quick recovery times relative to disturbance frequency (Figure 7.16). A landscape could also appear relatively stable, exhibiting low variance in p values, if disturbances were still relatively infrequent, but disturbance size increased. These conditions resulted in a stable system with low variance in which much of the landscape was still occupied by mature vegetation; this region of the state space may be comparable to the stochastic or relative constancy defined by Botkin and Sobel (1975). The landscape could also appear stable with low variance when disturbance sizes increase even further, although the early seral stages would then dominate. The landscape may be stable (sensu Loucks, 1970), but show very high variance with intermediate values of S and T and show extremely high variance when disturbance size exceeds 50% of the landscape and the disturbance interval is very long. Landscapes in this region of the state space would likely be characterized as nonequilibrium systems.

This model was extremely simple and certainly ignored the biological complexity that would characterize disturbance and succession in real landscapes, and many landscapes are affected by multiple disturbances that occur at different spatial and temporal scales and that may interact. Nonetheless, determination of S and T parameters for several known landscapes supported the general results (Turner et al., 1993b) and clearly demonstrated the strong influence of scale. For example, consider gap dynamics in eastern hardwood forests (e.g., Runkle, 1985). At the scale of the entire Great Smoky Mountains National Park, Runkle estimated that treefall gaps occur every year and affect approximately 1% of the landscape annually. The recovery time for a treefall gap was estimated at ~ 91 yr, the approximate time at which the trees reach the canopy (Runkle, 1985). Thus,

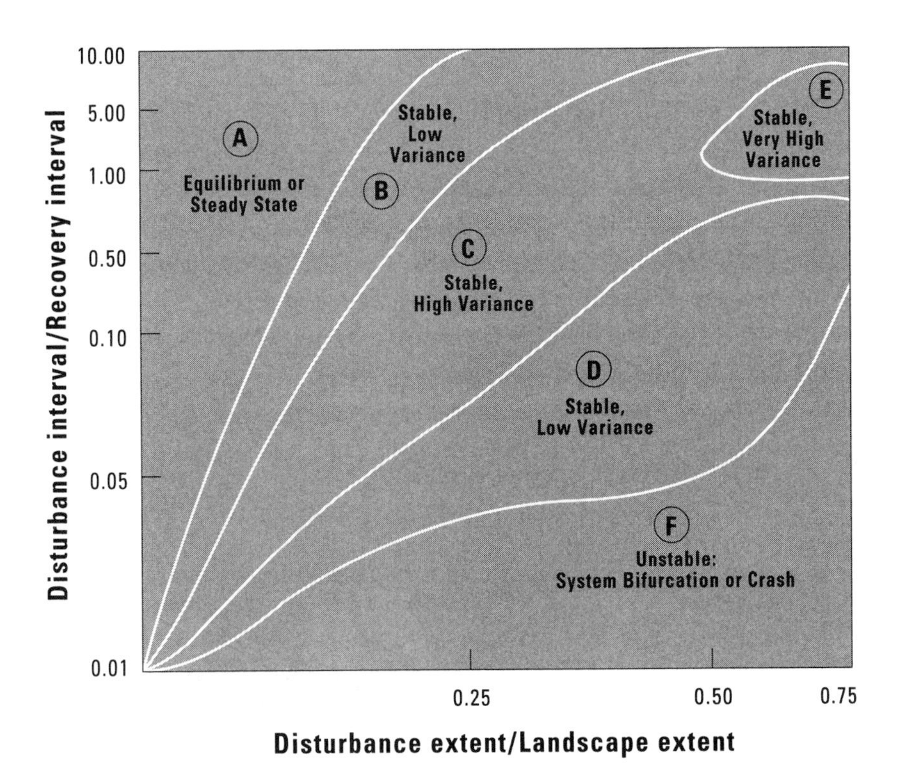

FIGURE 7.16.
State-space diagram of the temporal and spatial parameters that illustrates regions that display qualitatively different landscape dynamics.

REDRAWN FROM TURNER ET AL., 1993B.

for treefall gaps in the Great Smokies, the temporal parameter $T \sim 0.01$ and the spatial parameter $S \sim 0.01$. This disturbance regime occurs in the extreme lower-left corner of the diagram in the equilibrium region (Figure 7.16). Consider now what happens when the spatial scale is altered such that we consider treefall gaps only within a 100-m^2 plot within the park. The median gap size is ~75 m^2 (Runkle, 1982), and we can estimate that a treefall occurs within a 100-m^2 plot every 100 years or so. Recovery time is still ~50 yr. Under these conditions, the temporal parameter $T \sim 2.0$, and the spatial parameter $S \sim 0.75$, bringing this disturbance regime into the region of a stable system with very high variance. This example illustrates nicely the dependence of our observations of landscape dynamics on the spatial extent that is considered.

Much of the disagreement surrounding equilibrium versus nonequilibrium and stability versus instability can be attributed to several factors: the ambiguity in various definitions, different views of spatial heterogeneity and its effects, the lack of explicit specification of scales, and differences in theoretical foundations (Wu and Loucks, 1995). Landscapes can exhibit a variety of behaviors under different disturbance regimes, and the same landscape may shift among different regions of behavior. Landscapes that traditionally are considered as being in equilibrium are characterized by small and infrequent disturbance and rapid recovery. Stable systems with high variance are characterized by intermediate size and frequency of disturbance and intermediate rates of recovery. Potentially unstable systems are characterized by large and frequent disturbance and slow recovery. In these landscapes, a system crash or bifurcation to a qualitatively different system is possible (Paine et al., 1998; Romme et al., 1998). Conclusions regarding landscape equilibrium are appropriate only for a specified spatial and temporal scale. Failure to recognize this dependence can lead to sharply different interpretations of the same dynamics.

SUMMARY

Disturbance creates patterns and is an important and integral part of many ecosystems and landscapes. The causes, patterns, dynamics, and consequences of disturbances have been major research topics in landscape ecology. Disturbance and disturbance regime are characterized by a variety of attributes, including size, frequency, intensity, severity, and shape. The definition of disturbance is scale dependent.

Are various spatial locations in the landscape differentially susceptible to disturbance? If so, can we predict which areas are more or less susceptible to particular types of disturbance? Susceptibility to disturbance of sites located at particular landscape positions is evaluated by comparing the probability or frequency of occurrence of a particular disturbance at many places in a landscape. Results from a variety of studies suggest that landscape position appears to influence susceptibility to disturbance when the disturbance has a distinct directionality (e.g., hurricane tracks) such that some locations are usually exposed more than others. In addition, landscape position may influence susceptibility if the disturbance is of moderate intensity, such that its spread is influenced by subtle differences in the landscape. However, if the disturbance itself has no spatial directionality (e.g., downbursts) or is sufficiently intense that its spread is unaffected by differences in the landscape (e.g., high-intensity crown fire), then landscape position does not influence susceptibility to the disturbance.

Understanding how landscape heterogeneity can influence the spatial spread of disturbance has been a focus of landscape ecological research. Research from theoretical and empirical studies suggests that we cannot generalize as to whether landscape pattern always enhances or retards disturbance spread, but that its potential effects on disturbance spread may be substantial. If the disturbance spreads within the same cover type, then greater landscape heterogeneity may retard the spread of disturbance. If the disturbance spreads between cover types or is otherwise enhanced by edge effects, then increased greater landscape heterogeneity should enhance the spread of the disturbance. However, there may also be thresholds in environmental conditions beyond which landscape pattern will not affect the spread of a disturbance.

Disturbance is an important agent of pattern creation in landscapes. Disturbances create very complex heterogeneous patterns across the landscape, because the disturbance may affect some areas but not others, and severity of the disturbance often varies considerably within the affected area. These resulting mosaics may show considerable persistence through time. Recovery following disturbance can be very sensitive to spatial pattern created by disturbance and is strongly influenced by the spatial pattern of biotic residuals left behind. Disturbance and succession are inextricably linked when we consider landscape dynamics.

A thorough understanding of succession must include an understanding of how successional processes vary with respect to disturbance intensity, size, and frequency. The availability of propagules is a fundamental determinant of successional patterns and can be especially sensitive to the combination of high intensity and large size. If dispersal from outside the disturbed area is important, then the size, shape, and configuration of disturbed patches will influence propagule availability and thus vegetation composition. Spatial effects of disturbance (disturbance size, shape, and arrangement) become increasingly important when residuals are few or sparse and the disturbance is large; under these conditions, colonization and hence succession become slower and more difficult to predict.

Applying the theoretical and empirical advances in understanding of how disturbance and succession interact to large, heterogeneous landscapes is challenging because of the spatial interactions and the long time scales involved. Integrated models of disturbance and succession can reveal trends and dynamics in landscapes that cannot be easily observed empirically, and such models may be particularly useful for addressing questions of broad-scale disturbances or global change.

Because disturbances can be such a strong source of landscape structure, intentional or unintentional shifts in the disturbance regime may dramatically alter the landscape. For example, past climatic changes of small magnitude have caused significant changes in fire regimes in forested landscapes. Shifts in disturbance regime will lead to shifts in landscape structure, with varying time lags; the response of landscape pattern can lag substantially behind a change in the disturbance regime.

Questions of whether equilibrium can be detected on landscapes subject to disturbance and how large a landscape must be to incorporate a given disturbance regime have been important themes in landscape ecology. Much of the disagreement surrounding equilibrium versus nonequilibrium and stability versus instability can be attributed to several factors: the ambiguity in various definitions, different views of spatial heterogeneity and its effects, the lack of explicit specification of scales, and differences in theoretical foundations. Landscapes can exhibit a variety of behaviors under different disturbance regimes, and the same landscape may shift among different regions of behavior. Landscapes that traditionally are considered as being in equilibrium are characterized by small and infrequent disturbance and rapid recovery. Stable systems with high variance are characterized by intermediate size and frequency of disturbance and intermediate rates of recovery. Potentially unstable systems are characterized by large and frequent disturbance and slow recovery. In these landscapes, a system crash or bifurcation to a qualitatively different system is possible. Conclusions regarding landscape equilibrium are appropriate only for a specified spatial and temporal scale. Failure to recognize this dependence can lead to sharply different interpretations of the same dynamics.

DISCUSSION QUESTIONS

1. How can a disturbance both create landscape pattern and respond to landscape pattern? Use at least one example of a natural disturbance to illustrate your answer.
2. Succession is generally defined as change in vegetation through time; thus, temporal dynamics are explicit. Is succession also spatial? Why or why not? Under what conditions will the spatial pattern of disturbance influence succession?
3. How would you compare a natural and human-driven disturbance regime? What criteria would you suggest for determining whether a human-driven disturbance is comparable to a natural disturbance for a given landscape?

4. Explain how scale dependence is important in understanding disturbance dynamics and the effects of disturbances on a landscape. Consider both spatial and temporal scale.
5. You are charged with developing a strategy for monitoring the effects of disturbance on landscape structure for a large region over the coming century. What would you measure and why? How could you determine the sensitivity of your indicators to changes in the disturbance regime?
6. Are disturbance-driven landscapes stable? Justify your answer.

RECOMMENDED READINGS

Boose, E. R., D. R. Foster, and M. Fluet. 1994. Hurricane impacts to tropical and temperate forest landscapes. *Ecological Monographs* 64:369–400.

Franklin, J. F., and R. T. T. Forman. 1987. Creating landscape patterns by forest cutting: ecological consequences and principles. *Landscape Ecology* 1:5–18.

Hunter, M. L., Jr. 1993. Natural fire regimes as spatial models for managing boreal forests. *Biological Conservation* 65:115–120.

Knight, D. H., and L. L. Wallace. 1989. The Yellowstone fires: issues in landscape ecology. *BioScience* 39:700–706.

Pickett, S. T. A., S. C. Collins, and J. J. Armesto. 1987b. A hierarchical consideration of causes and mechanisms of succession. *Vegetatio* 69:109–114.

Romme, W. H., and D. H. Knight. 1982. Landscape diversity: the concept applied to Yellowstone Park. *BioScience* 32:664–670.

Turner, M. G., W. H. Romme, R. H. Gardner, R. V. O'Neill, and T. K. Kratz. 1993. A revised concept of landscape equilibrium: disturbance and stability on scaled landscapes. *Landscape Ecology* 8:213–227.

CHAPTER 8

Organisms and Landscape Pattern

THE SPATIAL DISTRIBUTION OF RESOURCES in heterogeneous landscapes can have important effects on the growth, reproduction, and dispersal of organisms. The changing distribution of resources due to land-use change is a critical issue for conservation design and management (Hobbs, 1993; Fahrig and Merriam, 1994; Noble and Dirzo, 1997). Development of the science in this area has certainly been encouraged by environmental questions that demanded answers grounded in basic understanding. For example, issues associated with *habitat fragmentation*, in which the amount of habitat is reduced and the remaining habitat apportioned into smaller, more isolated patches (Figure 8.1), have come to the forefront (Wilcove et al., 1986; Saunders et al., 1991; Meffe and Carroll, 1994; Opdam et al., 1996). Ecologists have been pressed to understand and to predict the prognosis for a variety of organisms that face increasingly fragmented habitat. It is now recognized that the maintenance of *biodiversity* (the abundance, variety, and genetic constitution of native animals and plants) requires a landscape perspective (Franklin, 1993) that complements population, community, and ecosystem considerations. In addition, there has been increasing recognition of the importance of animals (e.g., their movement and foraging patterns) affecting the

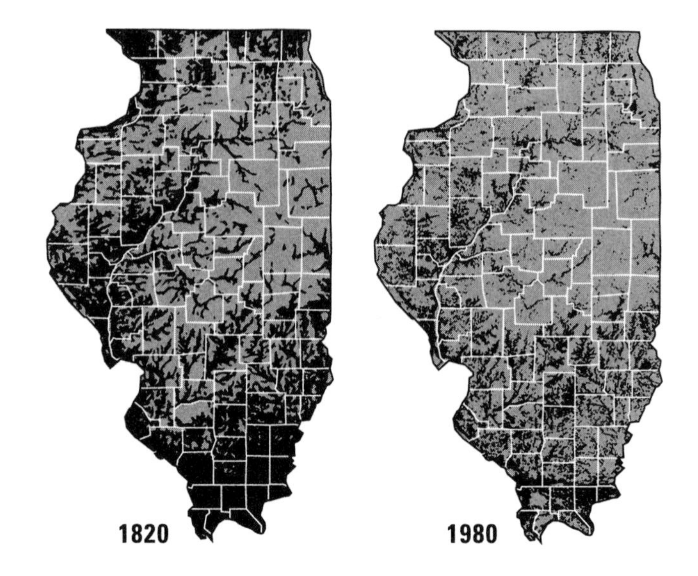

FIGURE 8.1.
Distribution of forest cover in Illinois in 1820 and 1980. Forest habitat was not continuous in this prairie–forest ecotone, but settlement resulted in reduced forest abundance and increased isolation of remaining forest fragments. Most remaining forests in Illinois now occur within 300 m of rivers and streams.

ADAPTED FROM IVERSON, 1991.

spatial heterogeneity of ecosystem processes (e.g., Pastor et al., 1999; see Chapter 9).

Population ecologists have also found it increasingly interesting, and often necessary, to incorporate spatial heterogeneity. Although spatial structure has always been a part of population and community studies, ecologists' view of the role of space has changed through time (Hanski and Simberloff, 1997). Consideration of the implications of spatial heterogeneity by population ecologists had focused primarily on issues other than space through the 1960s and 1970s (but see MacArthur, 1972). However, spatial structure is now considered an essential element of theories for processes involving genes, individuals, populations, and communities, although rigorous empirical work involving space remains a significant challenge (Kareiva, 1990, 1994; Wiens et al., 1993; Hanski and Simberloff, 1997). Wiens (1976) laid out many of these considerations in his review article, which still makes for excellent reading. In the introduction, Wiens (1976) wrote:

> In the real world, environments are patchy. Factors influencing the proximate physiological or behavioral state or the ultimate fitness of individuals exhibit discontinuities on many scales in time and space. The patterns of these discontinuities produce an environmental patchwork which exerts powerful influences on the distributions of organisms, their interactions, and their adaptations. (Wiens, 1976:81)

Obviously, interactions between spatial heterogeneity and organisms is a subject for which there is tremendous overlap among population ecology, landscape ecology, and conservation biology; even so, the needed synthesis among these fields has yet to emerge. The traditional cultures in these subdisciplines are very different and have often resulted in different emphases (see Ives et al., 1998). For instance, research in population ecology typically addresses how interactions within and among populations (e.g., competition and predation) generate spatial patterns and how these patterns then influence the outcomes of subsequent interactions. Unlike landscape ecologists, population ecologists usually do not begin with a map describing spatial patterns of resources, and more frequently employ more theoretical and analytical models that are not directly oriented toward a particular organism or management issue (Ives et al., 1998). In contrast, research in landscape ecology on populations typically addresses the effects of habitat abundance and spatial configuration on a population of interest and begins with an explicit map, which might also change through time. Questions of management relevance for specific plants or animals are often addressed. Relatively complex simulation models of organisms acting on real (or realistically complex) maps are more frequently used. Conservation biology may employ either of the two approaches described above, but the focus is often on threatened or endangered species whose long-term persistence in a given landscape is of concern.

In this chapter, we begin with a brief review of conceptual developments of interactions of organisms and spatial dynamics, focusing especially on the shift from the paradigm of island biogeography to that of metapopulation biology, and then discuss the scale-dependent nature of interactions between organisms and space. We next present a series of insights about the effect of spatial pattern on organisms that have emerged from landscape ecological studies, that is, a summary of current understanding, and suggest the conditions under which spatial pattern is important for organisms. Finally, we discuss the use of spatially explicit population models.

CONCEPTUAL DEVELOPMENT OF ORGANISM–SPACE INTERACTIONS

Ecologists have long observed that habitat can be isolated in patches, like islands in an inhospitable ocean of other land uses. Lack (1942), for example, noted that remote British islands had fewer bird species than nearer islands. Watt (1947) pointed out that the isolated patches of vegetation on the heterogeneous landscape were fundamental to understanding community structure (see also Chapter 7). Andrewartha and Birch (1954) discussed the importance of spatial relationships among largely isolated local populations, noting that local extinction of populations was a common phenomenon and that these sites may subsequently become reoccupied. As described in Chapter 1, Huffaker (1958) demonstrated how spatial pattern could create stable or unstable dynamics in a predator–prey system.

Some of the earliest theoretical work on spatial dispersal of organisms made analogies to physical diffusion (Skellam, 1951). The diffusion model, first applied to biological systems by population geneticists (e.g., Fisher, 1937; Dobzhansky and Wright, 1947), has been clearly presented by Andow et al. (1990) and Holmes et al. (1994). Texts by Okubo (1980) and Turchin (1998) also provide a comprehensive discussion of the theory, application, and measurement of diffusion. Our focus in this chapter is on concepts relevant to the movement of organisms in heterogeneous landscapes. The basic equation for the diffusion of a population of size N is given by

$$\frac{\partial N}{\partial t} = f(N) + D\nabla^2 N \tag{8.1}$$

This equation states that the change in number, N, with time, t, can be estimated by two functions: the description of local population growth $f(N)$, and the diffusion of organisms from the surrounding region. The description of population growth $f(N)$ depends on the needs and interests of the investigator and might be as simple as a linear function of net growth (e.g., birth–death) or a complex nonlinear function that can account for density-dependent or competitive effects on growth. The parameter D is the diffusion coefficient, which describes how rapidly the population moves in space, and ∇^2 is the diffusion operator, which describes the rate of change of N with distance (the density gradient). In spite of its simplicity, equation (8.1) has had remarkable success in explaining the rates at which species have invaded new environments (Lewis, 1997).

Diffusion theory shows that if organisms invade a uniform landscape, the rate of spread, V, will reach an asymptotes equal to

$$V \approx \sqrt{4rD} \tag{8.2}$$

where r is the intrinsic rate of population growth. Andow et al. (1990) tested the adequacy of equation (8.2) against observed rates of spread for three different species. The results showed that equation (8.2) adequately explained the invasion process of muskrats in Europe and the cabbage white butterfly in North America. However, in the third case, movements of the cereal leaf beetle, the estimation of D was made by observing fine-scale patterns of movements. These data were believed to produce an underestimation of the real value of D and a biased prediction of V. Thus, equation (8.2) gives a good approximation to spread across the landscape, provided that the data are gathered at sufficiently broad scales. The complexities of actual landscapes are included as the average value of D and, as long as spatial patterns remain relatively constant, provide an adequate description of invasion at landscape scales. When landscape patterns change dramatically, the values of D may need to be reestimated or alternative dispersal models considered.

Island Biogeography

The theory of *island biogeography* (MacArthur and Wilson, 1963, 1967; MacArthur, 1972) was an important influence on how ecologists think about organisms and spatial pattern, and for some time it was the prevailing paradigm guiding the design of conservation reserves. Island biogeography was developed as a general theory to predict the number of species found on oceanic islands. The theory predicts that the number of species on an island will reach an equilibrium that is positively related to island size (larger islands would contain more species) and negatively related to distance from the mainland (fewer species on islands far from the mainland and the source of new colonists). The number of species on an island depends on the immigration rate of species to the island and the extinction rate of species from the island. Immigration to the island is assumed to be a linear function of distance, d, and also depends on the size of the mainland source community, such that

$$I = d(P - R)^k$$

where P is the number of species in the mainland pool, R is the number of species on the island, and k is a parameter that would differ with communities of different organisms. The value of k is determined by fitting data to a specific island system. Once a species finds its way to the island, its rate of extinction depends on the available resources. If all islands are similar, the available resources should be proportional to island size.

$$E = nS^m$$

where S is island size and n and m are parameters fitted by regression from data.

Early field studies provided empirical support for the theory (Simberloff and Wilson, 1969, 1970). The basic concept was expanded to include alpine zones (Vuilleumier, 1970), which have communities isolated on the tops of mountains much like oceanic islands. The theory was also applied to cave communities (Culver, 1970; Vuilleumier, 1973). For these communities, the aboveground landscape separating the cave mouths functions very much like an inhospitable ocean.

With growing concern about habitat fragmentation, drawing the analogy between habitat fragments and islands was easy, and island biogeography theory was readily embraced by ecologists. The theory was applied to the design of nature preserves in terrestrial landscapes, generating a long debate among ecologists about whether a single large preserve would be better than having several smaller preserves spaced such that organisms could move among them. The argument centered on the fact that a single preserve might hold more total species, but it could be wiped out with a single catastrophic event. In contrast, the smaller preserves would each contain fewer species, but some preserves would be likely to survive any particular catastrophic event. If even a single small reserve escaped the catastrophe, it would provide a source for recolonization of the damaged areas (Burkey, 1989; Soule and Simberloff, 1986).

Island biogeography theory was subjected to a number of criticisms (Carlquist, 1974; Gilbert, 1980), and many modifications have been suggested. Perhaps the primary criticism has been the assumption of equilibrium (Diamond, 1972; Terborgh, 1975). An island system would require a very long period of time to reach such an equilibrium number of species (Simpson, 1974), perhaps best measured in geologic time units during which climate and many other factors change. In many ecosystems, chronic disturbance (Villa et al., 1992) would also invalidate the assumption, because the next disturbance would occur long before the system reached equilibrium. It has also been pointed out that islands close to the shore will experience very large immigration rates (Brown and Kodric-Brown, 1977). The immigration could overwhelm the extinction rate so that effects of island size would not be evident, although some models incorporated size-dependent immigration. In addition, island size and distance become less important as dispersal ability increases (Roff, 1974b).

Other studies have identified factors in addition to those considered in island biogeography theory that influence species richness in habitat fragments. Webb and Vermaat (1990) documented species diversity in isolated heathland remnants that form islands in an ocean of other vegetation. Counter to the prediction, they found that small islands had higher species diversity because they could not resist invasion by surrounding vegetation. Large islands, in contrast, had lower species diversity because they resisted the invaders. Habitat quality (Murphy et al., 1990) or interspecific competition (Hanski, 1981, 1983) on an island may sometimes be more important than size. Some island systems are dominated by catastrophic disturbances, such as hurricanes, that may cause extinction, irrespective of island size (Ehrlich et al., 1980). In disturbance-driven systems, population size and community diversity may be dominated by reinvasion processes. In fact, when natural disturbances are common, persistence may be higher on a landscape with several patches, because this spreads the risk of the entire population being wiped out (Goodman, 1987). Other ecosystem processes, such as trophic dynamics, that influence community composition must also be considered. Spiders on Bahamian islands largely followed island biogeography theory except when predatory lizards were present (Toft and Schoener, 1983). Then the relationships changed significantly, and predation tended to dominate the extinction rates.

The theory of island biogeography clearly dominated much of conservation biology through the 1970s and 1980s. Despite the criticisms, island biogeography theory has been important in highlighting effects of the size and isolation of natural areas on their effectiveness in meeting conservation objectives; indeed, these factors remain important considerations in conservation planning. There is overwhelming evidence that species richness increases with area, whether on islands or on the mainland. However, metapopulation models emerged in the late 1980s as a way of thinking about fragmented habitats and heterogeneous terrestrial environments in general; some authors have referred to this as a paradigm shift (Hanski, 1989; Merriam, 1991), but it can also be considered as a switch to questions at finer spatial scales than those considered by island biogeography.

Metapopulation Biology

Levins (1969, 1970) observed that most populations have a finite probability of extinction, m, that is measurably greater than zero, implying that populations will eventually go extinct. However, if the population is fragmented into a patchwork of subpopulations and the probability of extinction of the subpopulations remains small, local extinctions may be balanced by recolonization from neighboring populations. When these conditions exist, the process of extinction and colonization is locally dynamic but may be regionally stable. A simple equation summarizes this concept:

$$\frac{dp}{dt} = cp(1 - p) - mp \tag{8.3}$$

where c and m are the probabilities of colonization and extinction, respectively, and p is the proportion of available locations (patches) colonized at any point in time. The equilibrium solution to equation (8.3) provides an estimate of the expected proportion of colonized sites, p'.

$$p' = 1 - \frac{m}{c}$$

Populations persist regionally (p' is greater than zero) only if the probability of extinction, m, is less than the probability of colonization, c. It is a simple matter to rearrange the terms to calculate the expected number of empty sites, s'.

$$S' = 1 = p' = \frac{m}{c}$$

The interconnected set of subpopulations that function as a demographic unit was termed a *metapopulation* by Levins (1969). The concepts and terms used in metapopulation theory (Table 8.1) have been readily incorporated into landscape ecology, in part because metapopulation concepts seem to be an apt description of population dynamics in natural or artificially fragmented landscapes.

One interesting application of metapopulation theory has been the assessment of the effect of habitat destruction on regional population dynamics. Habitat destruction, in terms of metapopulations, is equivalent to the destruction of a habitat patch and the elimination of a site that may support a single subpopulation of the metapopulation. If the fraction of occupied sites, p, is assumed to decrease linearly as the fraction of potentially occupied sites destroyed increases, then equation (8.3) becomes

$$\frac{dp}{dt} = cp(1 - D - p) - mp \tag{8.4}$$

where D is the proportion of sites that have been destroyed. The equilibrium solution for equation (8.4) is

$$p' = 1 - D - \frac{m}{c}$$

This equilibrium equation predicts that the threshold for extinction of the entire metapopulation ($p' = 0$) will occur when the fraction of available sites, $(1 - D)$, is less than or equal to m/c. This relationship, illustrated in Figure 8.2, shows that a metapopulation within a region will disappear long before the final patches have been destroyed (in fact, for the parameters used in Figure 8.2, approximately one-third of the habitat patches still remain when the population goes extinct). There are many reasons to believe that the potential relationship between D and p' may not be linear. For instance, nonlinear relationships may occur when habitat destruction is nonrandom (spatially correlated) or when progeny must traverse increasingly hostile terrain to find unoccupied habitat sites. Under these circumstances, the linear relationship in Figure 8.2 becomes nonlinear, and precipitous declines in the survival of the metapopulations occur with even small amounts of habitat destruction near the extinction threshold (Lande, 1987). Tilman and colleagues (Tilman, 1994; Tilman et al., 1994; Tilman et al., 1997), in an interesting series of papers, have extended these concepts to the problem of predicting the combined effect of habitat destruction and competition between sympatric species. Time lags induced by competitive displacement and life-history characteristics result in a delay in the eventual extinction of species due to habitat loss. This effect, termed *extinction debt*, implies that a precipitous decline in species abundance and diversity may be the inevitable consequence of habitat loss but will not be observed until many years after the disturbances have occurred.

The Levins model of metapopulations is spatially implicit; that is, the process of colonization and extinction of suitable habitat patches is independent of their spatial locations. Sites that are a long distance from neighbors have the same probability of colonization and extinction as sites closer to neighbors, implying that

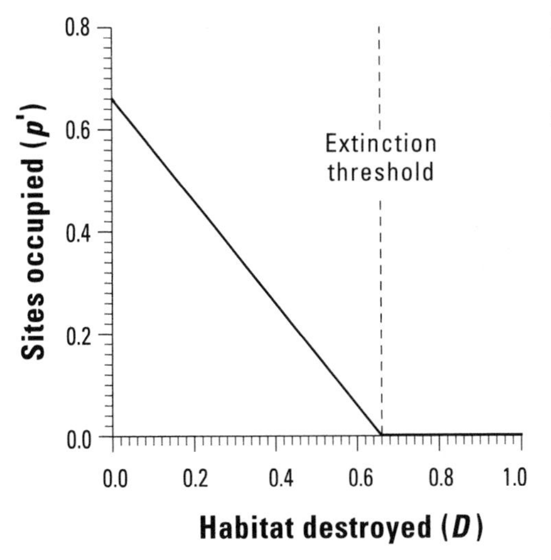

FIGURE 8.2.

Equilibrium percentage of occupied sites ($p' = 1 - D - m/c$) as a function of the fraction of habitat destroyed (D). Here $m = 0.2$ and $c = 0.6$. The extinction threshold occurs when the fraction of habitat destroyed is equal to $1 - m/c = 0.666$.

ADAPTED FROM BASCOMPTE AND SOLE, 1996.

Table 8.1.
Definitions and synonyms of terms used in metapopulation studies.

Term	Synonyms	Definition
Levins metapopulation	Classical metapopulation	Metapopulation structure assumed in the Levins model: a large network of similar small patches, with local dynamics occurring at a much faster time scale than metapopulation dynamics.
Local population	Population, subpopulation, deme	Set of individuals that live in the same habitat patch and therefore interact with each other; most naturally applied to populations living in such small patches that all individuals practically share a common environment.
Metapopulation	Composite, population, assemblage (of populations, when local populations are called subpopulations)	Set of local populations within some larger area, where typically migration from one local population to at least some other patches is possible.
Patch	Habitat patch, habitat island, site, locality	A continuous area of space with all necessary resources for the persistence of a local population and separated by unsuitable habitat from other patches (at any given time a patch may be occupied or empty).
Patch model	Occupancy model, presence–absence model	A metapopulation model in which local population size is ignored and the number (or fraction) of occupied habitat patches is modeled.

Source–sink metapopulation	—	Metapopulations in which there are patches in which the population growth rate at low density and in the absence of immigration is negative (sinks) and patches in which the growth rate at low density is positive (source).
Spatially explicit metapopulation model	Lattice model, grid model, cellular automata model, steppingstone model	Model in which migration is distance dependent, often restricted to the nearest habitat patches; the patches are typically identical cells on a regular grid, and only presence or absence of the species in a cell is considered (the model is called a coupled map lattice model if population size in a patch is a continuous variable).
Spatially implicit metapopulation model	Island model	Model in which all local populations are equally connected; patch models are spatially implicit models.
Spatially realistic metapopulation model	Spatially explicit model	Model that assigns particular areas, spatial locations, and possibly other attributes to habitat patches, in agreement with real patch networks; spatially realistic models include simulation models and incidence function models.
Turnover	Colonization–extinction events; dynamics	Extinction of local populations and establishment of new local populations in empty habitat patches by migrants from existing local populations.

Adapted from Hanski and Simberloff, 1997.

organisms can easily locate unoccupied patches no matter how far away they may be or how hostile the intervening landscape matrix. Bascompte and Sole (1996) used a spatially explicit metapopulation model to examine the effect of limited dispersal and habitat destruction on the persistence of metapopulations (Figure 8.3). The results indicate that predictions of the spatially explicit model differ from the spatially implicit model near the extinction threshold. The effect of limited dispersal exacerbated the effect of habitat destruction by increasing the probability of extinction. Not surprisingly, when little habitat loss has occurred, the two modeling approaches predict similar dynamics (Figure 8.3).

Sources and Sinks: A Special Case

Metapopulation theory considers that patches are of the same quality across the landscape and therefore that birth and death rates are the same in each patch. Pulliam (1988) proposed the special situation in which, in a mosaic of habitats, lo-

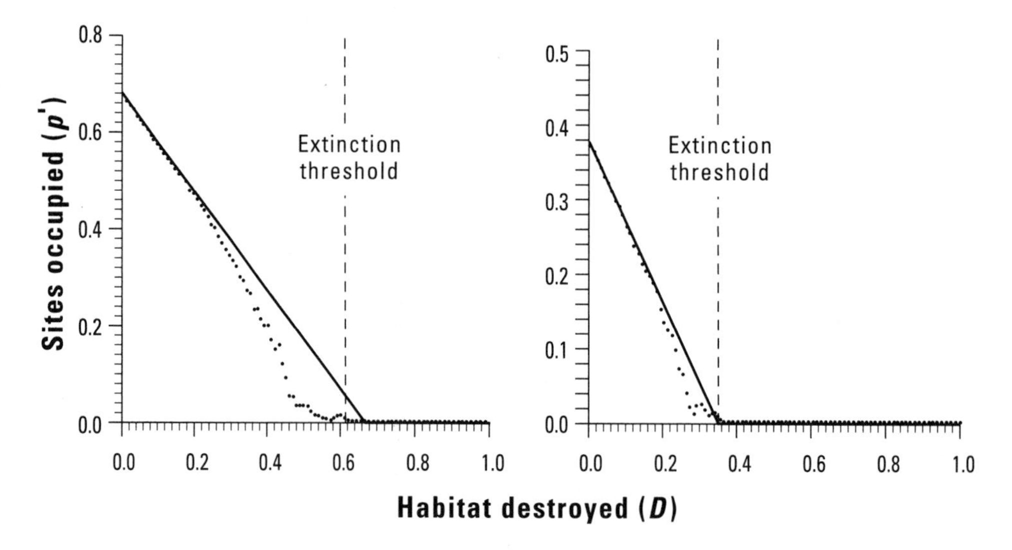

FIGURE 8.3.
Equilibrium percentage of occupied sites (p') as a function of the fraction of habitat destroyed (D). The solid line represents the value of p' predicted by equation (8.2), the dots represent realizations of a spatially explicit model, and the vertical line shows the extinction threshold for the spatially explicit model. For any given value of D, the fraction of occupied sites and extinction threshold is lower in the spatially explicit model.

ADAPTED FROM BASCOMPTE AND SOLE, 1996.

cal populations have unique demographic responses to local variation in habitat characteristics (Figure 8.4). When the demographics reflect the heterogeneity of the habitat, then source–sink concepts naturally emerge (Dias, 1996). *Sources* are habitat areas where local reproductive success is greater than local mortality. Populations in source patches produce an excess of individuals, who must disperse from where they were born to settle and breed. In contrast, *sinks* are poor habitats, that is, areas where local mortality exceeds reproductive success. Without immigration from sources, these populations would go extinct. A key insight from this work was that migration of the surplus organisms from the source to the sink could maintain the populations in an apparent demographic equilibrium. Even a small amount of source habitat added to a landscape can increase the total population size. Similarly, removal of patches that were serving as sources for a larger population could lead to catastrophic decline of the population. Although the sink patches do not ordinarily produce emigrants, their presence on the landscape can sometimes maximize population abundance. Kadmon and Schmida (1990), for example, showed that desert grasses produce most of their seeds in moist wadis. Although the grass spreads out over surrounding areas, it relies on the moist areas (source patches) to replace losses due to mortality. The heterogeneous mixture of habitat patches supports a larger population on the landscape. However, if sink

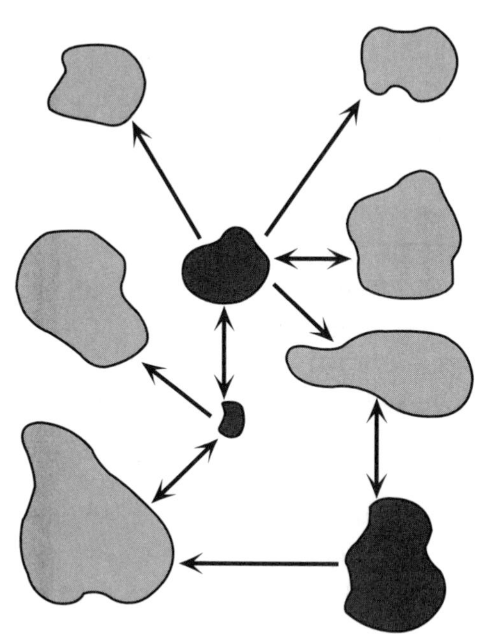

FIGURE 8.4.
A metapopulation with source (dark patches) and sink (light patches) subpopulations. A few source habitats provide excess individuals when they emigrate and colonize sink habitats. Arrows indicate the primary directions of movement between patches.

ADAPTED FROM PULLIAM AND DUNNING, 1994.

habitat is too abundant relative to source habitat, organisms with limited dispersal will be unable to find source habitat, and the landscape will be unable to support a viable population (Pulliam and Danielson, 1991). The maintenance of high-quality habitat on a landscape is important, and the effects of habitat loss cannot be mitigated by the preservation of sink habitat. In addition, a patch that is a source for one species may be a sink for other species (Danielson, 1991, 1992).

Identification of Suitable Habitat

Our discussion thus far has implicitly assumed that suitable habitat for a population is known and that its spatial distribution has been described. In a standard ecological metapopulation study, a key initial task is to distinguish between suitable and unsuitable habitat and delimit the patches of suitable habitat in a study area (Hanski and Simberloff, 1997). Suitable habitat includes those areas that have been defined as having the conditions required for a given species to meet its needs for resources, shelter, and reproduction. A variety of methods can be employed for identifying suitable habitat (Hansen et al., 1993), ranging from associations with seral stage or stand structure (Hansen et al., 1995), habitat suitability models (Van Horne and Wiens, 1991), logistic regression (Mladenoff et al., 1995; Mladenoff and Sickley, 1998), Bayesian rules (Miller et al., 1997), and multivariate statistical methods (e.g., Hansen et al., 1995; Knick and Dyer, 1997). However, it is important to recognize the complexities inherent in this task, and that it is difficult to distinguish among source, sink, and unsuitable habitats. In particular, population density is not a means of identifying suitable habitat, because poor habitats may be population sinks that have a temporary surplus of individuals relative to source habitats (e.g., following juvenile dispersal) (Van Horne, 1983).

In some instances, the simplifying binary assumption of suitable and unsuitable habitat, which is necessary to keep the models reasonably simple, is very clear. However, in other cases, it may be difficult to partition the landscape into patches embedded in an unsuitable matrix. Consider habitat perceived as suitable or favorable by eastern timber wolves (*Canis lupus lycaon*) in the upper midwestern United States. Although driven nearly to extinction during the early part of the century, wolves have been gradually expanding their range during the past 15 years, moving eastward from Minnesota into northern Wisconsin and the Upper Peninsula of Michigan. Analyses of data from radio-collared wolves revealed that suitable habitat was a function not only of vegetation type and deer density (deer are commonly preyed on by the wolves), but also of land ownership, road density, and human population density across the landscape (Mladenoff et al., 1995). Wolves were moving throughout the landscape

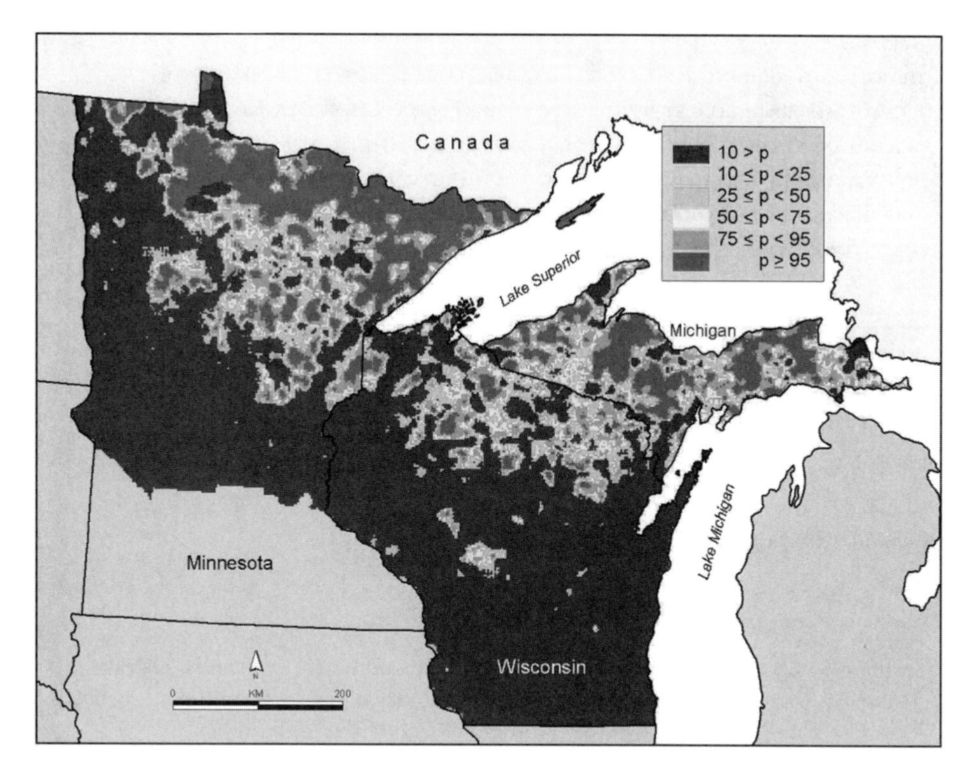

FIGURE 8.5.
Map of the probability of occurrence of suitable wolf habitat across the upper Midwest, USA, based on spatial extrapolation of a logistic regression model. Refer to the CD-ROM for a four-color reproduction of this figure.

ADAPTED FROM MLADENOFF ET AL., 1995.

and often crossing unsuitable areas. Although successful establishment of a wolf pack was restricted to the higher-quality habitat, suitable habitat was not simply a binary category, but rather a continuous probability surface (Figure 8.5). Although there are areas where the probability of suitable habitat is high, there are also extensive areas of moderate probability that may still serve as habitat patches. Landscape and population ecologists are only beginning to incorporate a more continuous representation of habitat suitability into their conceptual framework.

The same population may also require resources from different types of habitats, or patches, on the landscape (Dunning et al., 1992). For example, wintering birds may use some patches for foraging and others for shelter during storms (Pe-

TABLE 8.2.
RULES FOR DESCRIBING POTENTIAL SUITABLE HABITAT FOR SELECTED SPECIES IN TWO TEMPERATE FOREST WATERSHEDS, THE LITTLE TENNESSEE RIVER BASIN, LOCATED IN WESTERN NORTH CAROLINA, USA, AND THE HOH RIVER BASIN, LOCATED ON THE OLYMPIC PENINSULA, WASHINGTON, USA. RULES WERE IMPLEMENTED IN A GEOGRAPHIC INFORMATION SYSTEM DATABASE AND USED TO CONTRAST POTENTIAL EFFECTS OF LAND-USE CHANGE OF DIFFERENT COMPONENTS FOR THE BIOTIC COMMUNITIES.

Species	Habitat description
A. Little Tennessee River Basin	
Showy orchis (*Orchis spectabilis*)	Forest at elevations <1210 m on north-facing slopes
Catawba rhododendron (*Rhododendron catawbiense*)	Open areas or forest edges at elevations >900 m
Princess tree (*Paulownia tomentosa*)	Exotic species found in open areas or forest edges at elevations <762 m
Mountain dusky salamander (*Desmognathus ochrophaeus*)	Forested areas with northerly aspects; restricted to streamsides at elevations <950 m; found throughout mesic forest at higher elevations
Southeastern shrew (*Sorex longirostris*)	Open areas at elevations <760 m
Northern flying squirrel (*Glaucomys sabrinus*)	Forests at elevations above 1210 m
European starling (*Sturnus vulgaris*)	Exotic species found in open areas throughout the watershed
Wood thrush (*Hylocichla mustelina*)	Forest-interior habitat (areas >200 m from edge) at elevations <1370 m
B. Hoh River Basin	
Horsetail (*Equisetum telmateia*)	Early successional habitats next to streams in the Western Hemlock Zone
Cascade Oregon grape (*Berberis nervosa*)	Conifer forest in Montane Forest Zone
Mountain alder (*Alnus sinuata*)	Early successional habitats in the Montane Forest Zone
Mountain huckleberry (*Vaccinium membranaceum*)	Deciduous forest in the Montane Forest Zone

(*continued*)

TABLE 8.2. (*continued*)

Species	Habitat description
Licorice fern (*Polypodium glycyrhiza*)	Deciduous forest next to streams in the Western Hemlock Zone
Twinflower (*Linnaea borealis*)	Conifer forest in Western Hemlock Zone
Honeysuckle (*Lonicera ciliosa*)	Shrub areas with steep slopes in Western Hemlock Zone
Heather vole (*Phenacomys intermedius*)	Grassy habitats in the Montane Forest Zone
Red squirrel (*Tamiasciurus hudsonicus*)	Conifer forest in the Western Hemlock Zone

FROM PEARSON ET AL., 1999.

tit, 1989). The checkerspot butterfly needs cool slopes for prediapause larvae, but warmer slopes for postdiapause larvae and pupae (Weiss et al., 1988). Barred owls and pileated woodpeckers will supplement their diet from surrounding suboptimal patches (Whitcomb et al., 1977). In all these cases, the populations respond to the spatial patterning of different patches on the landscape, complicating the distinction between suitable and unsuitable habitat.

The suitability of a given landscape is species dependent, and the spatial description of land cover alone is often a misleading indicator of the location and amount of suitable habitat. Pearson et al. (1999) developed simple rules for describing species' habitat based on information easily obtained from natural history information. This information, combined with spatial data (e.g., land cover and topography) and simple geographic relationships (such as patch size or proximity to edge) (Table 8.2), allowed the potential effects of landscape change to be evaluated for a wide range of species (Figure 8.6). The results showed that potential habitat was more fragmented than the land-cover class that included the habitat. For example, there were about 2400 patches of suitable habitat identified for the showy orchis (*Orchis spectabilis*) in a Southern Appalachian watershed within 325 patches of forest cover (Pearson et al., 1999). Furthermore, a 12% increase in the size of the largest patch of forest in this watershed produced a >400% increase in the size of the largest patch of suitable habitat for showy orchis.

Habitat-based approaches in which the abundance and arrangement of habitat is the focus, rather than the demography of a population, may be important substitutes for direct estimates of population responses to landscape heterogeneity. Although maintenance of biodiversity is often desired in land planning and resource management, it is impossible to manage simultaneously for all species present at a site. Identification of suitable habitat for species and tracking changes in habitat abundance and pattern through time allow the modeling of risks of biodiversity changes at landscape scales and are a first step to anticipate potential effects of land-use changes (Riitters et al., 1997; White et al., 1997). Hansen et al. (1993) recommended that habitat suitability and life-history attributes be used as surrogates for detailed demographic data for vertebrate species, an approach that they believe is intermediate between the coarse-filter and fine-filter approaches described by Noss (1987) and Hunter (1991). Associating target species with specific habitat configurations and evaluating alternative management prescriptions using simulation models were among the key steps of their approach. If habitat is well defined, the method can be used to monitor habitat change through time, compare habitats among areas, or examine the effect of scale by developing GIS-based signatures (spatial pattern analyses) for varying scales surrounding the locations (Knick and Dyer, 1997; Riitters et al., 1997).

Habitat-based approaches have already offered insight into understanding the implications of variation in landscape pattern for a variety of taxa. A rule-based probabilistic model linked with land cover and physiographic data was used to predict the suitability of a landscape in northeast England for three species of birds

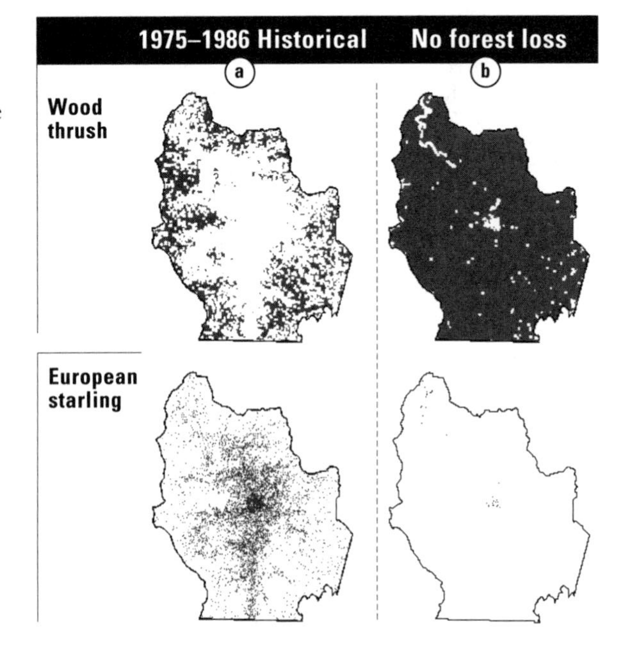

FIGURE 8.6.
Maps of suitable habitat for two bird species, European starling and wood thrush, in the Little Tennessee River Basin, western North Carolina, at the end of a 100-yr simulation using transition probabilities (a) extrapolated from the 1975–1986 historical period and (b) with no loss of forest permitted.

ADAPTED FROM PEARSON ET AL., 1999.

(Tucker et al., 1997). In low- to mid-elevation forests in the Pacific Northwest, risk to bird species under four different management scenarios was evaluated by quantifying suitable habitat for each species using habitat maps, species–habitat associations, and other life-history requirements (Hansen et al., 1995). Recently, White et al. (1997) used habitat and area requirements for a large sample of vertebrate species in Monroe County, Pennsylvania, to assess risks to biodiversity associated with land development. Six possible alternative versions of Monroe County that differed in the extent and spatial distribution of human impact were developed for the year 2020 based on projected current rates of population growth. Results demonstrated little variation in species richness among the six alternatives (extinction of species was not prevalent), although loss of habitat might have been substantial. The Gap Analysis approach (Scott et al., 1987, 1993), which analyzes existing preserves and land cover types to identify regions with high species diversity that remain unprotected, has been useful for allocating conservation resources to protect diversity over broad state-level scales. Riitters et al. (1997) derived habitat maps from digital land-cover maps by using a spatial filtering algorithm at various scales; results demonstrated that the amount and arrangement of suitable habitat changed with the home-range size of the organism. Collectively, these studies suggest that analyzing species habitat requirements may serve as a suitable starting point for relating land-use change to biodiversity. In such applications, however, it is important to remember that the availability of habitat implies potential rather than actual use of the habitat by the species of interest.

In sum, landscape ecologists tend to view the complex mosaic structure of real landscapes as the template on which populations operate. Patches of the same habitat may vary in quality, yet many studies simplify nature's complexity by describing habitat as either suitable or unsuitable and assuming a featureless background matrix (Wiens, 1997). Although empirical and modeling studies all require simplification, we must remember that the issue of how to identify suitable habitat is an important one, and its implementation influences the observations or predictions for a given landscape. The same landscape may be very different for different species, and an organism-centered view of the landscape is required to understand the response of populations to spatial patterning.

Contrast between Metapopulation Theory and Landscape Ecology

There is clearly much common ground between metapopulation biology and landscape ecology, yet Wiens (1997) identified several important differences. As we have seen, metapopulation models are typically focused on idealized habitat patches in a featureless landscape and emphasize local extinction, interpatch movement, and recolonization. Four features that characterize landscape ecology are largely missing from metapopulation models: (1) variation in patch quality, (2) variation in the quality of the surrounding environment, (3) boundary effects, and (4) how the landscape influences connectivity among patches (Wiens, 1997). The interpatch matrix becomes important in landscape ecology because dispersal between patches depends on the quality of the matrix. If the matrix is inhospitable, like the ocean, isolation becomes more important. For example, Bolger et al. (1997a) found that the urban matrix is essentially impervious to native rodents in southern California, because animals did not cross even short distances between remnant habitat fragments. If the matrix has relatively low contrast with the habitat patches, isolation may be much less important. Metapopulation models also implicitly assume a landscape in which the suitable habitat patches occupy a relatively low proportion of the landscape, whereas landscape ecologists are often focusing on a wide range of proportions of availability of suitable habitat. On landscapes, it is also important to keep in mind that the matrix and the edges of the patches are themselves habitat for other species (McCollin, 1993).

The main difference between the metapopulation view of nature and that embraced by landscape ecologists revolves around the degree of complexity that is considered: the metapopulation view of nature, complex as it is, seems much simpler in comparison to how landscape ecologists view reality (Hanski and Gilpin, 1997, Preface). As noted by several authors, we do not yet have a conceptual and practical synthesis of metapopulation biology and landscape ecology, yet establishing more common ground between these lines of inquiry is imperative (Turner et al., 1995b; Hanski and Gilpin, 1997; Preface; Wiens, 1997; With, 1997a). Work by With and King (1999b) in which a generalized metapopulation model is coupled with fractal neutral landscape models is an example of a step toward a theoretical synthesis between metapopulation and landscape ecology.

Synthesis: Conceptual Development

In spite of the simplifying assumptions of the theories of island biogeography and metapopulation biology, an impressive body of literature has confirmed the importance of patch isolation and patch size. The evidence ranges from pika on talus slopes (Smith, 1980), to birds in forest patches (Opdam et al, 1984, 1985; Whitcomb, 1977; Verboom et al., 1991), to frogs (Gulve, 1994) and snails (Aho, 1978) in freshwater ponds. Some studies stress the importance of isolation of the habitat islands (Helliwell, 1976; Hayden et al, 1985). Others stress island size (Williams, 1964; Freemark and Merriam, 1986) or effective island size in the presence of nest predators (Wilcove et al., 1986; Small and Hunter, 1988). Even with the complexities of the terrestrial landscape, basic theory explains a great deal of the variance in the spatial distribution of the biotic community (Saunders et al., 1991), as long as the assumptions are reasonably valid. Thus, many studies generally confirm insights expressed in island biogeography and metapopulation theory, and, along with population genetics, these models provide much of the theoretical foundation for conservation (With, 1997b). Landscape ecology introduces more spatial complexity than is considered in either theory. As ecologists continue to grapple with issues regarding how organisms respond to spatial pattern, there will be many opportunities for synthesis among population ecology, conservation biology, and landscape ecology.

SCALE-DEPENDENT NATURE OF ORGANISM RESPONSES

Important concepts relating to scale arise when the interactions between diverse organisms and the spatial pattern of suitable habitat are considered (Wiens, 1989a). A beetle does not relate to its environment on the same scales as a vulture, even though both are scavengers. Similarly, a resource patch for one species is not necessarily a resource patch for another, which implies that descriptions of patchiness are species and process specific (Figure 8.7). Furthermore, conclusions about how species respond to pattern at one scale are difficult to translate to another species at another scale. Returning once again to Wiens' (1976) review article, we read, "First, it is essential that the fabric of spatial scales on which patchiness is expressed be unraveled, and the structure of spatial heterogeneity be related to the variations in environmental states on diverse time scales. The key to achieving this is in shedding our own conceptions of environmental scale and instead concentrating on the perceptions of the organisms, attempting to view environmental structure through their senses" (Wiens, 1976:110). Thus, there is a strong imperative to focus on the scales that are appropriate for the organism, recognizing that our human-based perception of scale and pattern may not be the right one. So, why do appropriate scales differ among taxa?

Some differences in appropriate scales are due to various attributes of the species. For example, vagile species may be less sensitive to fine-scale patterns of adjacency than sedentary species or those that have very limited dispersal distances. In the species-rich mesic forests of the Southern Appalachians, studies of forest herbs have demonstrated that native herbaceous species with good dispersal (e.g., maidenhair

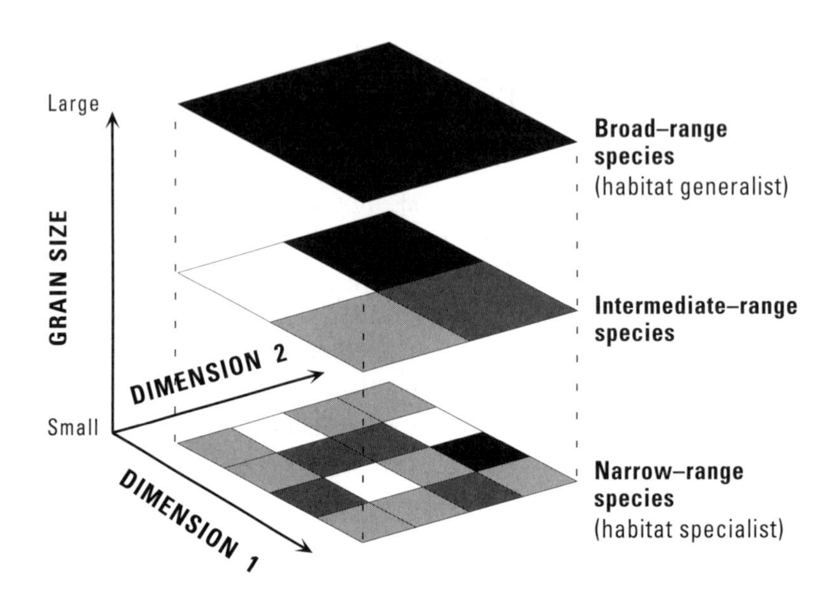

FIGURE 8.7.
Conceptual model of habitat structure in which homogeneity and heterogeneity depend on the species and the resolution at which species perceive their environment.

ADAPTED FROM KOLASA, 1989.

fern, *Adiantum pedatum*) are found in small isolated forest patches, but species with limited dispersal (e.g., the ant-dispersed bellwort, *Uvularia grandiflora*) are absent from small isolated forest patches (Pearson et al., 1998). Species that use sparse or clumped resources may need to operate over larger spatial scales (e.g., searching or dispersal distances), particularly if the resources are nonsubstitutable.

To some extent, differences in spatial scales among taxa may parallel differences in body mass. For example, a 20-g bird may occupy a home range of 4 ha, whereas a 200-g bird may occupy a home range of 92 ha. Such *allometric rules* for scaling ignore variations in diet, age, season, and the like, but they may still provide an approximation of organism-dependent scaling that is less arbitrary than those often used. Allometric relationships between habitat use and body size have been suggested. Foster and Gaines (1991) conducted an experimental study designed to answer this question: Is patch selection related to organism size? Their goal was to determine whether plant succession, patch size, or both affected the small mammal community in an old-field habitat. They tested three predictions: (1) Patch size affects densities of small mammals; larger animals, which require more resources, should remain on large patches. (2) Patch size affects persistence rates (length of time remaining on a patch), such that smaller patches should have shorter persistence rates. (3) Plant successional changes affect densities of resident small mammals.

Patches of three sizes (large, 5000 m^2; medium, 288 m^2; and small, 32 m^2) were created by mowing. The presence, density, and persistence rates of four small mammal species in the patches of different size were then monitored. The species included the cotton rat, ranging in mass from 90 to 160 g; prairie vole, 28 to 55 g; deer mouse, 15 to 25 g; and western harvest mouse, 8 to 15 g. Results demonstrated that patch size had a greater effect than successional stage, and this effect was most pronounced for the largest species, the cotton rat, which was never found in the smallest patches. It was also the large patches that were used to create territories. The densities of the smaller mammals were often highest on the smallest patches, because the animals avoided the interstitial areas. Persistence rates were greatest on the largest patches, and animals were more transient in the smaller. The researchers concluded that there were allometric relationships associated with patch size and that animals forced to move through a heterogeneous landscape may behave differently than conspecifics occupying the same habitat type.

Holling has proposed an empirically based general theory relating body mass to home range size (Holling, 1992). Holling accumulated a vast amount of data describing the occurrence of various species of different body sizes in a wide range of habitats and found differences in the frequency distribution of body masses in different habitat types (Figure 8.8). In addition, whole classes of body masses were rare or missing from the data. These holes in the frequency distributions were attributed to the scale of resources in these landscapes, suggesting that the scale of resource distribution constrains the types of species that occur there. It was also found that particular size classes were missing from prairies because fine-scale variability of resources did not exist, whereas the three-dimensional structure of the forests offered such fine-scale variability. Holling's ideas have stimulated much discussion and research, and they have implications for human-induced changes in landscapes, because humans alter both the structure and scale of landscape resources.

Although it is easy to acknowledge that interactions between species and spatial heterogeneity must be scale dependent, it is difficult to identify the right scales in practice. Without a reasonable means of scaling, comparisons across species and landscapes are more likely to be misleading than useful. Even different life stages within a species may operate at different scales on the landscape, especially if they differ in vagility as well as size [e.g., nymphal (flightless) and adult (flighted) grasshoppers; With, 1994b]. The concept of ecological neighborhoods proposed by Addicott et al. (1987) offers a practical way of identifying scales.

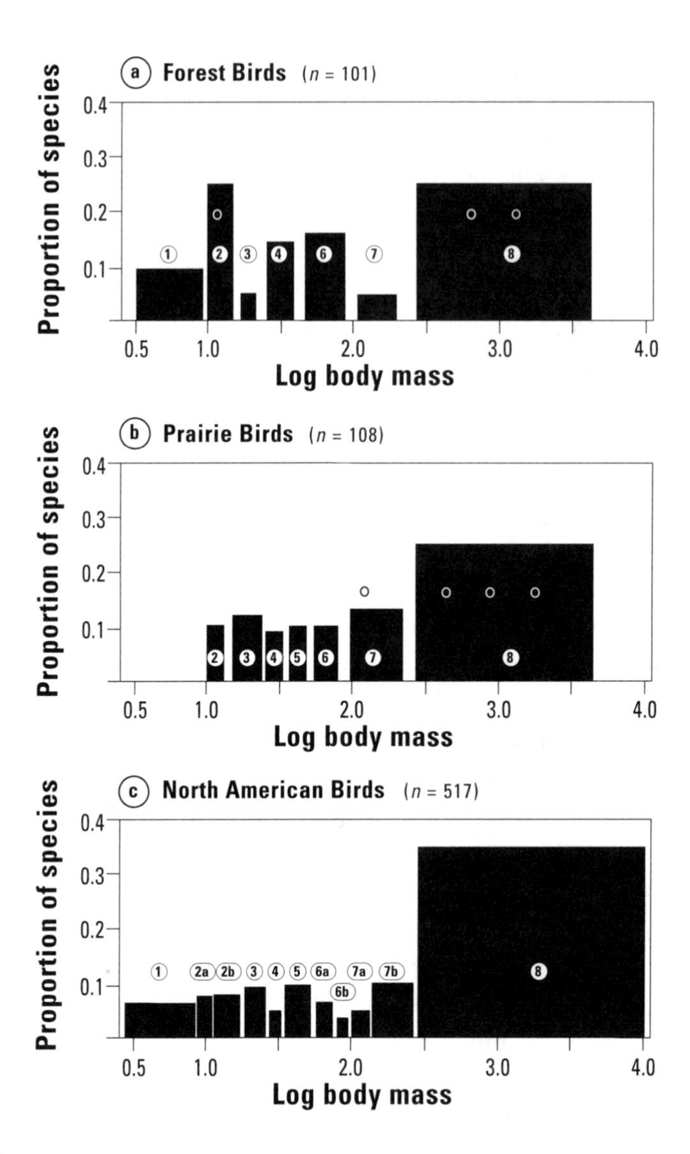

FIGURE 8.8.

Proportion of bird species in each body-mass clump category (labeled by circled numbers) in the (a) boreal region forest, (b) boreal region short grass prairie, and (c) for all remaining North American species. Body mass was measured in grams. Small open circles indicate additional breaks identified with the hierarchical clustering technique. The position and width of each bar represent the range of masses in a particular body-mass clump.

AFTER HOLLING, 1992.

Addicott et al. (1987) proposed that ecological neighborhoods for an organism be empirically defined by using three criteria: (1) a particular ecological process (e.g., foraging, reproduction), (2) a time scale appropriate to the process (e.g., day, week, season, year) and (3) the organism's activity or influence during this time period. By tracking the space that the organism uses during the time period and focusing on the process, the spatial extent used can be estimated by applying a criterion, for example, that 95% of the activity is included in a particular area (Figure 8.9). This then is the organism's neighborhood. Note that specifying a particular process is important, because the ecological neighborhood of an individual's daily foraging may be very different from that of its annual reproductive activities. The distribution of patches, their isolation, and their temporal duration can then be assessed relative to the size of the neighborhood by defining the following metrics: (1) *rp* = relative patch size = patch size (m^2)/neighborhood size (m^2); (2) *ri* = relative isolation = interpatch distance (m)/neighborhood radius (m); and (3) *rd* = relative patch duration = patch duration (*t*)/neighborhood duration (*t*). By using relative metrics, the effect of scale is removed (note that *rp, ri*, and *rd* are all dimensionless) and dynamics of different species may be compared. For example, a 10-ha habitat patch for a grizzly bear foraging over a 1000-ha land-

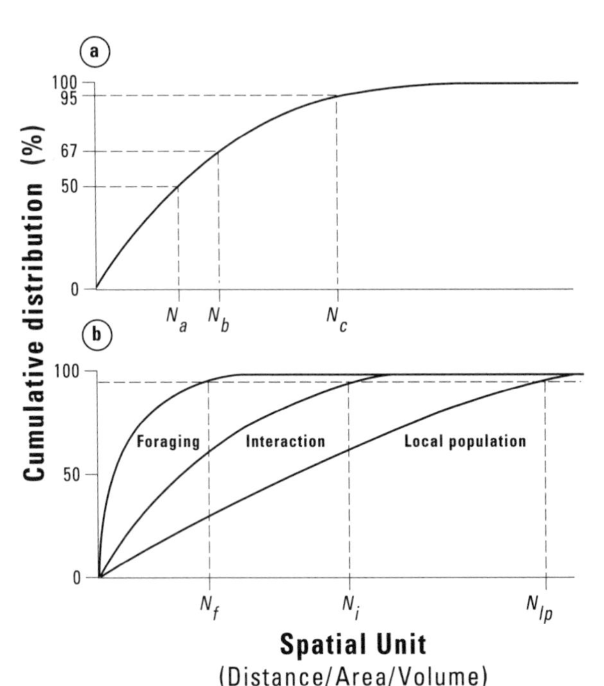

FIGURE 8.9.
Hypothetical examples of the relationship between cumulative distribution of movement and influence against spatial unit as a means of identifying the ecological neighborhood of an organism. (a) Dashed lines at arbitrary points on the cumulative distribution indicate different neighborhood sizes for different decision criteria. (b) Each curve represents a cumulative distribution with an associated neighborhood size for a different ecological process using a decision criterion of 95%.

MODIFIED FROM ADDICOTT ET AL., 1987.

scape may be functionally similar to a 0.10-habitat patch for a raccoon that forages over a 10-ha area; in both cases, the *rp* = 0.01, even though the resource patches themselves are very different. An organism's perception of heterogeneity clearly depends on its scale of activity, and this approach offers a practical way to identify and compare patchiness across a range of different species.

Related to the concept of ecological neighborhoods is the notion that organisms may respond to heterogeneity at multiple scales. Clearly, the spatial arrangement and density of vegetation influence the success of herbivores in finding food (Kareiva, 1983; Risch et al., 1983; Stanton, 1983; Cain, 1985; Bell, 1991). Indeed, a variety of authors (e.g., Wiens, 1976, 1989a; Addicott et al., 1987; Senft et al., 1987; Kotliar and Wiens, 1990; Orians and Wittenberger, 1991; Bergin, 1992; Russell et al., 1992; Crist and Wiens, 1994; Pearson et al., 1995) have suggested that animals make scale-dependent choices in habitat use and/or foraging. Studies focused on winter foraging dynamics of ungulates [elk (*Cervus elaphus*) and bison (*Bison bison*)] in northern Yellowstone National Park nicely illustrate this idea (Turner et al., 1997c).

During winter, foraging ungulates seek resources that are highly variable in space and time and that vary across multiple scales. Spatially, forage may vary in abundance and in quality across a landscape as a function of plant species' composition, moisture, soil fertility, and topography. This spatial variability ranges from meter by meter variation in forage quality or quantity to between-habitat variation over a scale of kilometers (O'Neill et al., 1989, 1991a, b; S. Turner et al., 1991). The spatial pattern of winter forage is also modified by the distribution of snow, which may reduce or even eliminate potential foraging sites. Temporally, forage resources change daily as the resource is depleted by grazing and as snow conditions change. Although depletion by grazing is gradual and patchy, a major snow event can rapidly change the distribution of forage availability across an entire landscape. Ungulates respond to this variability in space and time by making foraging decisions hierarchically (Senft et al., 1987; Kotliar and Wiens, 1990; Danell et al., 1991; Schmidt, 1993), but the scales at which decisions are made and the cues at each scale are not well understood.

Measurements of forage abundance and distribution of feeding stations (locations where ungulates dug or "cratered" through the snow to expose forage) within 30- by 30-m plots suggested that the spatial pattern of foraging by ungulates was random within feeding patches as long as biomass was present (Wallace et al., 1995). However, large differences in forage abundance across the landscape suggested that ungulates might make choices at broader spatial scales. Analyses of grazing intensity in a set of 15 large study areas encompassing 7500 ha of the northern range revealed that grazing intensity per hectare could best be predicted by environmental variability at broad scales (100 to 500 ha), rather than by local (per hectare) environmental variability (Pearson et al., 1995). A spatially explicit, individual-based simulation model of elk and bison foraging on the northern range (Turner et al., 1994a) was also used to explore ungulate responses to changes in the spatial scale of forage availability. Simulation experiments revealed strong responses in ungulate survival to removal of between-habitat variation in forage abundance, but not within-habitat variation, and also indicated that snow heterogeneity enhanced ungulate survival (Turner and O'Neill, 1995). Collectively, the field and modeling results suggest strong effects of resource heterogeneity and environmental cues at broad scales on ungulate habitat use and survival during winter. Interpreting or predicting ungulate grazing at any given location requires an understanding of the environmental heterogeneity of the surrounding landscape, not simply a description of local site attributes (Turner et al., 1997c).

The table can also be turned, so to speak, by asking how landscape pattern influences the scales at which organisms must operate. O'Neill et al. (1988b) used a neutral landscape model to ask this question: At what scale must an organism operate in order to find resources on a given landscape? The proportion of the landscape, p, occupied by resources was varied in simple random maps and in maps in which the resource cells were clumped by a simple contagion algorithm. The distance that an organism would need to travel on the map to be assured of locating a resource cell was then calculated. Results revealed a strong effect of spatial pattern (Table 8.3). When resources were above the critical threshold of connectivity ($p = 0.59$ for the four-neighbor rule used in this study; see Chapter 6), the spatial clustering of the resources did not influence the distance an organism would travel; a suitable resource site would be within one cell. When resources were rare, however (e.g., $p = 0.20$), clumping of resources increased the distance

TABLE 8.3.
SCALE AT WHICH ORGANISMS WOULD NEED TO OPERATE TO BE ASSURED OF LOCATING RESOURCES ON SIMPLE RANDOM MAPS AND MAPS IN WHICH RESOURCES WERE CLUMPED.

	Distance (in grid cells) that the organism must travel	
Proportion (p) of landscape occupied by resource cells	Random maps ($Q = 0.0$)	Maps with high contagion ($Q = 0.9$)
0.59	1	1
0.20	4	28
0.01	99	881

FROM O'NEILL ET AL., 1988B.

the organism must search by a factor of 7. When resources were very rare ($p = 0.01$), the organism needed to search a distance nearly nine times greater when resources were clumped than when they were distributed at random. Thus, the structure of the landscape itself may dictate the scales at which organisms must operate. In addition, the effectiveness of different foraging tactics may vary with the spatial distribution of resources (e.g., Cain, 1985; Roese et al., 1991).

To test empirically for different scales of interaction with patch structure among different species, With (1994b) studied three grasshopper species using experimental microlandscapes (Wiens and Milne, 1989; Johnson et al., 1992a). The study was conducted in grassland habitat, and 25-m^2 microlandscapes were established in which the heterogeneity of shortgrass cover was varied. With (1994b) recorded the movement patterns of the grasshopper species in different landscape mosaics and applied fractal analysis to compare the landscape perceptions of the different species in the same environments. Results demonstrated that the largest of the grasshopper species moved up to six times faster than the two smaller species, and the species responded differently to microlandscape structure in the 25-m^2 plots. The two smaller species also had more complex movement patterns than the larger species, suggesting that these species were interacting with patch structure at a finer scale of resolution than the larger species (With, 1994b). With concluded from these studies that the grasshopper species were scaling the landscape differently and suggested that the scale independence of fractal analysis provides a useful tool for identifying such differences among taxa. Thus, the scale of pattern will interact with the scale at which an organism operates to determine its dynamics on a given landscape.

Synthesis: Scale Dependence

Just as our ability to detect pattern depends on the scale at which we make measurements, the ability of organisms to detect and respond to heterogeneity depends on how they scale the environment (Wiens, 1989a). The multiple scales at which species perceive their environment and the fact that these scales often differ from our own must be recognized in any attempt to understand or predict the response of organisms to spatial heterogeneity. Because species differ in the scales at which they use resources or perceive the environment (their ecological neighborhoods), studies of the interactions among species may be particularly sensitive to scale (Wiens, 1989a). However, understanding the responses of organisms to spatial pattern at multiple scales is in its infancy (Kareiva, 1990; Kotliar and Wiens, 1990; Hyman et al., 1991; Ward and Saltz, 1994) and remains a high priority for ecology (Lubchenco et al., 1991; Levin, 1992).

EFFECTS OF SPATIAL PATTERN ON ORGANISMS

We now summarize a number of insights that have emerged from landscape ecological studies about organisms and space. Examples selected are illustrative rather than comprehensive. The focus is on effects of patch size and heterogeneity, ecotones and boundary shape, habitat connectivity and the role of corridors, and landscape context. This section ends with a synthesis, suggesting the conditions for which spatial pattern will be important for organisms.

Patch Size and Heterogeneity

In general, larger, more heterogeneous patches support more species. A well-studied aspect of population responses to habitat arrangement is the effect of patch size. In general, larger patches of habitat contain more species and often a greater number of individuals than smaller patches of the same habitat. This is not at all surprising based on the well-documented relationship between species and area, and it occurs for several reasons. First, the larger the habitat patch is the more local environmental variability is contained within it, such as differences in microclimate, structural variation in plants, and diversity of topographic positions. Even a seemingly uniform expanse of habitat such as forest or grassland is, at some scale, a mosaic of different habitats (Wilcove et al., 1986). This variability provides more opportunities for organisms with different requirements and tolerances to find suitable sites within the patch. In addition, the edges and interiors of patches may have different conditions that favor some species but not others, and the relative abundance of edge versus interior habitats varies with patch size. Smaller patches have a greater perimeter-to-area ratio than larger patches, which means that smaller patches will have a greater proportion of edge habitat and the larger patch will have a greater proportion of interior habitat.

Consider these differences for forested patches. At the edge of a forested patch, there is generally more light, a warmer, drier microclimate, and greater access for organisms that frequent open habitats (Chen et al., 1992, 1995, 1996). In contrast, the interior of the patch tends to be more shady, cooler and more moist, and often off limits to the organisms of open habitat (Ranney et al., 1981). When a patch gets sufficiently small or elongated in shape, all interior habitat is lost, leading to a loss of interior species and dominance by edge species. Large patches typically include both edge and interior species. An important point to keep in mind is that fragmentation may actually increase total species diversity (Opdam, 1991); although the fragmentation decreases the habitat of forest-interior species, the resulting heterogeneity increases opportunities for forest-edge and open-field populations. However, this increase in species diversity that may result from habitat fragmentation may not be desirable from a conservation perspective.

Many studies have examined the effects of patch size on bird communities (e.g., Van Dorp and Opdam, 1987; Verboom et al., 1991; McGarigal and McComb, 1995). Freemark and Merriam (1986) found that larger forest patches in Canada had more species than smaller forest patches, with forest interior birds showing the strongest effect. In addition, forest patches that had greater within-patch heterogeneity in terms of habitat structure and microenvironmental conditions also had more bird species and more total numbers of birds, with edge-related species showing the greatest effect. However, forest-edge birds may be nest predators and decrease the effective size of the patch for other species (Andren, 1992).

In Amazonian forests, a long-term study initiated by Tom Lovejoy of the Smithsonian Institution has examined the effects of patch size on a wide range of tropical biota. The biota of continuous unfragmented forest in the Amazon region were sampled during 1983 and 1984, after which time the forest was cut experimentally to create fragments of 1 and 10 ha. Responses of the biota were then measured during the subsequent 9 years following fragmentation. Results for insectivorous birds (Stouffer and Bierregaard, 1995) demonstrated that bird abundance and species richness within the forest fragments declined dramatically following isolation, despite the fact that the fragments were separated from continuous forest by only 70 to 650 m. Army ant colonies were eliminated soon following fragmentation. The bird species that were obligate army ant followers dropped out first; of the non-army ant obligates, only two species of edge specialists were not affected by fragmentation. The bird communities in the 1-ha fragments diverged from the preisolation community more than did the communities in the 10-ha fragments, indicating an effect of patch size. Interestingly, as secondary succession proceeded in the cleared areas, communities in the larger (10-ha) fragments became more like the preisolation communities, although communities in the smaller fragments continued to diverge.

In sum, there is overwhelming evidence from many taxa and geographic locations that larger patches support a greater number of species, and that an increase in within-patch heterogeneity (e.g., vertical complexity, microsite variety) will generally increase species richness.

Shape of boundary can influence species' relative abundances. Closely related to issues associated with patch size are those relating to the shape of the boundaries, or ecotone, between two cover types. Patch shape refers to the two-dimensional form of a given area as determined by its perimeter (edge or boundary may be used synonymously). Some of the effects of patch shape on organisms result from the relative amounts of interior and edge habitat that are provided by a given shape. For a given area, a circular shape will have the least edge habitat, whereas a very long narrow shape will have much more, and perhaps only, edge habitat, depending on the width. Because some organisms specialize on edges while others require interior habitats, patch shape has important implications for the biotic community.

Boundary shape can influence ecological processes, yet there is not a large amount of literature on the effects of varying edge shapes (Forman, 1995). Human activities often simplify boundary shapes, changing complex shapes that may follow topographic variability or result from natural disturbance into straight lines (Krummel et al., 1987). A study of revegetation on disturbed mine areas examined the influence of different shapes of the boundary between the undisturbed forest, which is a source of propagules for succession, and the reclaimed mine area. Hardt and Forman (1989) compared boundaries of three shapes, convex, concave, and straight, and found that the density of colonizing trees was 2.5 times greater in areas with concave boundaries. In addition, there was greater evidence of browsing on the vegetation adjacent to convex boundaries. These results suggested that the convex boundary shape would result in more rapid rates of succession. Another interesting consideration related to patch shape is the orientation, or general directionality, of a patch. Patch orientation influences bird species richness in forest fragments, presumably because patches that are oriented perpendicular to migratory pathways are more likely to attract birds than patches that are aligned with migratory pathways (Gutzwiller and Anderson, 1992).

Forman has been conducting an interesting long-term study of the effects of the shape of the boundary between pinyon–juniper (*Pinus–Juniperus*) and grassland (*Bouteloua–Artemisia*) in northern New Mexico on wildlife usage in and across the boundary. Results have demonstrated that use of the edge by both elk (*Cervus elaphus*) and deer (*Odocoileus hemionus*) increases with curvilinearity of the edge and movement along the boundaries decreases (Forman, 1995:106–107). Whereas movement across boundaries increases with curvilinearity, straight boundaries appear to act as partial boundaries. However, the influence of boundary shape on organisms remains a topic for which more research is needed.

The particular type of edge that surrounds a given habitat patch may also be an important influence on the species found within the habitat patch. For example, Watts (1996) determined that the species richness and total abundance of wintering sparrows in habitat patches were affected by the type of adjacent edge vegetation. Habitat patches supported more species and individuals when they were adjacent to dense edges (privet and brambles) than when adjacent to sparse edges (e.g., deciduous forest). Watts (1996) suggests that the occurrence of individual species may be independent of resource availability if there is a strong effect of edge quality on the species.

A review of theoretical and empirical studies of species interactions with habitat edges (Fagan et al., 1999) suggested four general classes of effects: (1) edges that may be barriers or filters to movement; (2) agents that alter mortality rates; (3) areas providing energetic subsidies or refuge; and (4) regions where novel interspecies interactions may occur. The wide differences in response of different species to edges can result in dramatically different effects. Fagan et al. (1999) have noted that our knowledge of "edge-mediated dynamics place(s) severe limitations" on our understanding of processes leading to species colonization or extinction, which comprises the conceptual core of island biogeography and species–area relations.

Habitat Connectivity

Connectivity is a scale-dependent threshold phenomenon. The connectivity of suitable habitat can constrain the spatial distribution of a species by making some areas accessible and others inaccessible. Plants and animals need suitable areas in which movement and dispersal can occur to maintain the populations. Both plants and animals have varying degrees of mobility, although plants, of course, usually move at the seed stage rather than as mature organisms. Once suitable habitat for a species of interest is characterized, determining whether the habitat is or is not spatially connected is often of interest.

A series of studies of small mammals in an agricultural landscape mosaic in Canada conducted by Merriam and his colleagues nicely illustrates important effects of habitat connectivity on small mammals, as well as the integration of field studies and modeling. Their study area is a landscape containing crop fields along with scattered woodlands. These woodlands are sometimes isolated from one another by being completely surrounded by crop fields, and sometimes the woodlands are connected by fencerows—narrow corridors of trees and shrubs that grow up along the borders of the crop fields. Early studies of this landscape demonstrated that chipmunks and white-footed mice traveled frequently along the fencerows, but seldom moved between the wood and the field or across open fields (Wegner and Merriam, 1979). Birds also seldom flew directly over the open fields between the woods and preferred to fly along the fencerows. Similar results have been observed in many locations in Europe, where fencerows between woodlands have developed over many centuries (Forman and Baudry, 1984).

Merriam and colleagues next explored what happened to the small mammal populations if they were extirpated within the woodlands (Henderson et al., 1985). Local extinctions are a frequent natural occurrence each year because relatively small numbers of animals survive the winter. Chipmunks were live-trapped and removed from woodland patches, and the rate of recolonization of these patches depended on their connectivity to other woods; that is, recolonization depended on whether the patches were joined or linked to one another. Recolonization occurred more rapidly in patches that were connected to other wooded areas by fencerows than in the isolated woodlands. This study also suggested that an area of at least 4 km^2 and containing at least five woodlands and interconnecting fencerows would be required for the populations in this mosaic to persist through the years (Henderson et al., 1985).

A simulation modeling study was used to explore further the effects of alternative patterns of connectivity among patches (Lefkovitch and Fahrig, 1985). Thirty-four different arrangements of connections among five patches were simulated to determine which spatial characteristics of groups of habitat patches were important predictors of the survival of a resident animal population, like the chipmunks or white-footed mice. Results indicated that populations in isolated patches died out much earlier and had lower population sizes than did the populations in the connected patches. These results were supported by field data in which four interconnected woodlands had higher mean population sizes of white-footed mice than two isolated woodlands of similar size. New organisms could disperse into the connected woodlands and augment or replenish the local populations, whereas the isolated woodland received fewer new mice.

Connectivity also influences the use of suitable habitat by larger organisms. Milne et al. (1989) examined the effects of habitat fragmentation on wintering white-tailed deer. Using a GIS-based model including 12 landscape variables, deer habitat was predicted independently at each of 22,750 contiguous 0.4-ha grid cells. When predicted habitat use was compared with empirical data, results demonstrated that deer did not use sites of suitable habitat that were isolated from other suitable sites.

Whether habitat is connected or not is a threshold phenomenon that depends on both the abundance and spatial arrangement of the habitat, as well as the movement or dispersal characteristics of the organism. A *threshold* refers to there being a point at which the habitat suddenly becomes either connected or disconnected. Recall from Chapter 6 how the use of neutral landscape models demonstrated connectivity across simple random or structured landscapes. When a four-neighbor rule is employed, suitable habitat is distributed in a number of relatively small disconnected patches when it occupies less than ~60% of a landscape; however, when suitable habitat occupies more than 60% of the landscape, the habitat is distributed in a few, large, well-connected patches. Under these conditions, the threshold of connectivity is ~60%. If the habitat is not distributed at random, but has a greater degree of clumping as observed in most landscapes, the threshold of connectivity usually moves down. In other words, the habitat would be connected at some lower fraction of occupancy on the landscape. Consider next the different movement abilities of an organism on a set of landscapes where the suitable habitat is again distributed at random. If the organism is a bit better at getting around, say it can move to its eight nearest neighbors, the adjacent and diagonal neighbors, then the threshold of connectivity decreases to about 40%. If the organism can jump across a single cell of unsuitable habitat and essentially ignore this interruption (albeit, a very simple view of the matrix), the threshold of connectivity decreases again, and a landscape containing only 25% suitable habitat could be traversed by the organism. Changes in the clumping of the habitat will also interact with the movement capability of the organism to determine the threshold of connectivity. Thus, connectivity is greatly influenced by the behavior of the organism.

The important idea about habitat connectivity is that habitat is either connected or disconnected and the change between these two states occurs at a threshold of habitat abundance. For an organism, this means the qualitative difference between being able to move about the landscape to locate suitable sites for foraging, nesting, and dispersal and being unable to do so. This has important implications for conservation, because suitable habitat might be lost for a while with no apparent negative effect on a plant or animal of interest until this threshold is passed. Then negative effects may occur suddenly as the organisms can no longer meet their needs on the fragmented landscape. Exactly where the threshold is depends on the organism, the amount of habitat, the spatial clustering of the habitat, and the nature of the matrix; different species might perceive different thresholds in the same landscape (Pearson et al., 1996; With, 1997a).

Do such thresholds in connectivity exist in real landscapes? Andren (1994) reviewed the empirical evidence for birds and mammals that could be used to test predictions derived from a neutral landscape model regarding habitat connectivity. He examined the process of habitat fragmentation, with its loss of habitat, reductions in patch sizes, and increased distances between patches. His results lead to the conclusion that the relative importance of these three aspects of fragmentation differs depending on abundance of suitable habitat in a landscape. When landscapes have >30% suitable habitat the primary effect of habitat fragmentation is habitat loss. This is because in landscapes with relatively high proportions of suitable habitat the habitat is reasonably well connected, and the configuration of habitat is less important. In experimental landscapes designed to study beetle movement, Wiens et al. (1997) found threshold effects when grassy habitat was less than 20%. Simulations conducted by Fahrig (1997) also demonstrated that the effects of habitat loss could outweigh the effects of habitat fragmentation on population extinction. In landscapes with low proportions of suitable habitat (10% to 30%), the spatial arrangement of patches is very important. Further reduction in habitat results in an exponential increase in distances between patches, that is, rapid decreases in connectivity; With and King (1999a) found evidence for a strong effect of gap structure on dispersal success. Moreover, the effect of patch size and isolation will not only depend on the proportion of original habitat in the landscape, but also on the suitability of the surrounding habitats (e.g., the matrix, areas between patches) for movement. Note that the results from both the theory and empirical studies suggest that conservation actions, such as adding habitat or protecting key locations, are most likely to have substantial effects on habitat connectivity when the suitable habitat is relatively low in abundance (Andren, 1994; Pearson et al., 1996; Fahrig, 1997). It is in this range where small changes in habitat abundance are likely to cause the threshold of connectivity to be passed. However, it is also important to recognize that the effects of habitat loss cannot be mitigated simply by connecting remaining habitat fragments.

What Is the Role of Corridors?

Corridors, narrow patches of land that connect similar patches but that differ from the surrounding matrix, have been proposed in conservation plans as a means of maintaining connectivity between otherwise isolated patches of habitat. As noted by With (1997a), corridors are appealing in a patch-based view of the world in which habitat is either suitable or unsuitable and corridors provide a physical bridge linking islands of habitat. The presence of corridors is assumed to increase population persistence by providing for an exchange of individuals among a population that was previously connected but that is now fragmented. However, there has been much controversy regarding their effectiveness (e.g., Simberloff and Cox, 1987; Saunders and Hobbs, 1991; Hobbs, 1992; Beier and Noss, 1998).

Functional corridors may not be discrete structures. Gustafson and Gardner (1996) simulated dispersal and patch colonization on heterogeneous landscapes and identified the regions of the landscape in which flows were funneled and that, therefore, functioned effectively as corridors. They found that these "corridors" were diffuse and difficult to identify. Reduced contrast between habitat patches and the intervening matrix may enhance connectivity more than would a discrete corridor. Although protecting naturally existing corridors probably benefits regional and local biodiversity, the creation of linear patches may not provide such benefits (Rosenberg et al., 1997).

Rosenberg et al. (1997) provided a useful distinction between two functions of linear landscape features: (1) they may themselves provide habitat, containing the resources needed for survival, reproduction, and movement; or (2) they may serve as biological corridors that provide for movement between habitat patches, but not necessarily for reproduction. Because corridors may be heavily influenced by edge effects, their interior habitat is reduced and may even be absent completely. By examining the literature on the use of corridors, Rosenberg et al. (1997) then synthesized a set of common patterns regarding the effectiveness of corridors. First, individual animals are likely to select pathways for movement that include components of their habitat when confronted by a choice, and this behavior is most pronounced for individuals moving within their home range. Second, the relative use of the matrix as movement habitat depends on its contrast with the organism's suitable habitat. Third, the behavior of animals may change in areas of less favorable habitat; for example, animals may move more rapidly when traversing low-quality habitat than in high-quality habitat. Evaluating the effectiveness of a linear habitat patch as a corridor requires a three-step evaluation of whether organisms can find, select, and successfully move through the patch; more empirical study is needed of a wider range of species in conditions that represent the complexity of real landscapes (Lidicker and Koenig, 1996; Rosenberg et al., 1997). Enhancing our understanding of how organisms move through heterogeneous landscapes (along corridors or through the matrix) is a key component of understanding the responses of organisms to spatial pattern.

Effects of Landscape Context

Characteristics of the surrounding landscape can strongly influence local populations. Whether the presence or abundance of organisms at a given location or sampling point is explained by characteristics of the immediate locale or by attributes of the surrounding landscape is an interesting question. Pearson (1993) studied wintering birds in powerline rights-of-way (ROW) in the Georgia piedmont. These ROWs are corridors in which the vegetation is maintained in an open state, usually by mowing, so that shrubs and herbaceous plants dominate. The areas surrounding the ROW may also be open, forested, or in cultivation. Pearson recorded the abundance of different bird species and the characteristics of the vegetation (such as height, density, and species composition) within each ROW and quantified the types of habitats in the surrounding landscape based on aerial photography (Figure 8.10). He found that variability in the presence and abundance of certain wintering birds [e.g., Parids (titmice) and Rufous-Sided Towhee (*Pipilo erythrophthalmus*)] was best explained by the habitats in the surrounding landscape (Table 8.4). Other species [e.g., Northern Cardinal (*Pyrrhuloxia cardinalis*) and White-Throated Sparrow (*Zonotrichia albicollis*)] responded only to the characteristics of the local habitat, and yet other species [e.g., Carolina Wren (*Thryotho-*

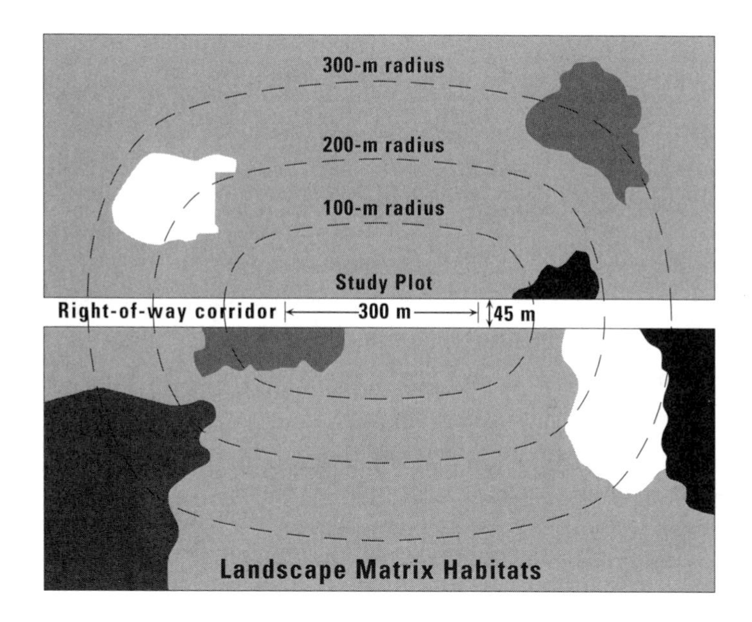

FIGURE 8.10.
Different spatial extents at which landscape patterns were characterized by Pearson (1993) and used in statistical models of bird species presence and abundance in powerline right-of-way sample plots.

ADAPTED FROM PEARSON, 1993.

TABLE 8.4.
RESULTS OF STEPWISE REGRESSION MODELS OF BIRD RESPONSE VARIABLES AS EXPLAINED BY WITHIN-HABITAT VARIABLES AND CHARACTERISTICS OF THE SURROUNDING LANDSCAPE.

Response variable	Matrix r^2	Model r^2
Community measures		
Total number of birds	0.24	0.73
Species richness	0.74	0.74
Shannon diversity	0.82	0.82
Functional groups		
Forest species	0.57	0.84
Early successional species	0.18	0.18
Generalists	0.13	0.73
Parids	0.65	0.76
Selected species		
Carolina wren	0.48	0.83
Field sparrow	0.24	0.24
Dark-eyed junco	0.24	0.24
Northern cardinal	0.31	0.83
Rufous-sided towhee	0.54	0.83
Song sparrow	0.19	0.19
White-throated sparrow	0.00	0.83

FROM PEARSON, 1993.

MATRIX r^2 REFLECTS THE AMOUNT OF VARIATION EXPLAINED ONLY BY THE LANDSCAPE MATRIX VARIABLES.

MODEL r^2 IS THE COEFFICIENT OF DETERMINATION FOR THE ENTIRE REGRESSION EQUATION. AMOUNT OF VARIATION EXPLAINED ONLY BY WITHIN-HABITAT VARIABLES CAN BE COMPUTED BY DIFFERENCE.

rus ludovicianus)] responded both to local conditions and to the landscape context (Table 8.4).

In a study that considered an even greater spatial extent, Flather and Sauer (1996) used data from the Breeding Bird Survey (BBS), annual censuses taken along fixed routes that are 39.4 km in length, and analyzed land-cover patterns in a circular scene of 19.7-km radius centered along each route, an area of 1200 km^2. They then tested the null hypothesis of no relation between landscape structure and neotropical migrant abundance by correlating the measures of landscape structure with breeding-bird abundance. Results demonstrated that neotropical migrants were more abundant in landscapes with a greater proportion of forest and wetland habitats, fewer edge habitats, larger forest patches, and forest habitats well dispersed throughout the scene. Landscapes with high proportions of human-dominated land uses (e.g., agriculture and urban) supported lower relative abundances of neotropical migrants. Neotropical migrants were also low in abundance where landscape diversity and edge were high and forest edges were more complex (as measured by fractal dimension). However, species that were permanent residents showed few correlations with landscape structure, and temperate migrants were associated with habitat diversity and edge attributes, rather than with the amount, size, and dispersion of forest habitats.

A number of other examples of the importance of landscape context can be found in the literature. Lindenmayer and Nix (1993) found that the occupancy of corridors by arboreal marsupials in Australia could not be predicted by habitat features within the corridor; information on the composition of the surrounding landscape was required. The winter foraging patterns of elk and bison in northern Yellowstone National Park, discussed earlier in this chapter, clearly differed with spatial scale and provide another example of the importance of landscape context (Pearson et al., 1995). Models explaining breeding-bird abundance in an urban southern California landscape were significantly improved when landscape variables were added (Bolger et al., 1997b).

Collectively, these examples demonstrate that it is often important to consider landscape context along with local site attributes when trying to explain local ecological processes. This insight has substantial implications for land management, because it suggests that what happens in small local areas may be influenced considerably by the surrounding landscape. Ecologists are conducting research to understand better when and over what scales the overall landscape influences are important or may even dominate and when the local conditions are more important.

Synthesis: When Is Spatial Pattern Important?

The influence of spatial heterogeneity on organisms, while interesting in its own right, certainly adds a substantial degree of complexity to population or community dynamics. But when is space really important, and when might spatial relationships be prudently ignored? Although the answer to this question depends on specific objectives and circumstances, there are instances where space (size, shape, and arrangement of habitat across the landscape) matters, and the answers to this question are clear. These include (1) when habitat is rare or fragmented, because theory and empirical data both point to the strong influence of spatial pattern in connecting or fragmenting habitat when it occupies <20% to 30% of a landscape, (2) if edge effects are important components of the process being studied, and (3) if dispersal limits movement between patches and metapopulation dynamics are likely to affect habitat use. Following from this, spatial considerations may not be needed if habitat is very abundant, because most suitable sites will likely be well connected and edge effects will be small; or if edge effects are unimportant for the process or species being studied; or if movement is relatively unlimited and organisms can reach all or most areas containing suitable habitat.

It is also important to recognize the limitations of a patch-based view of the world and to know when a more continuous or complex view of spatial heterogeneity is needed. Franklin (1993) argues that conservation considerations should include both a patch-based component, such as the design (size and spacing) of nature reserves, and the condition and management of the matrix, that is, the intervening areas between patches or reserves. Landscape context may be quite important for the organisms living in a given patch, and reducing the contrast across the edges of patches may enhance the suitability of the patches for organisms. The processes of movement and dispersal are also strongly influenced by the characteristics of the landscape between patches, yet quantitative data on this are sparse (and difficult to obtain).

SPATIALLY EXPLICIT POPULATION MODELS

The representation of spatial heterogeneity in population ecology varies considerably, ranging from assumptions that the world is subdivided into a large collection of patches that are all equally accessible to one another, to steppingstone models in which the patches have fixed locations in space (and so dispersal distances between patches will vary), to continuum models in which there is a continuous coordinate system in which populations interact and disperse (Kareiva, 1990). Landscape ecologists, however, typically utilize spatially explicit population models (SEPMs) to address questions about organisms in landscapes. SEPMs have a structure that specifies the location of each object of interest (e.g., organism, population, cell, or habitat patch) within a heterogeneous landscape, and therefore the spatial relationships among habitat patches and other landscape features are explicitly represented (Dunning et al., 1995). The models typically represent a landscape as a grid of cells and simulate the birth, death, foraging, and movement patterns of animals or plants across the landscape. Thus, SEPMs can be used to address questions of changing resource distributions, habitat fragmentation, isolation, habitat shape, and patch size on populations of interest. The models may be individual-based models, in which the attributes and movements of every individual in the population are represented in the model, or they may be cell-based models in which the occupancy of a cell, or site, is the focus (these models are also called cellular automata models). SEPMs have yielded interesting insights into the interactions between single populations and their habitats.

BACHMAP is an example of an individual-based SEPM model designed to elucidate the effects of habitat arrangement on the size and extinction probability of a Bachman's sparrow (*Aimophila aestivalis*) population in a region managed for timber production (Pulliam et al., 1992). Bachman's sparrow is a small bird found in pine woods of the southeastern United States, and the species is of concern to land managers because its population has declined since the 1930s over much of its range. The BACHMAP model includes life-history characteristics (such as dispersal, survivorship, and reproductive success) of Bachman's sparrow and the landscape characteristics of the Savannah River Site, South Carolina. An interesting aspect of the population is that the sparrows occur primarily in very young pine stands (<10 yr after planting) and in mature pine stands (>50 yr old), but they rarely use stands of intermediate age. Thus, as trees are harvested and regrow, the distribution of suitable habitat for Bachman's sparrow across the landscape changes. Simulations indicated that the demographic parameters (having to do with birth and death rates) had a greater impact on population size than the dispersal parameters (having to do with moving across the landscape to find suitable habitat); but many of the simulated population sizes were small, much less than the amount of suitable habitat might theoretically support. An extension of BACHMAP developed by Liu (1993) and called ECOLECON integrated economic considerations and harvesting scenarios, allowing future landscape patterns, economic returns, and sparrow population dynamics to be considered together.

Individual-based SEPMs have been developed for a variety of other species. In individual-based models, the location of each individual across the landscape is monitored, and the state of individuals (e.g., weight) may be tracked separately. An individual-based model for moose foraging in a boreal landscape (Moen et al., 1997, 1998) is discussed in Chapter 9, and individual-based models serve as components of a multilevel spatial model of the Everglades (DeAngelis et al., 1998), also discussed in Chapter 9. The spotted owl model (Noon and McKelvey, 1996a) used to assess the implications of old-growth forest harvesting in the Pacific Northwest and discussed in Chapter 10 also falls into this category, as does the elk and bison simulation model for northern Yellowstone National Park (Turner et al.,

1994a), discussed earlier in this chapter. Individual-based models represent a bottom-up approach to modeling population dynamics in space, but some researchers are interested in developing more formal protocols for linking these models with classical population ecology approaches (Fahse et al., 1998).

SEPMs may also be population based, in which each cell contains a population (Dunning et al., 1995). In these cell-based models, occupancy and even abundance are tracked on cells representing the landscape, but the fate of individuals is not tracked. These models have also yielded a variety of useful insights. With and Crist (1995) projected the dispersion patterns of grasshoppers across a larger landscape mosaic using a cell-based model. Lavorel et al. (1994b) used a spatially explicit simulation model to examine the interaction among dispersal strategies, seed dormancy, and disturbances on the pattern of coexistence. The results showed that when species differed only in mean dispersal distance the species with the shorter-range dispersal would always displace the other. When disturbances prevented reproduction, but activated dormant seeds, differences in dormancy strategy could result in coexistence of species with different dispersal capabilities. The predicted patterns of coexistence were consistent with conclusions of a field study, in species-rich Mediterranean old-fields, where coexistence of annual and perennial forbs depends on the interactions among dispersal and dormancy strategies, spatial heterogeneity, and small-scale disturbances (Lavorel et al., 1994a).

SEPMs can be used to address a wide range of questions in landscape ecology, population ecology, and conservation biology. They are particularly useful tools when questions require consideration of the amount, geometry, and rates of change in habitat. For management questions, SEPMs may be used to develop a robust relative ranking of management alternatives (Turner et al., 1995a). However, it is important to remember that these models are data hungry, and the lack of data for the scales or area of interest may limit their application. Validation and interpretation of the simulations can also be difficult. In addition, few SEPMs have been developed in which multiple species or interactions among species are considered. For example, Liu et al. (1995) demonstrated how the Bachman's Sparrow, described earlier, was affected by forest-management scenarios developed for the federally listed Red-Cockaded Woodpecker (*Picoides borealis*). Although the two species overlap in habitat requirements in these longleaf pine forests of the Southeast, the woodpecker requires stands older than 70 yr, whereas Bachman's Sparrow also occurs in younger stands. Thus, a shift to management of old-growth stands would actually reduce the habitat available for Bachman's sparrow. Management plans based on SEPMs developed for a single species may be insufficient for management of other species on the landscape, even those having similar habitat requirements. Again, a multispecies management perspective is sorely needed.

SUMMARY

Organisms exist in spatially heterogeneous landscapes. Landscape ecologists are particularly interested in how organisms utilize resources that are distributed across a heterogeneous landscape and how they live, reproduce, disperse, and interact with each other in space. Studies in population ecology have always been cognizant of spatial issues, but explicit consideration of spatial factors influencing populations has become much more prominent in the past three decades. Early studies focused on the diffusion of organisms across homogeneous landscapes, predicting rates of spread and number of sites that would be occupied under various conditions. However, the development of island biogeography and metapopulation theory fueled much research on the dynamics of populations in heterogeneous landscapes. Island biogeography was developed as a general theory to predict the number of species found on oceanic islands based on two attributes of an island: its size and its distance from the mainland. Island biogeography theory was subjected to a number of criticisms, one of which was its assumption of an equilibrium number of species. Despite the criticisms, island biogeography theory has been important in highlighting the effects of the size and isolation of natural areas on their effectiveness in meeting conservation objectives.

In the late 1980s, interest in Levins's metapopulation model revised and replaced equilibrium island biogeography as a way of thinking about fragmented habitats and heterogeneous terrestrial environments in general. A metapopulation was considered as a collection of subpopulations, each occupying a suitable patch of habitat in a landscape of otherwise unsuitable habitat, forming an interconnected set of subpopulations that function together as a demographic unit. The two key premises in the metapopulation approach are that (1) populations are spatially structured into assemblages of local breeding subpopulations and (2) migration among subpopulations results in a recolonization following local extinction, producing regionally stable metapopulation dynamics. The development of source–sink models resulted when differences between patches were considered. Patches with excess reproduction are source patches, and sink patches occur when local mortality exceeds reproductive success. Source–sink dynamics allow migration of the surplus organisms from the source to the sink patches, maintaining populations in an apparent demographic equilibrium.

Many approaches to studying populations in heterogeneous space require identification of suitable and unsuitable habitat. In some instances this simplifying is clear, but in other cases it is difficult to partition the landscape into patches embedded in an unsuitable matrix. Landscape and population ecologists are only beginning to incorporate a more continuous representation of habitat suitability into their conceptual framework. The same landscape will look very different to different organisms, and the spatial description of land cover alone may be misleading as a surrogate for suitable habitat. Thus, it is important to recognize the complexities inherent in habitat definition. However, habitat-based approaches may also be important surrogates for some estimates of population responses to landscape heterogeneity. Identification of suitable habitat for species and tracking changes in habitat abundance and pattern through time can contribute to modeling risks to biodiversity at landscape scales and serve as a first step toward anticipating potential ecological effects of land-use changes.

Landscape ecology adds several features to the consideration of how populations interact with spatial pattern: (1) variation in patch quality, (2) variation in the quality of the surrounding environment, (3) boundary effects, and (4) how the landscape influences connectivity among patches. The main difference between the metapopulation view of nature and that embraced by landscape ecologists revolves around the degree of complexity that is considered. We do not yet have a conceptual and practical synthesis of metapopulation biology and landscape ecology, yet establishing more common ground between these lines of inquiry is imperative.

There are important concepts relating to scale when the interactions between organisms and spatial pattern are considered, because the appropriate scales vary among taxa—and among questions. There is a strong imperative to focus on the scales that are appropriate for the organism, recognizing that our human-based perception of scale and pattern may not be the right one. Some differences in appropriate scales are due to various attributes of the organisms, such as differences in body mass. It is easy to acknowledge that interactions between organisms and spatial heterogeneity must be scale dependent, but it is difficult to identify the right scales in practice. The concept of ecological neighborhoods offers one practical approach to this thorny issue. Related to the concept of ecological neighborhoods is the notion that organisms may respond to heterogeneity at multiple scales. In addition, the structure of the landscape itself may dictate the scales at which organisms must operate.

A number of insights that have emerged from landscape ecological studies about organisms and space are presented, with examples:

- *Patch size and heterogeneity*: In general, larger more heterogeneous patches support more species.
- *Ecotones and effects of boundary shape*: Shape of boundary can influence species' relative abundances.
- *Habitat connectivity*: Connectivity is a scale-dependent, threshold phenomenon.
- *Effects of landscape context*: Characteristics of the surrounding landscape can strongly influence local populations.

The influence of spatial heterogeneity on organisms adds a substantial degree of complexity to any study of a population or community, and there remains the important question of when space really must be considered and when it might be prudently ignored. Instances where space (size, shape, and arrangement of habitat across the landscape) simply must be considered include (1) when habitat is rare or fragmented, especially when suitable habitat is <20% to 30%; (2) if edge effects are important to the process being studied; and (3) if dispersal limits movement between patches and metapopulation dynamics are likely to occur.

The representation of spatial heterogeneity in population ecology models varies considerably. Landscape ecologists typically utilize spatially explicit population models (SEPMs) to address questions about organisms in landscapes. SEPMs have a structure that specifies the location of each object of interest (e.g., organism, population, cell, or habitat patch) within a heterogeneous landscape, and therefore the spatial relationships between habitat patches and other landscape features are explicitly represented. SEPMs can be used to address a wide range of questions in landscape ecology, population ecology, and conservation biology. They are particularly useful tools when questions require consideration of the amount, geometry, and rates of change in habitat. Note, however, that while SEPMs are powerful tools they lack generality. In contrast, the patch-based models of island biogeography and metapopulation theory are general, but they often lack applicability to specific management questions. What is needed is a general, spatially explicit theory.

Field studies, experiments, and theory development have clearly demonstrated the important effects of spatial patterning on a variety of organisms. Patch size, shape, and arrangement; habitat connectivity; and landscape context may all have strong influences on the abundance and persistence of populations. When applied to conservation issues, landscape ecology encourages a broad-scale perspective that recognizes spatial complexity and dynamics. However, there remains both the opportunity and need for much greater synthesis among landscape ecology, population ecology, and conservation biology. A landscape perspective may also foster greater integration in approaches and understanding of the linkages between species and ecosystem function; this topic is covered in Chapter 9.

DISCUSSION QUESTIONS

1. Equation (8.2) predicts a rate of spread due to dependence on the estimate of the diffusion coefficient, D. Discuss the assumptions necessary to estimate the diffusion coefficient and the problems that may arise when comparing rates of spread among species within the same landscape or for the same species in different landscapes.
2. Discuss the concepts and principles of conservation design that have been affected by island biogeography theory. Are these concepts and principles dependent on the assumptions of the theory? Discuss how recent changes in our view of the applicability of island biogeography theory have affected the principles and practice of conservation design.
3. It is a simple process to convert equation (8.3) to its discrete form and produce a graph of the growth of a metapopulation. [*Hint*: the discrete form equation (8.3) will estimate the net change from one generation to the next.] Set $m = 0.2$, $c = 0.6$, and the initial value of $p = 0.2$ and graph the response of the system over 25 generations. What is this equilibrium value? Was equilibrium reached in 25 generations? What happens after 25 additional generations if $m = 0.62$?
4. Most source–sink metapopulation models are spatially explicit. How might equation (8.3) be modified to represent a spatially implicit source–sink metapopulation? Speculate on the conditions that would allow a spatially implicit model to accurately reflect source–sink dynamics in a heterogeneous landscape. What conditions will cause a spatially explicit model to fail?
5. What species attributes determine the spatial and temporal scales needed to characterize species abundance and distribution at landscape scales? Can allometric rules be derived for relationships among species groups or between diverse taxa?
6. Habitat connectivity is usually measured as a function of the geometry of land cover. Discuss the limitations and uncertainties that these methods pose for diverse species groups. How might measures of connectivity be improved?

RECOMMENDED READINGS

ANDREN, H. 1994. Effects of habitat fragmentation on birds and mammals in landscapes with different proportions of suitable habitat. *Oikos* 71:355–366.

DEN BOER, P. J. 1981. On the survival of populations in a heterogeneous and variable environment. *Oecologia* 50:39–53.

DUNNING, J. B., D. J. STEWART, B. J. DANIELSON, B. R. NOON, T. L. ROOT, R. H. LAMBERSON, AND E. E. STEVENS. 1995. Spatially explicit population models: current forms and future uses. *Ecological Applications* 5:3–11.

FRANKLIN, J. F. 1993. Preserving biodiversity: species, ecosystems, or landscapes? *Ecological Applications* 3:202–205.

HANSKI, I., AND M. E. GILPIN, editors. 1997. *Metapopulation Biology.* Academic Press, New York.

KAREIVA, P. 1990. Population dynamics in spatially complex environments: theory and data. *Philosophical Transactions of the Royal Society of London B* 330:175–190.

MCGARIGAL, K., AND W. C. MCCOMB. 1995. Relationship between landscape structure and breeding birds in the Oregon Coast Range. *Ecological Monographs* 65:235–260.

VERBOOM, J. A., P. SCHOTMAN, P. OPDAM, AND J. A. J. METZ. 1991. European nuthatch metapopulations in a fragmented agricultural landscape. *Oikos* 61:149–156.

WIENS, J. A. 1976. Population responses to patchy environments. *Annual Review of Ecology and Systematics* 7:81–120.

CHAPTER 9

Ecosystem Processes in the Landscape

DETERMINING THE PATTERNS, causes, and effects of ecosystem function across landscapes remains an important current topic in ecosystem and landscape ecology. Indeed, the interface between science and the management of ecosystems is one of the most dynamic fields of contemporary ecology (Christensen et al., 1996; Carpenter and Turner, 1998). The term *ecosystem* was first introduced by Tansley (1935) and refers to a spatially explicit unit of Earth that includes all the organisms, along with all components of the abiotic environment within its boundaries (Likens, 1995). Ecosystem ecology focuses on the flow of energy and matter through organisms and their environment. Ecosystem studies address questions about the capture of light energy by plants, its conversion into organic matter, and its transfer to other organisms and questions about nutrient cycling, in which essential elements such as phosphorus and nitrogen cycle repeatedly between the living and nonliving parts of ecosystems (Golley, 1993; Carpenter, 1998). Spatially, ecosystem science encompasses bounded systems like watersheds, spatially complex landscapes, and even the biosphere itself; temporally, ecosystem science crosses scales ranging from seconds to millennia (Carpenter and Turner, 1998). Ecosystem ecology is a well-developed field and will not be exhaustively reviewed here; interested readers are referred to Golley's review of the history of ecosystem research (1993), and the recent synthesis by Pace and Groffman (1998).

During the past several decades, studies of the effects of forest disturbances on biogeochemical processes produced a wealth of information about watershed dynamics, energy flow, and nutrient cycling in natural and managed systems in the United States. Watershed studies during the past several decades at a variety of research sites, including Hubbard Brook (New Hampshire), H. J. Andrews (Oregon), Coweeta (North Carolina), and Walker Branch Watershed (Tennessee), provided ecologists with new insights into the interactions between terrestrial and aquatic systems (e.g., Bormann and Likens, 1979; Swank and Crossley, 1988; Johnson and Van Hook, 1989; Likens and Bormann, 1995). Nutrient budgets were developed for whole watersheds, and the effects of experimental manipulations, especially forest clear-cutting, on rates of nutrient cycling, storage, and loss were determined. More recently, ecosystem-level research has expanded to consider spatial heterogeneity, for example, topographic or regional variation in nutrient cycling processes (e.g., Zak et al., 1987, 1989; Groffman et al., 1993; Benning and Seastedt, 1995; Burke et al., 1995; Gross et al., 1995; C. L. Turner et al., 1997); spatial variation across landscapes in evapotranspiration leaf area (Running et al., 1987; Running et al., 1989; Spanner et al., 1990; Band et al., 1991; Nemani et al., 1993; Pierce et al., 1994), biomass and productivity (Host et al., 1988; Sala et al., 1988); the transport of materials between ecosystems (Peterjohn and Correll, 1984; Shaver et al., 1991; Soranno et al., 1996); and spatial interactions among the plant community, large herbivores, and nutrient cycling (Coughenour, 1991; Pastor and Naiman, 1992; Pastor and Cohen, 1997). Understanding the implications of landscape heterogeneity and disturbance dynamics for ecosystem processes is among the most important challenges facing contemporary ecosystem ecologists (Schimel et al., 1997).

When landscape ecologists study ecosystem processes, they emphasize the causes and consequences of spatial heterogeneity in the rates of ecosystem processes (e.g., net primary productivity, nitrogen mineralization), the influence of landscape position on ecosystem function; and the horizontal movement of materials (such as water, nutrients, or sediments) and how these movements might differ with alternative spatial arrangements of land cover. We consider each of these topics here. We include land–water interactions in this chapter because many important questions and land-management issues revolve around functional interactions between terrestrial and aquatic systems, especially the transport of sediment and excess nutrients to streams, lakes, estuaries, and coastal oceans. When a watershed is considered homogeneous and point measurements are made at the outflow, spatial heterogeneity is not relevant. However, when different watersheds are compared or the heterogeneity within a watershed is of interest, such studies overlap with landscape ecology. Finally, the landscape provides an ideal template for developing much needed linkages between populations and ecosystem processes (Jones and Lawton, 1995), and we consider these interactions in this chapter.

SPATIAL HETEROGENEITY IN ECOSYSTEM PROCESSES

Ecosystem processes vary spatially in response to many factors. For example, temperature gradients, precipitation patterns, and topographic variation produce differences in the rates of processes such as productivity, decomposition, and nitrogen cycling across landscapes. Soils and topography exert a powerful influence over ecosystem processes, and this is a common theme throughout this chapter. Another common theme is that relevant spatial and temporal scales, and the appropriate model formulations, vary considerably with the process and question of interest. For example, the relevant landscape for studying microbial processes in the soil will be much smaller than the landscape used to study regional patterns of primary production. Similarly, a very detailed process model may work well for a relatively small extent, but a simpler model may be needed for large regions (see also Chapter 3).

Patterns of Biomass, Productivity, and Carbon

Global and regional patterns of variation in net primary production in terrestrial and aquatic systems have been recognized by biologists for many decades (Leith and Whittaker, 1975; Box, 1978). More recently, ecologists have gained tremendous insights into the spatial and temporal dynamics of net primary production (NPP) with the advent of remote imagery from which rates of NPP can be inferred. Since the first Landsat satellite was launched in 1972, estimation of terrestrial plant production has been one of the most important applications of satellite remote sensing (Running, 1990). Understanding and predicting such patterns became more urgent as scientists worked to understand the global carbon cycle, tried to quantify carbon sinks, sources, and fluxes; and began to grapple with predicting possible consequences of global warming (Bolin, 1977; Emmanuel et al., 1984).

New views of spatiotemporal patterns of vegetation and NPP that were produced in the 1980s by remote sensing scientists caught the attention of ecologists. Data from the AVHRR (advanced very high resolution radiometer) satellite, which provides daily images of Earth at a resolution of 1.1 km, were used to generate vegetation maps for Africa (Tucker et al., 1985), North America (Goward et al., 1985), and the globe (Justice et al., 1985). These maps relied on the dimensionless *normalized difference vegetation index* (NDVI), a ratio of the difference between near-infrared (NIR, 0.725 to 1.1 μm) and red (RED, 0.58 to 0.68 μm) portions of the spectrum: NDVI = (NIR − RED)/(NIR + RED). The value of the NDVI is directly related to the presence of photosynthesizing vegetation, providing an indirect measure of the health and growth of vegetation (Jensen, 1996). Recent global change research has validated the importance of coarse-scale NDVI measurements as inputs into global climate models (Los et al., 1994). In fact, the measurement of NDVI represents a critical input for general circulation models of the atmosphere by providing regional- to global-scale land surface parameters such as LAI (leaf area index), albedo, and FPAR (fraction of absorbed photosynthetically active radiation) (Sellers et al., 1994).

NDVI has proved to be quite useful, as it relates well to *leaf area index (LAI)* (Sellers, 1985, 1987) and, in turn, to NPP (Goward et al., 1987). LAI, the ratio of leaf area to ground area, usually reported as m^2/m^2, is a useful index for a variety of ecosystem processes, including the interception of light and water by the vegetation, attenuation of light through the canopy, transpiration, photosynthesis, and nitrogen content. Ecologists estimating ecosystem processes across landscapes now routinely measure LAI, in part because it can be well estimated for large areas by satellite imagery (Running, 1990).

Vegetation indexes derived from satellite data offer new ways of exploring spatial variation in ecosystem structure and function at broad scales. Riera et al. (1998) hypothesized that variability in vegetation cover and biomass should be related to topographic relief and to land use–land cover at the spatial extent of full Landsat Thematic Mapper (TM) images (~185-km swath). The *simple ratio vegetation index (SR)* [defined as the ratio between TM3 (the reflectance in band 3, 0.63 to 0.69 μm, corresponding to the red portion of the spectrum) and TM4 (the reflectance in band 4, 0.76 to 0.90 μm, the near-infrared) of the Landsat TM sensor] and the NDVI were compared across 13 study sites representing a wide range of biomes in North America. Marked differences in landscape heterogeneity were observed among the landscapes. Desert and grassland landscapes had low mean NDVI and low overall heterogeneity, whereas forested landscapes had high mean NDVI, but also low overall heterogeneity. Spatial heterogeneity was greatest for those landscapes that had intermediate values of the vegetation indexes. Furthermore, the heterogeneity of the vegetation, as measured by the standard deviation of NDVI, increased as topographic complexity increased (Riera et al., 1998), but only for areas with relatively high topographic relief. In landscapes with low topographic relief, the proportion of agriculture influenced the heterogeneity of NDVI.

Ecologists began combining remote imagery and other spatial data with ecosystem simulation models to predict spatiotemporal patterns of productivity in the late 1980s. Early developments were driven by the need to link global models, such as the general circulation models (GCMs) used to simulate the global climate, with changes in the vegetation cover and state. Running et al. (1989) were among the first to integrate biophysical data obtained from many sources and use these data to execute an ecosystem model over a large landscape. They used a 28- by 55-km coniferous forest landscape in western Montana, USA, and built a simple GIS in which climate and soil data were stored. The model required an estimate of LAI for each grid cell, and this estimate was obtained from satellite imagery. In addition, the model required soil water-holding capacity for each grid cell and daily meteorological data. The ecosystem model FOREST-BGC (Running and Coughlan, 1988) was then run in each of the 1200 grid cells representing the landscape to predict spatial patterns of annual evapotranspiration and net photosynthesis. The resulting estimates of LAI, evapotranspiration, and photosynthesis (Figure 9.1) demonstrated the power of these new integrative methods for producing spatially explicit projections of variation in ecosystem processes and offered insights into interactions among the controls on these processes (Running et al., 1989).

Extensive empirical studies were also conducted during the 1980s and 1990s on regional patterns of primary production, the accumulation of soil organic matter, and biogeochemical cycling; studies from the Great Plains region of North America nicely illustrate this approach. Using an extensive data set containing measurements of aboveground net primary production (ANPP) from 9500 sites throughout the Central Grassland region of the United States, Sala et al. (1988) evaluated (1) the spatial and temporal pattern of annual production for the region and (2) the importance of climatic variables as determinants of the pattern of ANPP when the site-level data were aggregated to major land resource areas. Results demonstrated that general trends in processes like net primary productivity and decomposition could be predicted reasonably well by broad-scale variability in temperature, precipitation, and soils (Sala et al., 1988).

The analyses by Sala et al. (1988) confirmed the importance of water availability as a control on ANPP, with the regional spatial pattern of production reflecting the east–west gradient in annual precipitation. ANPP was lower in the drier western part of the region and higher in the more moist eastern areas, but the spatial pattern shifted eastward during dry years and westward during wet years (Figure 9.2). Maximum variation between favorable and unfavorable years was observed in a wedge-shaped area centered on southwest Kansas (Figure 9.2d), which corresponded to a region that has a characteristic spring and summer rainfall deficiency (Sala et al., 1988). The analyses also revealed an interaction between precipitation and soil water-holding capacity. When annual precipitation was <370 mm, sandy soils with low soil water-holding capacity were predicted to be more productive than loamy soils with high water-holding capacity. The opposite pattern was predicted when precipitation was >370 mm. This occurs because bare-soil evaporation is lower in sandy soils than in loamy soils, because water penetrates more deeply into the soil; runoff is also lower in the sandy soils. Sala et al. (1988) also observed from their empirical analysis that a model will need to include a larger number of variables to account for the spatial pattern of

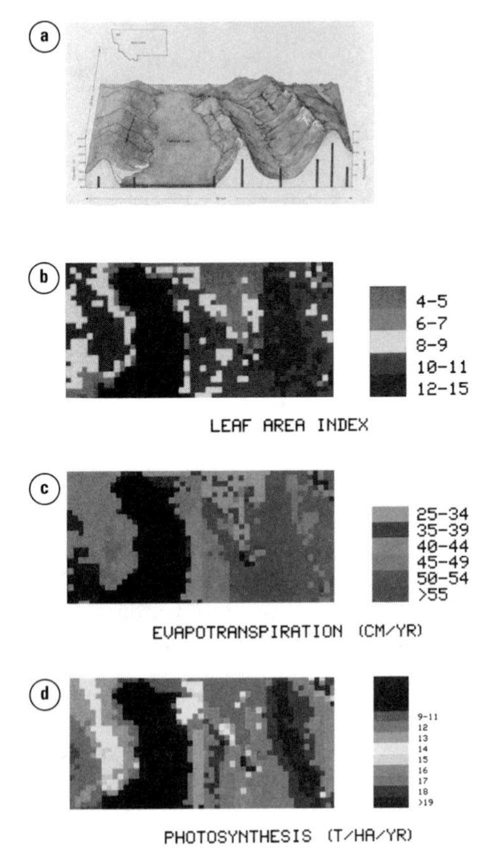

FIGURE 9.1.
(a) Schematic diagram showing prominent physiographic features of a 1540-km^2 study area in western Montana in which Running et al. (1989) combined satellite imagery, GIS, and an ecosystem simulation model to predict patterns of ecosystem processes. Maps of (b) leaf area index (LAI), (c) annual evapotranspiration, and (d) annual net photosynthesis for the 28- by 55-km study area using 1.1-km grid cells. LAI was estimated by satellite, microclimate data were extrapolated from a model, and ecosystem processes were simulated with the FOREST-BGC model. (Refer to the CD-ROM for a four-color reproduction of this figure.)

the same process as the scale of analysis becomes finer. The pattern of the process at the coarse scale constrained the pattern at the finer scale; thus, variability at the finer scale will be accounted for by factors at that scale plus the factors that determine the pattern at the coarse scale.

Spatial variation in soil organic matter and carbon has also been studied in the Great Plains region, where the accumulation of organic matter in soils depends on temperature, moisture, soil texture, and plant lignin content (Parton et al., 1987). Lignin is a structural compound that is difficult for most microbes to decompose, and thus plant tissues with higher lignin content tend to decompose more slowly. Across the dry landscape of the Great Plains, soil carbon increases from southwest to northeast and, additionally, is greater on fine-textured soils than on sandy soils. General patterns across the landscape are reasonably well known, but additional variables like plant species composition, past land use, and

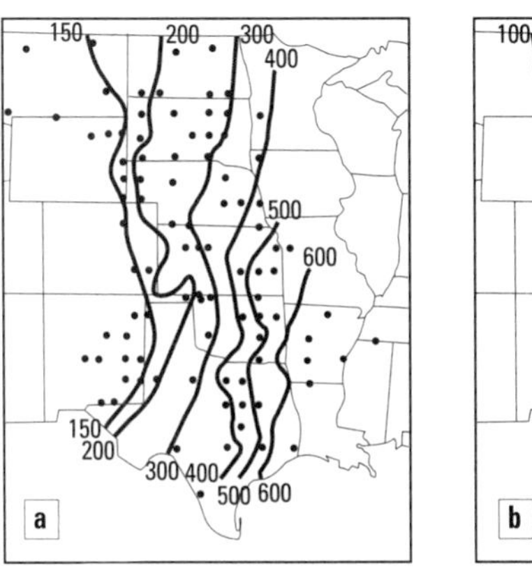

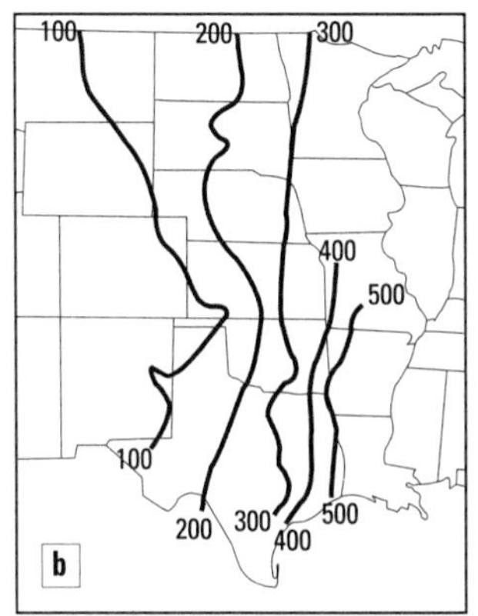

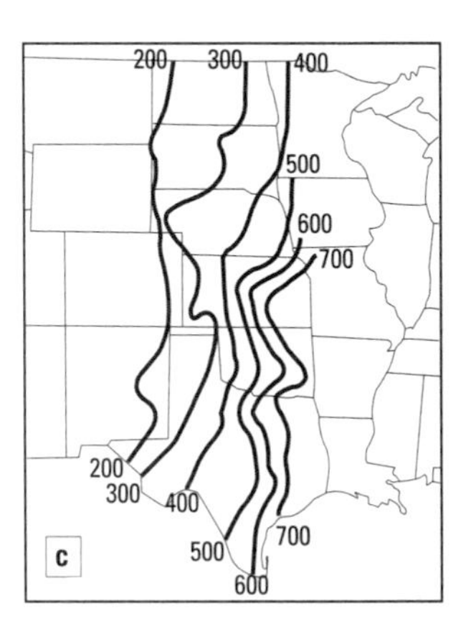

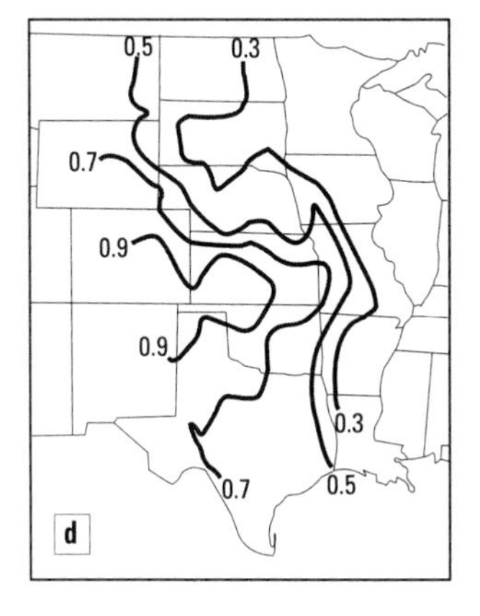

FIGURE 9.2.
Isopleths of aboveground net primary production (ANPP, g/m^2) for the Central Grassland region of the United States (a) during years of average precipition, (b) during relatively dry years, and (c) during relatively wet years. (d) Isopleths show the relative variability in production between favoral and unfavorable years, estimated as $(\text{ANPP}_{wet} - \text{ANPP}_{dry})/\text{ANPP}_{average}$.

ADAPTED FROM SALA ET AL., 1988.

landscape position are required to predict ecosystem processes at finer scales. A more recent study conducted in agroecosystems illustrates some of these patterns.

Burke et al. (1995) studied the patterns of soil organic matter (SOM), microbial biomass, and the availability of carbon (C) and nitrogen (N) in soils in two geographic locations, then at different landscape positions (at the bottom, or toe, of a slope; on a slope; or at the top of a ridge), and under different types of land management. They found that these indicators of ecosystem functioning (SOM, microbial biomass, and soil C and N) were highest at the more northern of two study sites in eastern Colorado, and at toeslope landscape positions, the bases of slopes where materials tend to accumulate. The geographic differences between sites reflected broad-scale influences of temperature and soil texture on the accumulation of organic matter; the northern site had cooler temperatures and finer soils. The effects of landscape position reflected the long-term transport of fine organic matter downslope, as well as gradients in available moisture and production. Ridgetops tend to be exposed and dry, toeslopes more sheltered and moist. Thus, there are natural gradients in soil organic matter and nutrient supply capacity across the landscape. Nutrient fluxes by litter redistribution can also be extensive in other landscapes. For example, redistribution of leaf litter in a topographically and edaphically complex landscape on the Allegheny Plateau in Ohio was found to be substantial, with some landscape positions (e.g., ridgetops and upper slopes) serving as a net source of litter and others (side and lower slopes) serving as a net sink (Boerner and Kooser, 1989).

While abiotic gradients exert a powerful influence, rates and spatial patterns of NPP can also vary with changes in land use and with natural disturbances. Examining land-use changes and NPP in Georgia, USA, Turner (1987c) found an overall increase in annual NPP from 2.5 to 6.4 tonnes/ha from 1935 to 1982. These changes were associated with a period of widespread abandonment of croplands, followed by natural succession to pine and increased urbanization in some areas. Production of croplands varied widely among physiographic regions in the state. Low NPP through the 1960s reflected persistent effects of poor agricultural practices that had caused the fertility of the land to decline; in the piedmont region, nearly all the topsoil had been lost from nearly half of the land area. Higher NPP after the 1960s was due, in part, to improvements in agriculture and silviculture, as well as large increases in irrigated area and inputs of nitrogen fertilizers. By 1982, the average NPP of forests in Georgia was 8.3 tonnes/ha, approximately two-thirds of the predicted natural rate.

Spatial and temporal variation in NPP has important implications for regional and global patterns of carbon dynamics. Secondary forests in regions that experienced widespread cropland abandonment may serve as important terrestrial sinks for global carbon (Delcourt and Harris, 1980). Much of the eastern United States may be serving this important function. Disturbance-initiated changes in the landscape mosaic also alter patterns of productivity and carbon accumulation. For example, wildfires release carbon dioxide (CO_2), carbon monoxide (CO), and particulate matter into the atmosphere (Crutzen and Goldhammer, 1993). In the conterminous United States, wildland fires affected about ten times more of the landscape annually during presettlement than at present and consumed about eight times more biomass and produced seven time more emissions (Clark and Royall, 1994; Leenhouts, 1998). However, long-term effects of increased fire frequency and extent on net carbon storage remain unresolved. Much of the carbon released by fire is recovered by new vegetation growth, with little net effect on atmospheric carbon. Moreover, the significant amounts of stable carbon (e.g., charcoal) left behind following fire can persist for centuries, suggesting that wildland fires could even be a significant long-term sink for carbon that would otherwise be cycled to the atmosphere.

Prediction of the future distribution of vegetation as a result of global change is a significant challenge and concern (Starfield and Chapin, 1996). The potential for major changes in spatial patterns of biomes and productivity at regional and global scales as a consequence of climate change and potential feedback interactions between vegetation and climate may alter the response of both systems to a doubling of atmospheric CO_2 (Neilson and Drapek, 1998). In addition, the inherent time lags of response of terrestrial biota means that recommendations for conservation management must be implemented long before observed changes are realized. Estimates of potential patterns of species redistribution must consider species physiological tolerances, competition, dispersal mechanisms, and the interaction of these factors with local disturbance regimes and landscape heterogeneity (Halpin, 1997).

Landscape Biogeochemistry

Interest in obtaining broad-scale estimates of biogeochemical processes and their spatial variability has increased in recent years. This increase is motivated in part by concern about human intrusions into global nutrient cycles, such as making more of a given element available in a biologically active form (Mooney et al., 1987; Groffman et al., 1992; Vitousek et al., 1997a). Nitrogen is a useful indicator of ecosystem function for several reasons. Nitrogen usually limits primary productivity in most temperate ecosystems (Vitousek and Howarth, 1991; Reich et al., 1997), and the presence of nitrate in soil water and streamwater can be used as an indicator of disturbances that lead to N leaching (Bormann and Likens, 1979; Vitousek and Melillo, 1979; Parsons et al., 1994). Nitrogen influences the quality of both water and air, and anthropogenic intrusions into the global nitrogen cycle may have profound implications for terrestrial and aquatic ecosystems (Vitousek et al., 1997a). The unprecedented production of industrially fixed nitrogen has resulted in massive global inputs of nitrogen, and nitrogen may no longer be the limiting nutrient on NPP in some regions.

Nitrogen dynamics are influenced by abiotic gradients and biotic interactions. For example, temperature and soil type explain a considerable amount of variation in both nitrogen mineralization rates and ANPP in cool temperate forests (Reich et al., 1997). Nitrogen mineralization (the production of ammonium by aerobic soil bacteria) and nitrification (conversion of ammonium to nitrate by aerobic soil bacteria) produce the forms of nitrogen available for plant uptake. The interesting question of whether nitrogen availability is an effect or a cause of variability in ANPP remains unresolved (Reich et al., 1997). Nitrogen mineralization rates may differ with vegetation composition, in part due to differences in litter quality among different plant groups. For example, nitrogen mineralization rates were higher in Canadian boreal forest stands with deciduous species than in conifer stands located on similar sites (Pare et al., 1993), and large differences in net N mineralization rates have been associated with successional changes from poplar to white spruce in an Alaskan chronosequence (Van Cleve et al., 1993).

Although spatially explicit studies of biogeochemistry are relatively few, they provide important insights into scale-dependent relationships between ecosystem pattern and process. For instance, Morris and Boerner (1998) quantified nitrogen mineralization and nitrification potentials in soils of hardwood forests in southern Ohio at three spatial scales: (1) the regional scale, represented by four study areas of 90 to 120 ha separated by 3 to 65 km; (2) the local scale, represented by three contiguous watersheds within each study area, and (3) the topographic scale, represented by different landscape positions within each watershed. Their results underscored the importance of understanding the patterns of variation manifested at different spatial scales. They observed no effect of spatial scale for nitrification potential in their study area, suggesting extrapolation from plot to region should be relatively easy. However, this was not the case for N mineralization potential or storage of organic carbon. These processes varied with topographic position, and thus stratification would be needed for extrapolation from plots to regions. Studies in other locations have also found strong topographic influences on soil nitrogen dynamics (e.g., Garten et al., 1994). Future research should reduce, define, and refine the scales at which microbially mediated ecosystem processes are measured (Morris and Boerner, 1998), considering both top-down and bottom-up approaches to scaling. In a study of the water-limited and discontinuous plant cover typical of shortgrass steppe ecosystems, Epstein et al. (1998) found that plant function types (C-3 versus C-4 plants) can affect both the spatial and temporal patterns of N cycling.

A regional study in southern Michigan, USA, used soil texture and natural drainage class to extrapolate rates of denitrification obtained from a landscape study (Groffman and Tiedje, 1989) to an even larger area using a GIS (Groffman et al., 1992). Denitrification is the production of gaseous nitrogen from nitrate by soil bacteria in the absence of oxygen. Denitrification is important in reducing nitrate pollution in groundwater. Results revealed that spatial patterns of soil texture strongly influenced regional patterns of denitrification. Loam-textured soils occurred under 47% of the forests in the region, but accounted for 73% of the denitrification. Sandy soils occurred under 44% of the regional forest, but produced only 5% of the regional denitrification, and clay loam soils, which underlie 9% of the regional forest, produced 22% of the denitrification (Groffman et al., 1992).

Disturbances can have substantial influences on biogeochemical cycles, but little is known about the long-term implications of a disturbance-generated landscape mosaic for nutrient cycling. The failure to adequately represent the effects of fire disturbance can lead to significant discrepancies between model predictions and observed data (Peng et al., 1998). Numerous studies report increased nutrient loss and runoff from forests following fire or clear-cutting (Vitousek et al., 1979; Vitousek and Matson, 1985; Binkley et al., 1992; Parsons et al., 1994; Jewett et al., 1995; Likens and Bormann, 1995), and others show how reductions in leaf area reduce evapotranspiration and increase nutrient outflow (e.g., Knight et al., 1985). Disturbances lead to short-term declines in nitrogen uptake by vascular plants and increased nitrification by soil microbes, thereby enhancing nitrate production. The ability to predict broad-scale patterns of ecosystem processes requires understanding the variability within and among ecosystems (Zak et al., 1989; Walley et al., 1996), as well as the consequences of disturbance for ecosystem function (Schimel et al., 1997). This challenge remains a research frontier in ecosystem and landscape ecology.

Spatial Models of Energy Flow and Nutrient Dynamics

Ecosystem science has had a close connection to simulation modeling since its development in the 1960s and 1970s (see also Chapter 3). The recent focus on landscape dynamics and ecosystem processes encouraged development of spatially explicit models of energy flow and nutrient dynamics (Sklar and Costanza, 1990). In one early integrated regional-scale model used to predict spatial heterogeneity in ecosystem processes, Burke et al. (1990) combined an ecosystem model, Century, with GIS data to simulate spatial variability in storage and flux of carbon and nitrogen for the northeastern quarter of Colorado in the U.S. Central Grasslands (Burke et al., 1990). Century is a widely used model developed to simulate grassland ecosystems (Parton et al., 1987, 1988). Using spatial data for climate and soils as inputs, the model simulated spatial patterns of NPP, soil organic carbon, net nitrogen mineralization, and oxidized nitrogen emissions. The study identified important scale-dependent effects. For example, climate data could be aggregated to a relatively coarse scale and still produce reasonable estimates of NPP. However, soil texture had to be represented at finer spatial scales (and without averaging) because of nonlinear relationships between soil texture and SOM (Burke et al., 1990). This study made broad-scale predictions based on the attributes of each cell, which could then be aggregated across the landscape, but exchanges between grid cells were not considered.

Another early spatially explicit ecosystem model was developed by Costanza et al. (1990) for the Atchafalaya Basin, a marsh–estuary landscape in southern Louisiana. This model was designed to evaluate a variety of alternative management strategies designed to reduce coastal erosion, and it considered the attributes of cells along with the movement of materials between cells. The model was structured as a spatial array of ecosystem models running in each of 2479 1-km^2 grid cells. Cells were then connected to one another by simulated fluxes of water, nutrients, and sediments. Emerging from this line of research, Fitz et al. (1996) subsequently proposed a more general ecosystem model that simulates a variety of ecosystem types using a fixed model structure. Other spatial models of landscape biogeochemistry focus on land–water interactions and are discussed later in this chapter.

Simulation modeling should continue as one of the tools used to study spatial variation in ecosystem processes, but such models are time consuming to develop and demanding of data needed to initialize and test the model. Current models are typically driven by basic questions, require considerable collaboration among scientists from many disciplines, and reflect a lengthy period of development, testing, and modification. For example, the FOREST-BGC simulator had its origins in the 1980s as a single-tree daily water balance model for 1 year, and it developed into an integrated carbon, nitrogen, and water cycle model with dual time-step resolution and run for 100 years (Running and Hunt, 1993). Further modifications have incorporated the effects of fire disturbances and succession by linking a forest gap model with BGC (FIRE-BGC, Keane et al., 1996a, b). The spatial FOREST-BGC model that predicts photosynthesis, respiration, evapotranspiration, decomposition, and nitrogen mineralization over broad landscapes is then used to calibrate simple models for implementation at the global scale (Hunt et al., 1991; Running, 1994). The calibration of simple ecosystem models for use on larger spatial and temporal scales also offers a powerful approach for scaling (Running and Hunt, 1993) that remains a topic of active research. The suggestion first made by Overton (1975) to use multiscale models that contain submodels operating at different scales and degrees of complexity now promises new insight into simulating ecosystem pattern and process (e.g., DeAngelis et al., 1998, discussed later in this chapter).

Synthesis

Empirical and modeling studies of the spatial heterogeneity in ecosystem processes have demonstrated several important points. First, spatial variations in abiotic variables (temperature, precipitation, soils, and topographic position) often produce substantial spatial variation in ecosystem processes. Thus, as ecosystem ecologists have long recognized, the abiotic template is a powerful constraint on ecosystem function. Second, abiotic factors vary over multiple spatial scales, and ecologists are still striving to determine the scales that are appropriate for developing predictive relationships. Considering these factors hierarchically may enhance our understanding of how they vary. Third, understanding the implications of the dynamic landscape mosaic for ecosystem processes remains a frontier in ecosystem and landscape ecology. We simply do not have a well-developed theory of ecosystem function that is spatially explicit, nor do we have a wealth of empirical studies from which to infer general conclusions.

EFFECTS OF LANDSCAPE POSITION ON LAKE ECOSYSTEMS

Our focus on the spatial variability in ecosystem processes has, thus far, emphasized processes in terrestrial systems. However, spatial variability is also observed in the aquatic systems embedded in landscapes. Limnologists have long considered lakes as discrete units of study, owing in part to the natural boundary of the lake shoreline (Forbes, 1887). Understanding the spatial patterns within individual lakes (e.g., patterns and processes in stratified lakes) occupied early limnologists (Forel, 1892; Birge and Juday, 1911; Soranno et al., 1999). Recognition of the interactions between lakes and their watershed and airshed led to a broader view of lake ecosystems in which atmospheric inputs and catchment characteristics such as geology, land use–land cover, and topography were recognized as important influences on the chemical and trophic status of lakes (Likens, 1985). In recent years, lake ecologists have begun to recognize spatial variation among lakes at landscape scales (Kratz et al., 1997).

When and where does the spatial arrangement of lakes within a landscape matter to lake functioning? Neighboring lakes generally lie within the same geologic setting, experience the same weather, and might be expected to be similar in their biological and chemical properties. However, lakes within such a *lake district* (regions of similar geomorphology and climate that contain many lakes) often show remarkable differences from one another, even though they are in close proximity. Many factors can contribute to differences between lakes, including lake size and depth, internal processes such as nutrient cycling or trophic dynamics, and the strength of interactions with the surrounding landscape. A lake's *landscape posi-*

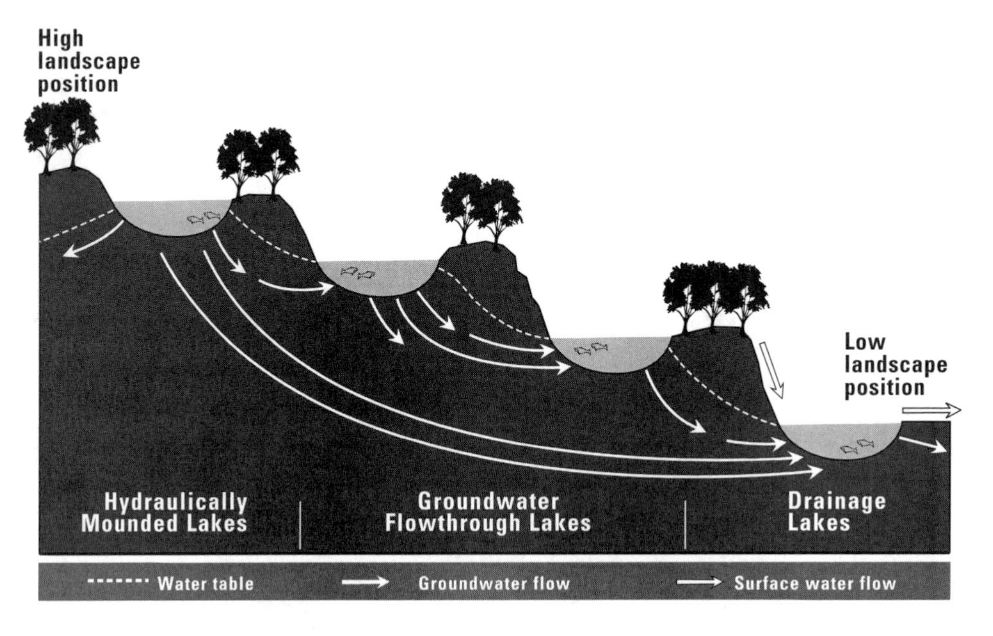

FIGURE 9.3.

The concept of landscape position applied to lakes.

MODIFIED FROM WEBSTER ET AL., 1996.

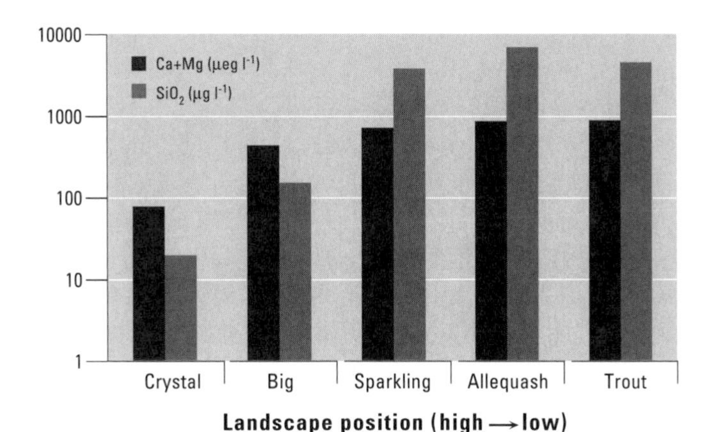

FIGURE 9.4.
The relationship between landscape position and the Ca + Mg and dissolved reactive silica concentrations in the five clearwater lakes of the North Temperate Long-term Ecological Research site in northern Wisconsin, USA. These five lakes lie within the same groundwater flow system. Crystal Lake is highest (the headwater) and Trout Lake is lowest in the flow system.

ADAPTED FROM KRATZ ET AL., 1997.

tion is described by its hydrologic position within the local to regional flow system and the relative spatial placement of neighboring lakes within a landscape (Kratz et al., 1997) (Figure 9.3). Many hydrologic properties of a lake are determined directly by landscape position. In northern Wisconsin, groundwater is an important component of the water balance of lakes; some lakes have no inflow of surface water. The amount of groundwater that enters a lake is directly influenced by the position of the lake in the landscape. Lakes higher in the flow system (which may differ in elevation by only a few meters in northern Wisconsin) have different relative sources of water than lakes lower in the flow system (Webster et al., 1996). Precipitation comprises a greater proportion of the water input to lakes higher in the landscape than to the lower lakes, which receive a greater proportion of their water input from groundwater. Groundwater typically has greater ionic strength than precipitation because of its contact with the soils and substrate; thus, landscape position influences ionic concentrations in lakes (Figure 9.4).

The effect of landscape position on a number of physical, biological, and chemical properties and processes in lakes has been demonstrated by studies conducted in the Northern Highlands Lake District of northern Wisconsin, an area containing about 2500 lakes (Magnuson and Kratz, in press). The response of lake chemistry to drought differs by lake position (Webster et al., 1996). There is a common tendency for lakes to increase in cation concentration during drought periods because chemical concentrations increase as water evaporates. However, lakes low in the landscape tended to increase in the total mass of cations in the lake, whereas lakes high in the landscape tended to have no change in mass, or even a decrease in cation mass in the lake (Webster et al., 1996). Changes in cation concentrations can also lead to changes in the biota; for example, crayfish and snails occur only in lakes with adequate calcium, and a 10% increase or decrease in calcium concentration would add or remove about 5% of the lakes above 4 ha in area as crayfish and snail habitat (Kratz et al., 1997). Silica concentrations, which vary with landscape position and are much higher in lakes with more groundwater input, may influence the robustness of freshwater sponges in the littoral zone of lakes and, in turn, influence the rest of the littoral zone food web.

Landscape processes that influence the amount of dissolved organic carbon (DOC) entering a lake can affect the magnitude and vertical distribution of primary production within the lake. Lakes with high concentrations of DOC tend to be tea colored, and the brown color reduces the clarity of the water and hence the light penetration. Colored DOC is derived mainly from the soils or wetlands in the landscape surrounding a lake. Therefore, the position of the lake relative to sources of allochthonous (from external sources) DOC inputs can influence net primary productivity within the lake (Kratz et al., 1997; Gergel et al., 1999).

An extensive analysis of 556 lakes in northern Wisconsin demonstrated that landscape position explained a significant fraction of the variance in 21 of 25 variables tested (Riera et al., 2000). Variables most strongly related to landscape position were measures of lake size and shape, concentrations of major ions (except for sulfate) and silica, biological variables (chlorophyll concentration, crayfish abundance, and fish species richness), and human variables (density of cottages). Lake depth, water optical properties, and concentrations of nutrients other than silica were poorly explained by lake order (Riera et al., 2000).

Recent work has extended these analysis to other lake districts. Soranno et al. (1999) analyzed long-term data from nine *lake chains* (lakes in a series, connected through surface or groundwater flow) from seven lake districts of diverse hydrogeomorphic setting in North America. The study asked (1) if there were predictable spatial patterns in chemical, algal, and water quality variables along lake chains and (2) if lakes that are closer together in a chain behave more similarly through time. Results indicated that spatial and temporal patterns of lakes within a lake district were organized along gradients of geology (depth of glacial till and spatial heterogeneity in soil characteristics), hydrology (water residence time and whether lakes were dominated by surface or groundwater flow), and some measure of landscape influence (e.g., ratio of watershed area to lake area). The spatial patterns along lake chains for a wide range of variables were surprisingly similar across lake districts. For example, weathering variables, alkalinity, conductivity, and calcium generally increased along lake chains, but these patterns were weaker in regions situated in calcium-rich tills or having high local heterogeneity in geologic substrate. Concentrations of particulate nutrients and measures of algal biomass increased along lake chains in surface-water lakes, but not in the groundwater-dominated lakes. Regarding temporal patterns, landscape position was important in determining *synchrony* (a measure of the degree to which lake pairs within a lake district behave similarly through time [Magnuson et al., 1990]) between lake pairs only for variables related to weathering. For most variables, synchronous behavior in lakes within a lake chain was unrelated to lake spatial position. Soranno et al. (1999) measured synchrony at the annual time scale, and they note the need for further analyses of synchrony at other temporal scales.

Synthesis

Results from these studies suggest that a landscape perspective for lakes based on landscape position is relatively robust (Kratz et al., 1997; Magnuson et al., 1998; Baines et al., 1999; Soranno et al., 1999; Riera et al. 2000; Webster et al., 2000). This perspective argues for a view of lake ecosystems embedded in a landscape matrix, with lakes interacting with one another and with the terrestrial environment. Soranno et al. (1999) argued further for an expansion of this view to encompass the set of lakes, streams, and wetlands that occur within a landscape; these aquatic systems are often treated separately (and as independent entities), yet they are often connected spatially and functionally. A landscape perspective may help to foster such an integration.

LAND–WATER INTERACTIONS

A common theme underlying many studies of land–water interactions is the degree to which land uses in the uplands and the spatial arrangement of these land uses influence water quality in streams and lakes. Freshwater and estuarine ecosystems act as integrators and centers of organization within the landscape, touching nearly all aspects of the natural environment and human culture (Correll et al., 1992; Boynton et al., 1995; Naiman et al., 1995; Naiman, 1996). Hynes, widely regarded as the father of modern stream ecology, stated, "We must not divorce the stream from its valley in our thoughts at any time. If we do, we lose touch with reality" (Hynes, 1975). As demonstrated by the effect of landscape position on lake characteristics, land–water interactions are apparent even in relatively undisturbed landscapes. In another example, studies along a toposequence of tundra, sedge, and shrub communities along a slope in Alaska have revealed the importance of ecosystem adjacencies to nutrient transformation and movement. Shaver et al. (1991) addressed the question of how heterogeneity among ecosystems influenced the outputs of nutrients into surface waters in an undisturbed tundra landscape (Figure 9.5). The entire sequence of community types occurred along a few hundred meters, but large differences were observed in the rates of plant uptake, mineralization, and transport between ecosystems.

Improving our understanding of the complex relationships between the land and water is an important goal of both basic and applied research in landscape ecology. Freshwaters are degraded by increasing inputs of silt, nutrients, and pollutants from agriculture, forest harvest, and cities (Carpenter et al., 1995, 1998). Consider a watershed containing croplands or pastures. Farmers often apply fer-

FIGURE 9.5.

(a) Toposequence studied by Shaver et al. (1991) in the Sagavanirktok River valley in northern Alaska. Six ecosystem types extend from the uplands down to a series of floodplain terraces and to the river. The entire toposequence is underlain by permafrost, and soil water flows downslope over the permafrost during summer. Total vertical drop is ~10 m, and horizontal distance ranges from 100 to 200 m. (b) Conceptual model of element cycling along the toposequence at the study site. The sequence is viewed in one dimension as a series of individual square meters linked by element transport in soil water. (c) The processes represented within and between cells. In each square meter, there may be internal recycling of elements by plant and soil (A and B), exchange between plants and soil through uptake or litterfall (C), and exchange with the atmosphere through processes such as nitrogen fixation or denitrification (D). Elements may also be weathered from parent materials (E). Transport of elements downslope in soil water is indicated by (F).

ADAPTED FROM SHAVER ET AL., 1991.

tilizers high in nitrogen and phosphorus (P) to their fields, but not all the added N and P is taken up by the plants. When it rains, some of these nutrients are leached from the soil and transported through the watershed and into the stream by both surface and subsurface water flow. Like agricultural areas, cities and suburbs are important contributors to such *nonpoint source pollution,* that is, pollution that does not come from a single source, like a pipe, but rather is delivered from

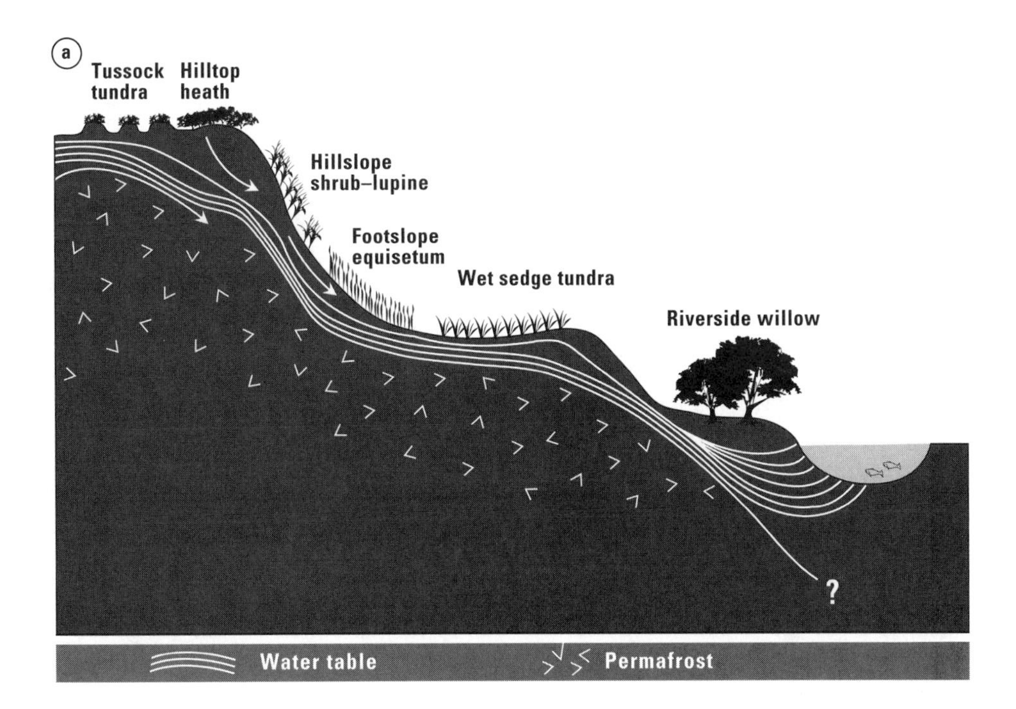

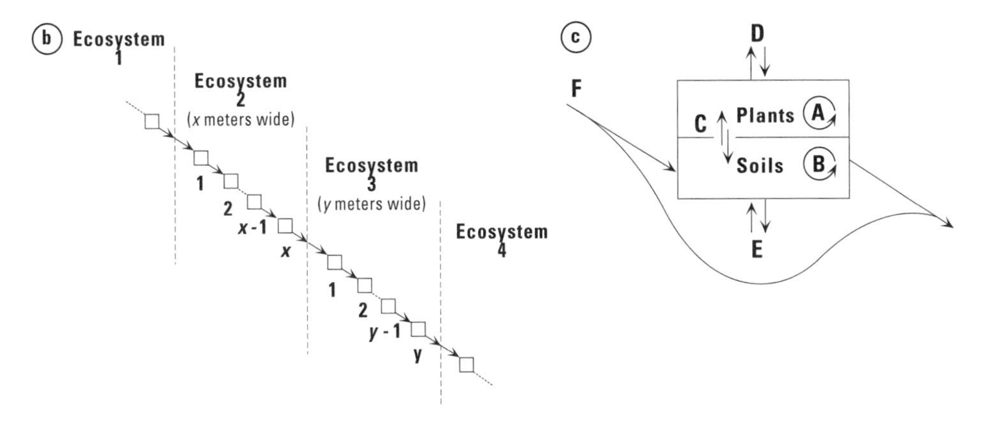

widespread areas of the landscape. Homeowners often apply as much fertilizer and pesticides per unit area to their lawns as farmers do to their crop fields, and a portion of these nutrients ends up in nearby aquatic systems. Lakes and reservoirs fill more rapidly with mud from agricultural and urban land uses, and the growth of nuisance plants, including toxic blue-green algae, is promoted by the increased silt and nutrients. Regional changes in land use cause widespread *eutrophication* of many lakes on the landscape. Eutrophication makes lakes more similar, because the lakes are all dominated by the same set of species that can tolerate eutrophic conditions. Therefore, the diversity of lake types within a landscape is reduced as all lakes become eutrophic and converge to have similar species (Carpenter et al., 1995).

The nutrients most often considered in studies of land–water interactions are nitrogen and phosphorus. Economic and health concerns about excess nitrogen inputs into aquatic ecosystems are growing throughout the world (e.g., Cole et al., 1993; Mueller and Helsel, 1996; Vitousek et al., 1997a). In rivers, nitrogen biogeochemistry is sensitive to land-use patterns, the structure of the riparian zone, and river flow regimes (Cirmo and McDonnell, 1997). Accumulation of excess phosphorus in lake and stream systems has long been recognized as a driver of eutrophication (e.g., Carpenter et al., 1998). However, understanding the effects of land-use patterns on water quality and the spatial scales over which these effects are manifest has become an important goal of landscape ecological studies, particularly since the mid 1980s.

Osborne and Wiley (1988) analyzed the nitrogen and phosphorus concentration of streams in the Salt River Basin, Illinois, and used regression analysis to determine whether there was a relationship to land-use patterns mapped from aerial photos. Results demonstrated that the amount of urban land cover and its distance from the stream were the most important variables in predicting nutrient concentrations in the stream water. In the Saginaw River, Michigan, watershed, 62 catchments were studied to relate seasonal stream-water chemistry to landscape characteristics (Johnson et al., 1997). Results demonstrated a strong influence of landscape characteristics, but the predictive power of particular variables differed among seasons (Table 9.1). For example, land use and the interaction of land use and geology were the best predictors of variance in nitrogen during summer, but geology and the interaction between land use and geology were the best predictors during autumn. Other studies have also found significant relationships between land use and concentrations of nutrients in lakes and streams (e.g., Geier et al., 1994; Hunsaker and Levine, 1995; Johnes et al., 1996; Soranno et al., 1996; Bolstad and Swank, 1997; Lowrance, 1998; Bennett et al., 1999).

Landscape patterns are also important in urban areas, as illustrated for Min-

Table 9.1.
Percentage of variance in stream-water chemistry accounted for by all landscape factors, partitioned by land use, geology and structure, and shared land use plus geology and structure, as well as unexplained variance (based on redundancy analysis).

Variance component	Summer	Autumn
Land use	20	7
Geology and structure	15	17
Shared: land use + geology and structure	22	16
Total variance explained	56	39

From Johnson et al., 1997.

neapolis–St. Paul by Detenbeck et al. (1993). In 33 lake watersheds in the Minneapolis–St. Paul area, landscape and vegetation patterns were obtained from aerial photographs and then compared with measured lake water quality. When lakes were dominated by forested lands in the surrounding watershed, they tended to be less eutrophic and have lower levels of chloride and lead. In contrast, lakes with substantial agricultural land uses in their watersheds were more eutrophic. When wetlands remained intact in the watersheds, less lead was present in the lake water. Percent urban land use has also been found to be positively correlated with the export of phosphorus to lakes in the Minneapolis–St. Paul area.

The impacts of human activities on sediment and nutrient budgets of watersheds are well-documented (Carpenter et al., 1998). Historically, broad-scale forest clearing and conversion to agriculture or residential land use have led to increased erosion and transport of sediments and nutrients into estuaries and the lower portions of river systems (Wolman, 1967; Trimble, 1974; Meade, 1982; Meybeck, 1982). In the Chesapeake Bay basin, such impacts have been documented for postsettlement times (1700s to the present) both broadly for the entire bay (Brush, 1984, 1986, 1989, 1997; Cooper and Brush, 1991; Boynton et al., 1995) and specifically for selected tributaries (Costa, 1975; Jacobson and Coleman, 1986; Jordan et al, 1997a, b). In addition to sediment inputs, metals and organics are also exported from land to water as a consequence of land-use change. Comeleo et al. (1996) found that the area of developed land within 10 km of a sediment sampling station was a major contributing factor in the concentrations of both metals and organics in the sediments. In freshwater systems in Wisconsin, land use influenced mercury concentrations in rivers (Hurley et al., 1995) and lakes (Watras et al., 1995).

Problems associated with nonpoint pollution have stimulated a variety of modeling studies designed to relate runoff and nutrient loading in aquatic systems to upland dynamics (e.g., Cluis et al., 1979; Berry et al., 1987; Gilliland and Baxter-Potter, 1987; Bartell and Brenkert, 1991; Freeman and Fox, 1995; Soranno et al., 1996; Zhang et al., 1996). In an early modeling study examining spatial variability in the loss, gain, and storage of total nitrogen, Kesner and Meentemeyer (1989) combined a simple mass-balance model with a GIS database to study N dynamics in an 11,490-ha agricultural watershed in southern Georgia, USA. Results demonstrated that it was possible to quantify and map source and sink regions of N in a watershed and that the riparian zone was critically important in buffering this watershed against excessive losses of N.

A simple model of phosphorus transformation and transport for the Lake Mendota watershed, Wisconsin, has provided useful insights into the effects of land use on water quality (Soranno et al., 1996). The watershed of Lake Mendota is dominated by agricultural and urban land uses, and the lake itself has a long history of limnological study (Brock, 1985; Kitchell, 1992). Soranno et al. (1996) developed a GIS-based model of phosphorus loading in which phosphorus-export coefficients varied among land uses (Figure 9.6). Phosphorus is usually bound to sediments, and its delivery to the lake is attenuated by movement across the terrestrial landscape. Soranno et al. (1996) accounted for this by weighting the con-

FIGURE 9.6.

Land-use data for the Lake Mendota, Wisconsin, watershed for presettlement (a), baseline (b), and urbanization (c) land-use scenarios used by Soranno et al. (1996) to simulate phosphorus loading. Effective land-area maps for the high-runoff model calibration set for three scenarios (d, e, f) and for the low-runoff model calibration set for three scenarios (g, h, i). The effective land area is the area that contributes to P loading, and it is shown as the black and gray areas in d–i. Tan areas are classified as noncontributing. In the light gray areas, all P produced on each 1-ha cell is transported to the water. The gray scales represent the proportion of a cell's P that reaches water. (Refer to the CD-ROM for a four-color reproduciton of this figure.)

FROM SORANNO ET AL., 1996.

LAND USE

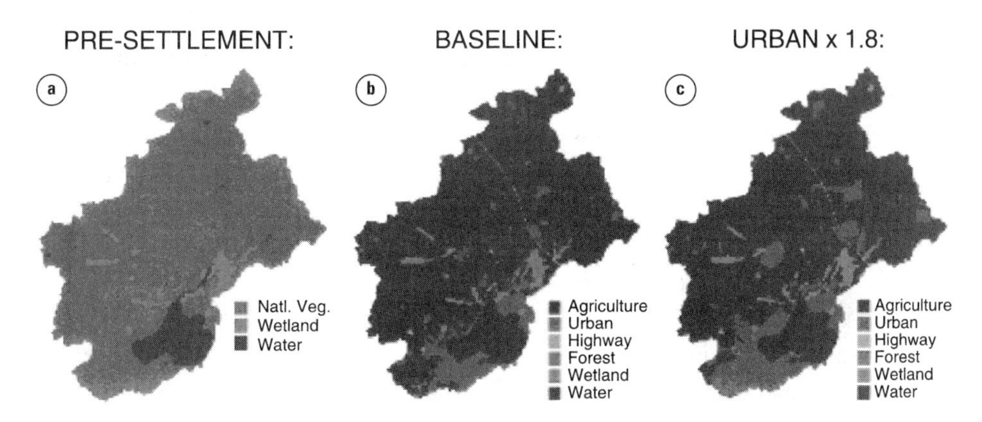

EFFECTIVE LAND AREA

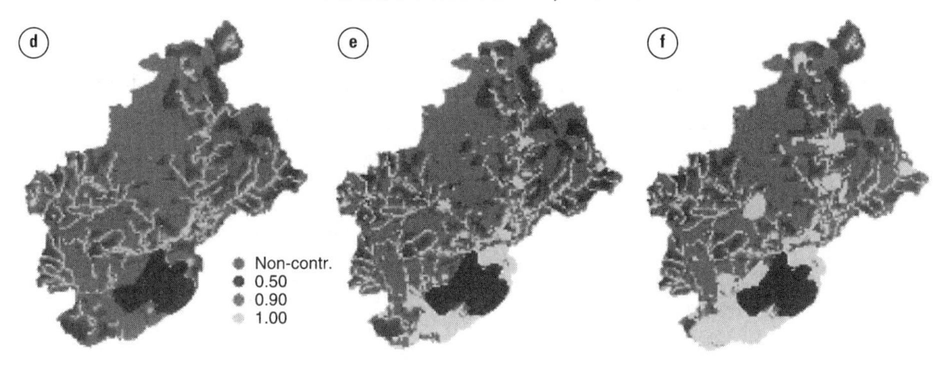

LOW RUNOFF, 1977

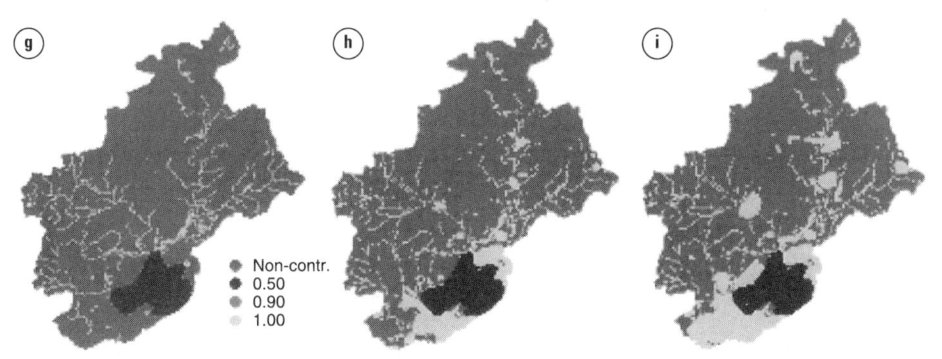

tribution of phosphorus to the lake by a given grid cell by its distance from the lake. Because of the network of storm sewers serving the urban areas of Madison, Wisconsin, and surrounding communities, urban areas were treated as though they were adjacent to streams. The model was then used to compare phosphorus loadings in Lake Mendota under current patterns of land use, presettlement land use, and projected future land use in which the urban area increased nearly twofold. Because rainfall events drive runoff, simulations were conducted for both high- and low-precipitation years. Results demonstrated that most of the watershed did not contribute phosphorus loading to the lake; most P came from a relatively small proportion of the watershed, ranging from 17% of the watershed contributing during low-precipitation years to 50% during high-precipitation years. A sixfold increase in phosphorus loading was estimated to have occurred since settlement. Results also demonstrated the importance of riparian vegetation in attenuating phosphorus runoff.

Role of Riparian Buffers

Riparian vegetation zones, including wetlands and floodplain forests, are conspicuous elements of many landscapes and important mediators of land–water interactions (Naiman and DeCamps, 1997). Freshwaters are especially sensitive to changes in these adjacent lands (Correll et al., 1992, Osborne and Kovacic, 1993; Correll, 1997; Lowrance et al., 1997). *Riparian buffers,* areas of relatively undisturbed vegetation along streams or adjacent to lakes, influence transport of nutrients and sediments from upland agricultural–urban areas to aquatic ecosystems. Riparian vegetation and microbial communities can take up large amounts of water, sediment, and nutrients from surfacewater and groundwater draining agricultural areas within a catchment, often substantially reducing the discharges of nutrients to aquatic ecosystems. The increased surface roughness created by vegetation can also trap particles. Therefore, landscape ecologists have taken particular interest in characterizing and understanding the function of patches or corridors of riparian vegetation, because their functional importance is large relative to their size.

Wetlands, floodplains, and riparian vegetation zones have often been altered by agricultural and urban development (Turner et al., 1998a) (Figure 9.7). Woody riparian vegetation once covered an estimated 30 to 40 million ha in the contiguous United States (Swift, 1974). At least two-thirds of that area has been converted to nonforest land uses, and only 10 to 14 million ha remained in the early 1970s. Floodplain clearing for agriculture, urbanization, and water resource development has been responsible for much of the loss of riparian forests. Riparian forests have been reduced by more than 80% in many portions of the United States, including the arid West, the Midwest, and the lower Mississippi Valley (Swift, 1974). A classic example of the loss of riparian forest has been described for the Willamette River, Oregon (Sedell and Froggatt, 1984). Prior to 1850, the streamside forest extended up to 3 km on either side of a river characterized by multiple channels, sloughs, and backwaters. By 1967, government-sponsored programs for forest clearing, snag removal, and channelization (channel deepening and straightening) reduced the Willamette River to a single uniform channel that had lost over 80% of its forest and land–water edge habitats.

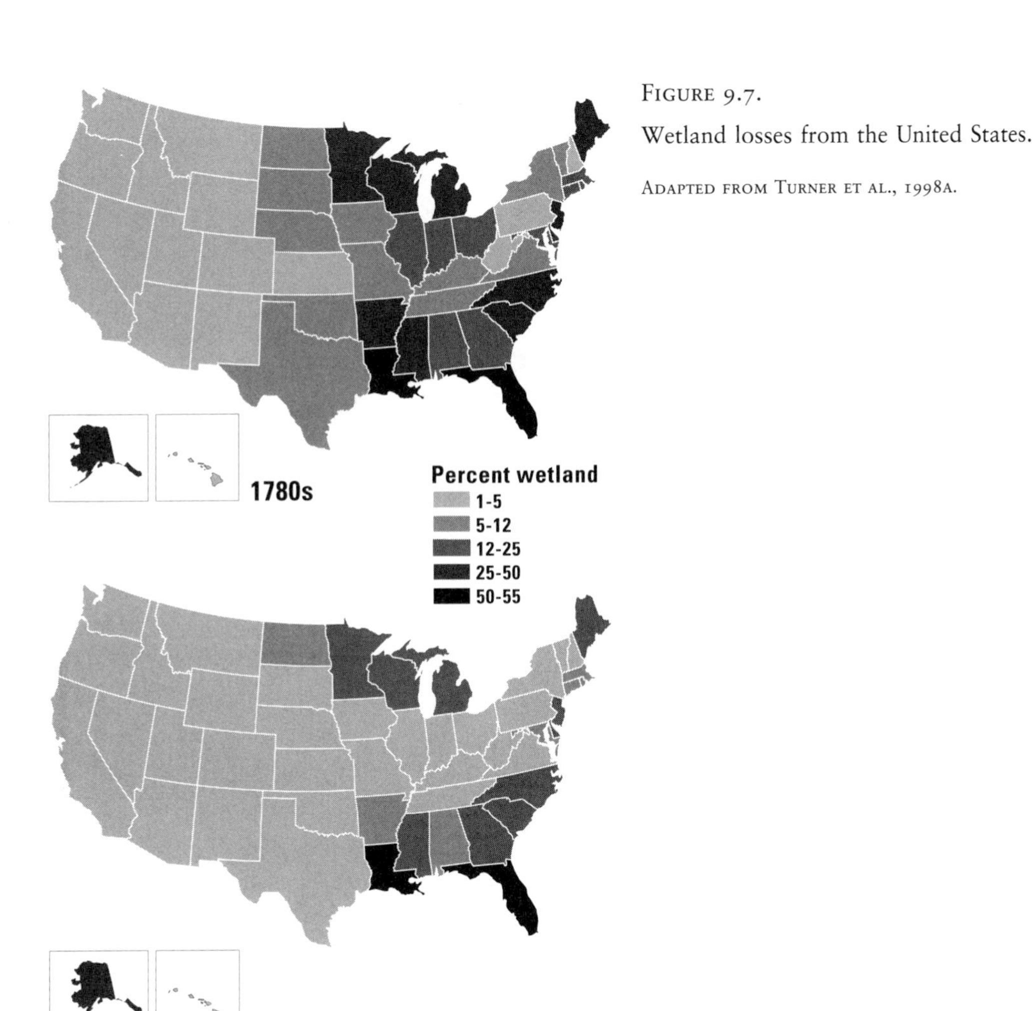

FIGURE 9.7.

Wetland losses from the United States.

ADAPTED FROM TURNER ET AL., 1998A.

What exactly is the functional role of the riparian zone? For nitrogen, the process of interest is denitrification. High rates of denitrification are fostered by inputs of nitrate and organic carbon to an anoxic environment. Nitrate is commonly supplied from agricultural areas, whereas persistent anoxia is associated with saturated wetland and riparian soils. Consequently, conditions for denitrification are often maximized in relatively narrow bands between disparate patch types (Holmes et al., 1996). Riparian vegetation along rivers can also be effective in retaining suspended matter, as well as attenuating the effects of flooding (Brunet and Astin, 1997). Because phosphorus tends to be adsorbed to soil particles, sediment trapping in the riparian zone reduces the transport of P from the land to the water. Riparian vegetation also provides important habitat for many plants and animals (Kelsey and West, 1998; Pollock, 1998). For an excellent review of the function of riparian zones, interested readers are referred to Naiman and DeCamps (1997).

In the mid-Atlantic region of the United States, studies of nutrient dynamics in mixed agricultural watersheds have nicely demonstrated the nutrient removal function of riparian vegetation. Substantial quantities of particulate materials, organic nitrogen, ammonium–N, nitrate–N, and particulate phosphorus were removed in an agricultural watershed when waters flowing from a cornfield passed across approximately 50 m of riparian forest (Peterjohn and Correll, 1984) (Figure 9.8). The effectiveness of vegetated riparian buffer strips (forest and grass) in retaining nutrients moving from adjacent agricultural lands was also examined by Osborne and Kovacic (1993). Results demonstrated that nitrogen runoff was reduced by 90% for both forest and grassy riparian buffers, but that forest vegetation retained more nitrogen, whereas grassy vegetation retained more phosphorus. This process of nutrient removal is ecologically important because it can substantially reduce cultural eutrophication. Thus, the presence and location of particular vegetation types can strongly affect the movements of materials across the landscape and help to regulate the quality of surface waters within the landscape.

The spatial pattern of riparian vegetation (variation in length, width, and gaps) influences its effectiveness as a nutrient sink. Weller et al. (1998) developed and

FIGURE 9.8.

(a) Total N flux and cycling and (b) total P flux and cycling in a study watershed from March 1981 to March 1982. All values are in kilograms per hectare of the respective habitats.

ADAPTED FROM PETERJOHN AND CORRELL, 1984.

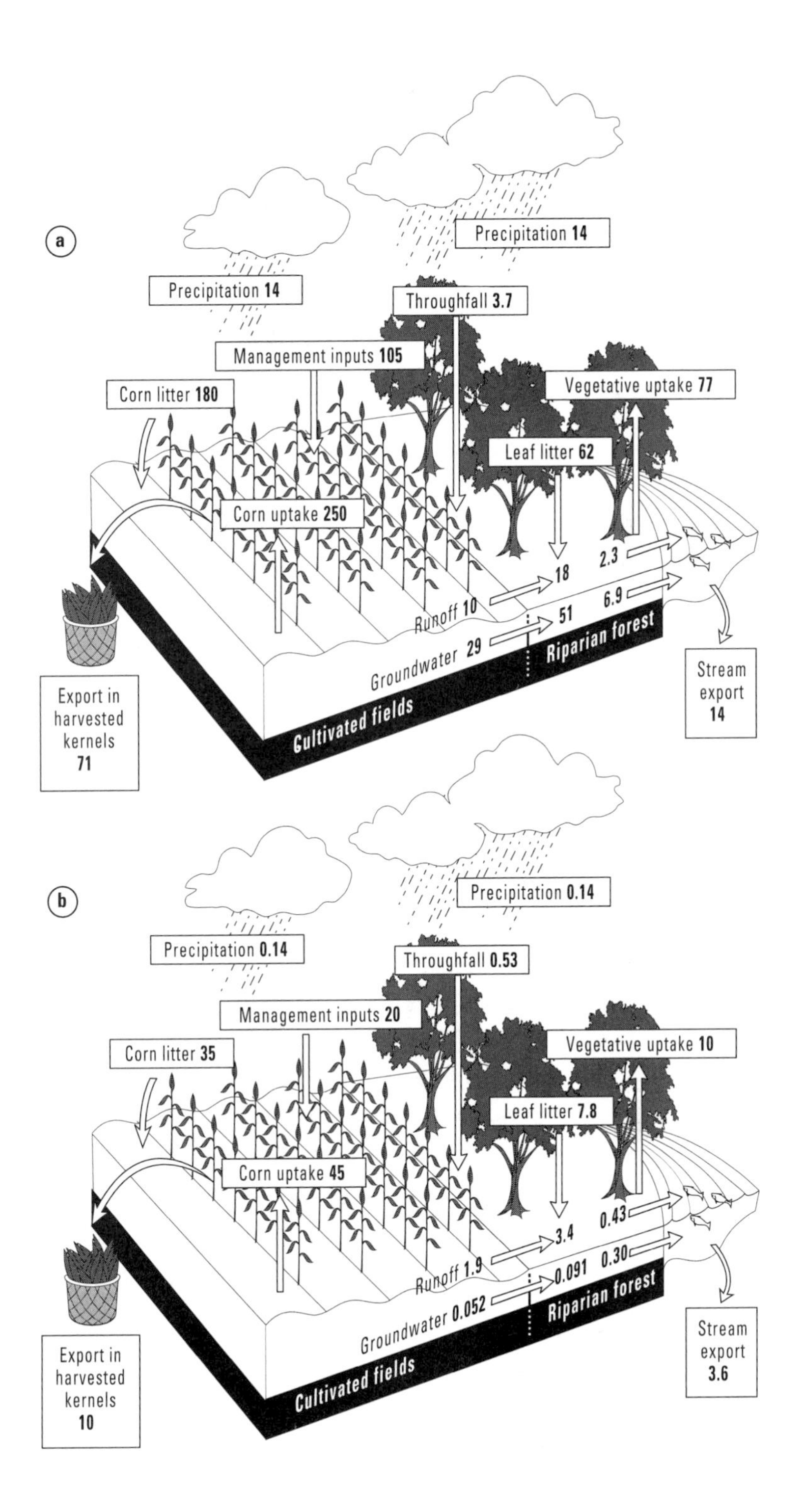
a
Precipitation 14
Precipitation 14
Throughfall 3.7
Management inputs 105
Corn litter 180
Vegetative uptake 77
Leaf litter 62
Corn uptake 250
2.3
18
6.9
Runoff 10
51
Groundwater 29
Riparian forest
Stream export 14
Export in harvested kernels 71
Cultivated fields
b
Precipitation 0.14
Precipitation 0.14
Throughfall 0.53
Management inputs 20
Corn litter 35
Vegetative uptake 10
Leaf litter 7.8
Corn uptake 45
0.43
3.4
0.30
Runoff 1.9
0.091
Groundwater 0.052
Riparian forest
Stream export 3.6
Export in harvested kernels 10
Cultivated fields

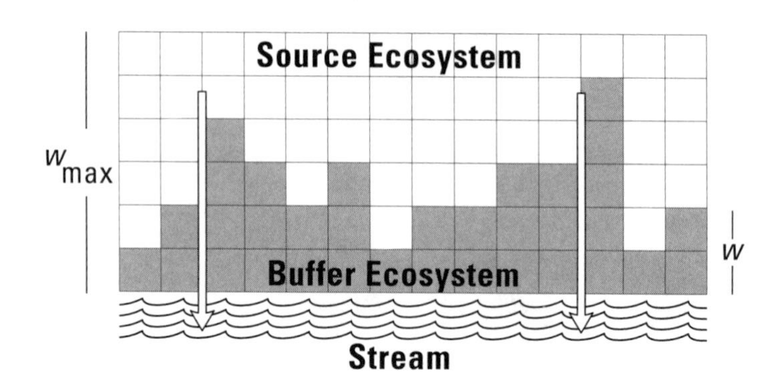

FIGURE 9.9.
Conceptual model of a landscape with a riparian buffer. The landscape is divided into a grid, and cells along the stream are occupied by the buffer ecosystem. Water and materials flow downhill from the source ecosystem, through the buffer, and to the stream. Weller et al. (1998) developed models in which the width (w) and length of the riparian buffer were varied, along with the width (w_{max}) of the entire simulated landscape, to evaluate the effectiveness of the buffer at retaining nutrients.

ADAPTED FROM WELLER ET AL., 1998.

analyzed models predicting landscape discharge based on material release by an uphill source area, the spatial distribution of riparian buffer along a stream, and retention of material within the buffer (Figure 9.9). The buffer was modeled as a grid of cells, with each cell transmitting a fixed fraction of the material received. Variability in the riparian buffer width reduced total buffer retention and increased the width needed to meet a management goal (Weller et al., 1998). Variable-width buffers were less efficient than uniform-width buffers, because transport through gaps dominated discharge, especially when buffers were narrow; average buffer width was the best predictor of landscape discharge for unretentive buffers, whereas the frequency of gaps was best for narrow, retentive buffers (Weller et al., 1998). This heuristic model offers predictions that are amenable for testing in a variety of riparian systems.

Spatial Scales at Which Landscape Pattern Influences Water Quality

A variety of investigators has asked questions about the spatial extent, or distance from a water body, over which landscape patterns influence water quality (Figure

9.10). That is, is it only the riparian zone that is important in maintaining water quality, or must land uses in a greater proportion, or even the entire watershed, be considered? Studies of such scale-dependent relationships between landscape characteristics and water chemistry have yielded mixed results (Omernik et al., 1981; Wilkin and Jackson, 1983; Cooper et al., 1987; Osborne and Wiley, 1988; Hunsaker et al., 1992; Hunsaker and Levine, 1995; Comeleo et al., 1996; Richards et al., 1996; Johnson et al., 1997; Gergel et al., 1999). The demonstrated importance of riparian buffers, described previously, is a testament to the importance of landscape attributes over smaller landscape scales. For example, Johnson et al. (1997) found that total phosphorus in stream water was better explained by land-use patterns within a 100-m buffer of a stream than by land use or other variables at the extent of the catchment. However, other studies have demonstrated that more distant upland land uses were as important as riparian land uses in larger watersheds (e.g., Omernik et al., 1981). Responses may also differ between lotic

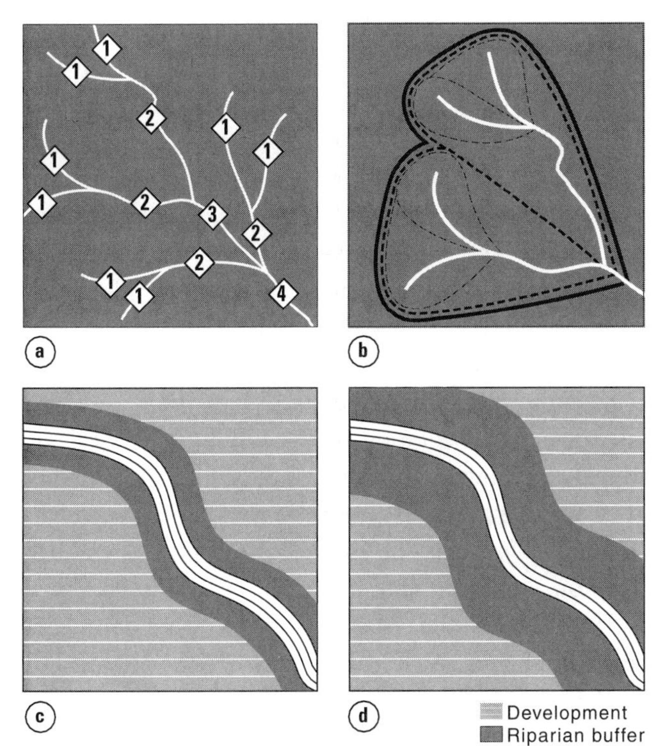

FIGURE 9.10.
Different spatial extents considered in studies of land–water interactions: (a) hypothetical hierarchical drainage network with subwatersheds (numbers refer to stream order); (b) subwatersheds considered separately; (c) fixed-width buffer; (d) larger fixed-width riparian buffer.

and lentic systems. Gergel et al. (1999) found that landscape characteristics (especially proportion of wetlands) within 50 m of lakes in northern Wisconsin explained significant variability in concentrations of dissolved organic carbon (DOC); in contrast, measurements from the whole watershed always explained more variability for DOC in rivers than did measurements from the nearshore area.

Thus, there has not been general agreement among studies that have explicitly examined the effects of the specific locations of particular land uses or the extent of the watershed considered. However, comparisons among studies is complicated because of differences in the choice of response variables (e.g., concentration or loading, and of which nutrient), spatial extent of the watersheds considered, and the temporal scale of measurements (e.g., monthly, seasonal, or annual). There remains a strong imperative for ecologists to better understand the interactions between land and water and the scales over which they are manifest.

Reciprocal Interchanges between Water and Land

Most of the emphasis on land–water interactions has focused on how the movement of materials from the terrestrial components of the landscape influences the aquatic components. However, recent studies have identified significant movements of materials and nutrients from the water into terrestrial communities. Willson et al. (1998) described an expanded perspective on interactions between fish and wildlife in the Pacific coastal region of North America. Each year millions of anadromous fish (e.g., salmon, char, and smelt) move from the ocean into numerous freshwater streams to spawn. These fishes provide an important seasonal resource base for a variety of terrestrial predators and scavengers, including Bald Eagles (*Haliaeetus leucocephalus*) and brown and black bears (*Ursus arctos* and *Ursus americanus*, respectively). These predators congregate along the spawning streams in great numbers, suggesting that the availability of spawning fish is important to the predators, although documentation of the specific ecological importance of this resource is only beginning (Willson et al., 1998). The anadromous fishes typically die after spawning, and the nutrient subsidies provided by their carcasses to the streams are well recognized. What has been surprising, however, is the potential fertilization effects of salmon carcasses on the terrestrial ecosystems (Willson et al., 1998). Bears and other carnivores commonly carry salmon, living or dead, onto stream banks and tens of meters into the forests (Figure 9.11). Marine-derived nutrients, which can be identified by isotopic markers, pass from the bodies of the salmon into the soil and then into the riparian and upland vegetation, with the nutrients probably then moving up the terrestrial food

FIGURE 9.11.
Movement of nutrients from a stream to fish, bears, and eagles; nutrients are then deposited in the uplands and absorbed into the terrestrial biota.

chain. Willson et al. (1998) report potential additions of P from bear-carried fishes of approximately 6.7 kg/ha, which is similar to the P application rate of commercial fertilizers for evergreens and trees! The consequences of this water-to-land fertilizer effect for terrestrial food webs could have substantial implications for the spatial patterns of ecosystem processes in these forested landscapes.

Synthesis

Components of the landscape surrounding a lake, stream, or river can strongly influence water quality. Elements of the landscape may serve as sources, sinks, or transformers for nutrient, sediment, and pollution loads. Land cover, such as agricultural or urban, is only part of the equation, because the actual practices employed within a land-cover type can have very strong effects. In addition, the topography of the region influences the rate of delivery from landscape components to water bodies. When watersheds are steeply sloped and soils are highly erodable, the flux or export of nutrients and sediments to surface waters will increase. In both urban and agricultural landscapes, native riparian vegetation, whether wetland or forest, can reduce nonpoint pollution and help to maintain satisfactory quality of surface waters. However, more research is needed to elucidate more fully the importance of the riparian zone within whole watersheds. Furthermore, recently identified movements of nutrients from the water into some upland communities point to important reciprocal interchanges between water and land that may have substantial implications for spatial heterogeneity in ecosystem processes in terrestrial landscapes.

LINKING SPECIES AND ECOSYSTEMS

When we seek to understand the functional dynamics of entire landscapes, interactions between species and ecosystem processes must often be considered. Grazers, for example, can enhance mineral availability by increasing nutrient cycling in patches of their waste (McNaughton et al., 1988; Day and Detling, 1990; Holland et al., 1992). In dealing with the management or restoration of landscapes, it is again important to account for both species and ecosystem processes. Landscape ecology offers both a conceptual framework and methods for making substantial progress in this area and may perhaps help to dissolve some fences between the traditionally distinct subdisciplines of population and ecosystem ecologies. In an introductory chapter in the book *Linking Species and Ecosystems* (Jones and Lawton, 1995), Grimm (1995) writes:

> Interactions between population/community and ecosystem ecologists would be facilitated by adopting, *as a starting point*, a spatially based conception of units of study. . . . Whatever the scale of the investigation, a spatially based perspective places species interactions (the traditional focus of community ecology) into a context in which their effect on ecosystem processes may be assessed. Interactions between patches may be critical to larger-scale processes and include biotic interactions that occur within component subsystems.

Species and ecosystems are inherently linked, but population ecology and ecosystem ecology often break this linkage conceptually. In this section, we highlight two studies that illustrate insights from and approaches for considering species–ecosystem interactions in a landscape context.

Vegetation, Large Herbivores, and Nutrient Cycling

Studies in the boreal forest landscape have demonstrated fascinating links among spatial patterns of plant species distributions and biomass, the foraging dynamics of moose (*Alces alces*), and rates of nutrient cycling (McInnes et al., 1992; Jeffries et al., 1994; Pastor et al., 1997). Studies on Isle Royale, an island located in Lake Superior and well known for long-term studies of moose and wolves, demonstrated how selective foraging by moose on hardwood species allows unbrowsed or lightly browsed conifers to dominate the boreal landscape (McInnes et al., 1992). Moose prefer to browse upon deciduous tree species such as birch (*Betula lutea*) and aspen (*Populus tremuloides*), as well as balsam fir (*Abies balsamea*), rather than on white spruce (*Picea glauca*). In areas of Isle Royale where fences (exclosures) were built to prevent moose from browsing, the deciduous trees have persisted and grown larger. However, outside the exclosures, where moose were allowed to browse, white spruce was the only tree species that could grow above the browsing height of a moose. Moose browsing on balsam fir and the deciduous trees prevented saplings of these preferred forage species from growing into full-sized trees. The browsing of moose also opens up the forest canopy and reduces tree biomass, allowing more light to reach the forest floor and stimulating more production of shrubs and herbaceous species.

Understanding the spatial heterogeneity of ecosystem processes in this boreal landscape requires forging a linkage between the feeding ecology and population dynamics of moose and the function of the ecosystem, all within the context of a landscape. By selectively foraging on specific plant species, moose and other large herbivores influence ecosystem dynamics, changing plant community composition, biomass, production, and nutrient cycling (McInnes et al., 1992). Soils in areas dominated by spruce received less litter, and the nutritional quality of the litter, especially its nitrogen content, declined for the decomposers. This decrease in litter quantity and quality leads to a decline in microbial processes that in turn determine nitrogen availability for the living plants. Conifer litter depresses the availability of soil nitrogen, which limits net primary production in boreal forests.

Moen et al. (1997, 1998) developed a simulation model to predict how alternative moose foraging strategies affect the net annual energy balance and density of moose and the spatial distribution of browse across the landscape. Simulations were conducted at fine resolution (grid cells of 1 m^2 over an 8-ha landscape), and results have demonstrated how moose create their own landscape by their patterns of foraging and the feedbacks of these patterns on vegetation structure and composition. Because moose are highly mobile and can forage all around the landscape, interactions between moose and vegetation create a mosaic of nutrient cycling regimes in these boreal forests, resulting in complex spatial and temporal patterns of browsing, conifer density, and soil nitrogen distribution across the landscape (Pastor et al., 1999).

Species and Ecosystem Dynamics in the Everglades

The basic and applied issues associated with restoration of the Everglades, Florida, provide an excellent example of the benefit gained by linking species and ecosystems in a landscape context (Davis and Ogden, 1994; DeAngelis et al., 1998). The Everglades is a huge freshwater marsh that extends from Lake Okeechobee in central Florida southward toward Florida Bay and the Gulf of Mexico. Water is the prevailing abiotic influence on this landscape, flowing south and southwest from Lake Okeechobee at a very slow rate across this flat region. Human modification of the natural flow began early in the 20th century and involved levees, canals, pumping stations, and water-control structures to drain large areas of the Everglades. Agriculture and urban land uses quickly occupied these former wetlands. Water flow to the southern Everglades, which was protected as a national park in 1947, is delivered through water-control structures, and both the volume and timing of flow have been altered. The net effects of these hydrological changes were (1) a loss of uninterrupted sheet flow, (2) pronounced fluctuations in water levels, (3) a higher frequency of major drydown events, and (4) a disproportionate loss of high-elevation short-hydroperiod wetlands (DeAngelis et al., 1998).

Before these modifications occurred, a broad area (up to 90 km wide) of the Everglades was inundated up to a depth of 0.5 m during part of the year, creating a continuous spectrum of hydroperiods (average annual inundation time for a particular area), which has important ecological consequences. During flooding, populations of small fish, crayfish, and the like, are nourished by detritus and algae growth, are relatively protected from large predatory fish, and reach high population sizes. During the dry period, the fishes are concentrated by the receding waters into pools and depressions. Wading birds then rely on this concentrated resource base to provide food for their young during the nesting period (Fleming et al., 1994). Wading bird populations have declined by >90% since the 1940s, and a variety of other species (e.g., Cape Sable Seaside Sparrow, *Ammodramus maritimus*; American alligator, *Alligator mississipiensis*) have also suffered negative consequences from the modification of the natural hydrology and disruption of trophic dynamics across the landscape (DeAngelis et al., 1998). Recent cooperative efforts by an array of federal and state agencies are striving to restore the Everglades and other ecosystems in southern Florida.

DeAngelis et al. (1998) developed a spatial model (Across Trophic Level System Simulation, or ATLSS) of the Everglades that explores the effects of changes in hydrology on a variety of species, including those at the top of the food web. The species differ in their use of the landscape and resources and include the

Florida panther (*Felis concolor coryi*), white-tailed deer (*Odocoileus virginianus*), Cape Sable Seaside Sparrow, Snail Kite (*Rostrhamus sociabilis*), White Ibis (*Eudocimus albus*), Wood Stork (*Mycteria americana*), Great Egret (*Ardea alba*), Great Blue Heron (*Ardea herodias*), American alligator, and American crocodile (*Crocodylus acutus*).

ATLSS includes four levels of modeling that integrate population, ecosystem, and landscape dynamics (Figure 9.12). The population models for the focal species

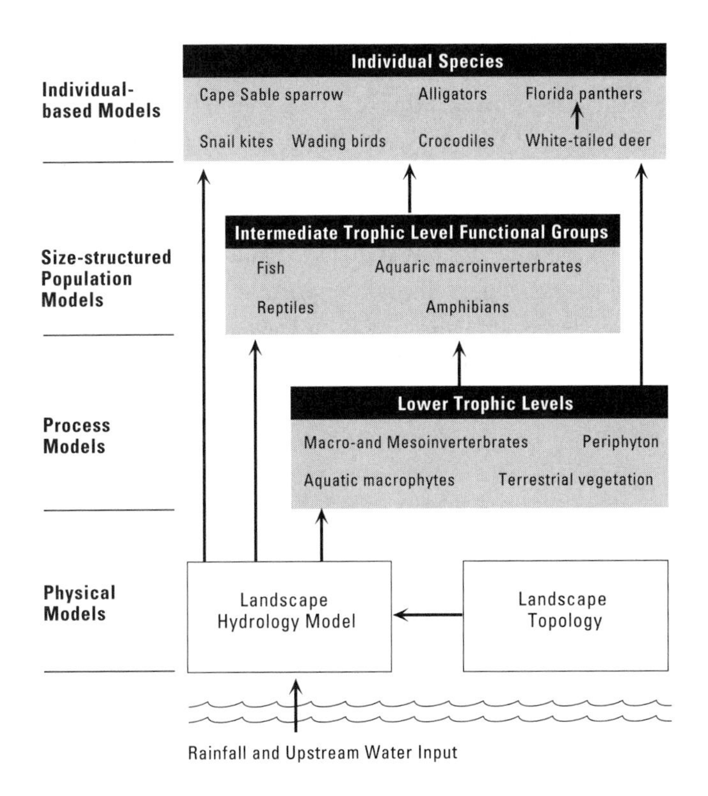

FIGURE 9.12.

Schematic of the Across Trophic Level System Simulation (ATLSS) approach for modeling ecosystems of South Florida. This shows the four levels of modeling used and the direction of effects from abiotic forces and lower and intermediate trophic levels to the higher trophic levels. Feedback effects of higher trophic levels on lower and intermediate levels also occur in the models.

ADAPTED FROM DEANGELIS ET AL., 1998.

are individual-based models and include the characteristics of each organism (age, size, spatial location, sex, health, social status, etc.). Immediately below the individual-based models are the intermediate trophic levels (fish, aquatic macroinvertebrates, and several groups of reptiles and amphibians), which are modeled as size-structured populations. Below this level and driving their biomass are the lower trophic levels, simulated by process models and consisting of macrophytes, periphyton, aquatic mesofauna, and macroinvertebrates. The macrophytes include both aquatic and terrestrial plants that provide forage for the white-tailed deer. Finally, there is a landscape structure level that includes the standard static data layers (e.g., surface elevation, vegetation types, soil types, road networks), as well as dynamic layers, such as changing water levels across the landscape.

The ATLSS modeling approach thus uses the spatially explicit view of the landscape to simulate effects of the ecosystem directly on individuals. The output from this model is expected to provide information to managers that would help them to rank the effectiveness of various water-management options against specific criteria for indicator species. DeAngelis et al. (1998) also used the Everglades restoration and modeling effort to highlight two important theoretical issues that are current challenges to ecologists: (1) uniting or at least connecting population ecology with the behavioral and physiological condition of individuals and (2) uniting population ecology with ecosystem ecology.

SEARCHING FOR GENERAL PRINCIPLES

Modeling will continue to be an important tool for understanding the causes and consequences of variation in ecosystem processes, although the logistics of intensively measuring such processes at fine resolution but broad spatial extent are daunting (not to mention very, very expensive). Many spatial models of ecosystem processes have focused on management issues, particularly nonpoint-source pollution (identifying critical source areas, relating land use to sediment and nutrient loads, predicting movements of materials during storm events, and mitigating contamination). Basic questions addressed by spatial ecosystem models focus on the problem of what happens when you consider space. For example, how does spatial variability in climate or soils affect ecosystem processes such as NPP or evapotranspiration? In general, modeling studies have elucidated the controls on ecosystem processes of spatial variation in the environment, served as informative tools for exploring effects of alternative landscape patterns on aquatic ecosystems, and are just beginning to delve into the spatial interactions between species and ecosystem processes. Modelers continue to grapple with a number of fundamental issues associated with spatial models of ecosystem processes. Parameterization of spatial models is difficult and can be highly uncertain, and highly parameterized models are less portable from one system to another. Therefore, it is not surprising that investigators continue to experiment with techniques for incorporating spatial dynamics while keeping the data demands of the model reasonable.

Furthering our knowledge of the patterns and causes of spatial heterogeneity in ecosystem function remains at the frontier of ecosystem and landscape ecology. The library of empirical data from a variety of landscapes is slowly building, yet we have no general answers to such questions as these: Just how spatially variable are ecosystem processes? How do the controls on processes and rates operate across space? Are there critical thresholds in spatial variability that are important for ecosystem processes? Despite tremendous advances in understanding ecosystem processes over relatively small spatial extents, there exists very little theory for predicting variability in ecosystem processes across larger areas. A more synthetic understanding of spatial heterogeneity in ecosystem processes remains an important research need, one that should include both theoretical development and empirical study.

SUMMARY

Determining the patterns, causes, and effects of ecosystem function across landscapes is an important current topic in ecosystem and landscape ecology. When landscape ecologists study ecosystem processes, they emphasize the causes and consequences of spatial heterogeneity in the rates of ecosystem processes (e.g., net primary productivity, nitrogen mineralization); the influence of landscape position on ecosystem function, and the horizontal movement of materials (such as water, nutrients, or sediments) and how these movements might differ with alternative spatial arrangements of land cover.

Spatial variation in abiotic variables (temperature, precipitation, soils, and topographic position) often produces substantial spatial variation in ecosystem processes (e.g., productivity, decomposition, and nitrogen cycling) across landscapes. For example, nutrient accumulation and cycling may vary between ridgetops, which tend to be exposed and dry, and the more sheltered and moist bases of slopes. Thus, as ecosystem ecologists have long recognized, the abiotic template is a powerful constraint on ecosystem function. However, changes in land-use patterns and natural disturbances also influence the rates and patterns of ecosystem processes. Disturbances can have substantial influences on biogeochemical cycles, but little is known about the long-term implications of a disturbance-generated landscape mosaic for nutrient cycling.

Abiotic factors vary over multiple spatial scales, and ecologists are still striving to determine the scales that are appropriate for developing predictive relationships. Empirical studies of ecosystem processes have demonstrated that a model will need to include a larger number of variables to account for the spatial pattern of the same process as the scale of analysis (either extent or grain) becomes finer.

The recent focus on landscape dynamics and ecosystem processes encouraged development of spatially explicit models of energy flow and nutrient dynamics. Simulation modeling should continue as one of the tools used to study spatial variation in ecosystem processes, but such models are time consuming to develop and demanding of data with which to initialize and test the model.

Aquatic ecologists have been exploring landscape-level characteristics that describe the state of lake ecosystems. Many hydrologic properties of a lake are determined directly by landscape position—its hydrologic position within the local to regional flow system and the relative spatial placement of neighboring lakes within a landscape. Results from a variety of studies suggest that a landscape perspective for lakes based on landscape position is relatively robust. This perspective argues for a view of lake ecosystems embedded in a landscape matrix, with lakes interacting with one another and with the terrestrial environment. A landscape perspective also may help to foster a view of land–water interactions that encompasses the set of lakes, streams, and wetlands that occur within a landscape.

Components of the landscape surrounding a lake, stream, or river can strongly influence water quality. Elements of the landscape may serve as sources, sinks, or transformers for nutrient, sediment, and pollution loads. A common theme underlying many studies of land–water interactions is the degree to which land uses in the uplands and the spatial arrangement of these land uses influence water quality in streams and lakes. Riparian vegetation zones, including wetlands and floodplain forests, are conspicuous elements of many landscapes and important mediators of land–water interactions. Most of the emphasis on land–water interactions has focused on how the movement of materials from the terrestrial components of the landscape influences the aquatic components. However, recent studies have also identified significant movements of materials and nutrients from the water into terrestrial communities. There remains a strong imperative for ecologists to better understand the reciprocal interactions between land and water and the scales over which they are manifest.

Understanding the implications of the dynamic landscape mosaic for ecosystem processes remains a frontier in ecosystem and landscape ecology. We simply do not have a well-developed theory of ecosystem function that is spatially explicit, nor do we have a wealth of empirical studies from which to infer general conclusions. Landscape ecology also offers promise as a conceptual framework within which interactions between species and ecosystem processes can be linked.

DISCUSSION QUESTIONS

1. A daunting challenge in studying the spatial variation in ecosystem function and the factors that control the rates of ecosystem processes is to balance data needs (e.g., spatial extent of the study and the number of measurements needed) with logistical difficulties and actual cost (e.g., person-hours required for collecting and processing samples and the costs of running laboratory analyses). Consider an extensive landscape of your choice. Develop a field sampling design that attempts to describe the spatial variation of an ecosystem attribute or process rate (e.g., NPP, LAI, nitrogen mineralization, denitrification, phosphorus accumulation) of your choice.
   a. Did you need to stratify sampling in your design? Why or why not? If yes, by what variables did you stratify, and why?
   b. How did you consider issues of spatial autocorrelation in your design?
   c. Estimate the cost of implementing your design in terms of person-hours and laboratory costs (if any). Could this cost be reduced by combining remote sensing with field measurements? Why or why not?
2. Describe how the processes associated with the release, uptake, and storage of carbon would change in the following scenarios over a period of 100 years (graphical representations may be helpful):
   a. A temperate deciduous forest is affected by small-gap disturbances that affect 1% of the landscape each year and initiate succession within the gaps.
   b. A temperate deciduous forest is cleared for agriculture and farmed continuously for 50 years. Farming is then abandoned, and the land undergoes natural succession for the next 50 years.
   c. A mature boreal forest landscape experiences wildfire that burns 60% of the landscape and initiates forest succession. Ten years later, a second fire burns 50% of the previously burned area and eliminates the newly established trees, resulting in the area of double-burn being dominated by herbaceous vegetation for the next 80 years.

3. Consider an agricultural watershed in which fertilizers are applied to upland crop fields. A management goal for this watershed is to maintain acceptable water quality while producing agricultural products. Under what conditions might you expect a riparian buffer to be effective at maintaining water quality? Under what conditions would a reduction in the source of nutrients (e.g., reduce or eliminate some fertilizer inputs) be needed to maintain water quality?
4. How do you expect the statistical relationships between a measurement of ecosystem function and a variable that controls that function to change with spatial scale? Are the changes with scale linear? Why or why not?

RECOMMENDED READINGS

BURKE, I. C., D. S. SCHIMEL, C. M. YONKER, W. J. PARTON, L. A. JOYCE, AND W. K. LAUENROTH. 1990. Regional modeling of grassland biogeochemistry using GIS. *Landscape Ecology* 4:45–54.

DEANGELIS, D. L., L. J. GROSS, M. A. HUSTON, W. F. WOLFF, D. M. FLEMING, E. J. COMISKEY, AND S. M. SYLVESTER. 1998. Landscape modeling for Everglades ecosystem restoration. *Ecosystems* 1:64–75.

KRATZ, T. K., K. E. WEBSTER, C. J. BOWSER, J. J. MAGNUSON, AND B. J. BENSON. 1997. The influence of landscape position on lakes in northern Wisconsin. *Freshwater Biology* 37:209–217.

NAIMAN, R. J., AND H. DECAMPS. 1997. The ecology of interfaces: riparian zones. *Annual Review of Ecology and Systematics* 28:621–658.

PASTOR, J., Y. COHEN, AND R. MOEN. 1999. Generation of spatial patterns in boreal forest landscapes. *Ecosystems* 2:439–450.

PETERJOHN, W. T., AND D. L. CORRELL. 1984. Nutrient dynamics in an agricultural watershed: observations on the role of a riparian forest. *Ecology* 65:1466–75.

RUNNING, S. W., R. R. NEMANI, D. L. PETERSON, L. E. BAND, D. F. POTTS, L. L. PIERCE, AND M. A. SPANNER. 1989. Mapping regional forest evapotranspiration and photosynthesis by coupling satellite data with ecosystem simulation. *Ecology* 70: 1090–1101.

SORANNO, P. A., S. L. HUBLER, S. R. CARPENTER, AND R. C. LATHROP. 1996. Phosphorus loads to surface waters: a simple model to account for spatial pattern of land use. *Ecological Applications* 6:865–878.

CHAPTER 10

APPLIED LANDSCAPE ECOLOGY

The practical application of landscape ecology is an exciting and rapidly expanding field. Landscape ecology contributes to many different aspects of applied ecology and is playing a significant role in the development and application of methods of ecosystem management (Agee and Johnson, 1988; Slocombe, 1993; Grumbine, 1994; Christensen et al., 1996). The challenges facing natural resource managers increasingly occur over entire landscapes and involve spatial interdependencies among landscape components at many scales. Demand for the scientific underpinnings of managing landscapes and incorporating the consequences of spatial heterogeneity into land-management decisions is substantial. Consequently, many resource managers are shifting their management from an approach that focuses on specific resources, such as fish, wildlife, and water, to one focused on the integrity of entire systems. An awareness is growing that ecological effects of resource management are sensitive to the temporal scales and spatial configuration of the activity (e.g., timber harvesting or land development). Indeed, nearly all land-management agencies in the United States have recognized that informed resource-management decisions cannot be made exclusively at the level of habitat units or local sites. This shift has caused an increased demand for applications and management recommendations that often outpaces the basic science (Turner et al., in press).

Applications of landscape ecology require integration and synthesis of the many facets of the discipline that have been considered separately in this book. For example, in managing a particular landscape, consideration must be given to the configuration of the landscape mosaic and its change with time, the disturbance regime and its likely consequences for patterns and processes, the responses of many different species that operate at a variety of scales, and the effects of change on ecosystem function. Many applications of landscape ecology depend on establishing a cause–effect relationship between landscape pattern and an environmental variable of interest. These relationships can then be used to plan the spatial configuration of landscape mosaics to minimize undesirable impacts. Ultimately, analyses of changes in landscape pattern may prove to be a practical and efficient approach to broad-scale environmental analysis (O'Neill et al., 1997).

Applied problems and resource management were instrumental in catalyzing the development and emergence of landscape ecology, and they continue to stimulate both basic and applied research in landscape ecology. For example, understanding the effects of landscape pattern on stream and lake ecosystems has been driven by the practical problem of how to maintain healthy aquatic ecosystems. Similarly, the need for forest managers to schedule timber harvests in space and time across landscapes stimulated much basic research on the effects of alternative harvesting regimes for landscape patterns and processes. In fact, the distinction between basic and applied research is often arbitrary; most applied problems have basic components. The theory and methods of landscape ecology clearly have implications for many applied problems and resource-management questions, which in turn will continue to promote research that contributes to our basic understanding of the interaction between pattern and process. In this chapter, we survey how principles of landscape ecology have proved useful and where the challenges lie for land-use decisions, forest management, regional impact assessment, and broad-scale environmental monitoring.

LAND USE

Understanding land-use changes and their ecological implications presents a fundamental challenge to ecologists, one in which landscape ecology clearly must play a role. Throughout the world, land cover today is altered principally by direct human use—by agriculture, raising of livestock, forest harvesting, settlement, construction, mining, and the like (Meyer, 1995; Dale et al., 2000). Over the centuries, two important trends are evident: the total land area dedicated to human uses has grown dramatically, and increasing production of goods and services has intensified both use and control of the land (Richards, 1990). The rate of land-cover alteration is accelerating worldwide, particularly in regions with rapid population growth. Forests and grasslands have undergone especially large changes (Houghton, 1995). Between 1700 and 1980, it has been estimated that the area of forests and woodlands decreased globally by 19% and grasslands and pastures diminished by 8%, while world croplands increased by 466% (Richards, 1990). The pace of change has accelerated, with greater loss of forests and grasslands during the 30 years from 1950 to 1980 than in the 150 years between 1700 and 1850.

Land-use activities change landscape structure by altering the relative abundances of natural habitats and introducing new land-cover types. Introduction of new cover types can increase biodiversity by providing unique habitats, but natural habitats are often reduced, leaving less area available for native species. Land-use activities may alter the spatial pattern of habitats, often resulting in fragmentation of once continuous habitat and reduction in the biodiversity of native species. Natural patterns of environmental variation can also be altered by land use, especially if disturbance regimes are changed. For example, the environment may be changed directly when fire control and logging alter the frequency and extent of natural fires. Thus, effects on landscape structure should be considered when decisions about development locations, densities, and uses of the land are made.

A recent report from a committee of the Ecological Society of America addressed the ecological implications of land use (Dale et al., 2000). Six principles of ecology with particular implications for land use were identified. These ecological principles deal with time, species, place, productivity, disturbances, and the landscape. The recognition that ecological processes occur within a *temporal* setting means that change over time is fundamental to analyzing the effects of land use. In addition, individual *species* and networks of interacting species have strong and far-reaching effects on ecological processes. Furthermore, each *site* or region has a unique set of organisms and abiotic conditions influencing and constraining ecological processes. *Productivity* is a complex expression of climate, available resources, and characteristics of the species present in an area. *Disturbances* are important and ubiquitous ecological events whose effects may strongly influence population, community, and ecosystem dynamics. Finally, the size, shape, and spatial relationships of patches on the *landscape* influence the structure and function of ecosystems. The importance of incorporating a landscape ecological perspective in considering issues associated with land use is thus stated explicitly (Dale et al., 2000).

The consequences of land-use change have received reasonable attention from ecologists, particularly with regard to effects on aquatic ecosystems and biodiversity. The cause–effect connection between water quality and land use is well established (see Omernick, 1977), but the relationship with landscape pattern is relatively recent (see Chapter 9 for an extended discussion of land–water interactions). For example, Hunsaker et al. (1992) found that 80% to 90% of the variance in water quality in 36 watersheds in southern Illinois could be explained with landscape metrics. In addition to the known relationship with the amounts of land use on the watershed, indicators of spatial configuration, such as dominance and contagion (see Chapter 5), contributed to the explanatory power of the statistical model.

The consequences of land-use change for biodiversity have also received much attention (e.g., see Turner et al., 1998a, for a review of land-use change and biodiversity in the United States, and see Chapter 8 for a discussion of habitat fragmentation and metapopulations). The interaction of land use and disturbance regimes needs more study, however. If land-use practices change the frequency, size, and intensity of natural disturbances, then altered sequences of vegetation development may lead to completely different plant communities (see Pickett, 1998). In addition, continued land development in regions subject to severe disturbances (e.g., hurricanes, crown fires, and floods) will produce continued conflict between human needs and desires and natural processes.

In the remainder of this section, we present examples that illustrate the integration of landscape ecology with land-use planning. We do not attempt to review the well-developed fields of land planning or landscape design. Rather, we use examples of case studies that consider the landscape and humans in an integrated manner and illustrate the range of approaches currently being used and the insights emerging from them. A unifying hypothesis that links the ecological and social sciences is that humans respond to cues both from the physical environment and from sociocultural contexts (Riebsame et al., 1994). Given the extensive influence of human land use on landscape structure and function, it is folly to consider the future of any landscape in isolation from the humans that inhabit or manage it.

Integrated Landscape Planning in the Front Range of Colorado

Protecting habitat for native plants and animals while making room for development remains a challenging task. Duerksen et al. (1997) made practical recommendations for making local decisions about habitat protection, and this group also developed an interactive decision support framework for the Front Range of Colorado (Cooperrider et al., 1999; Theobald et al., 2000; see also http://ndis.nrel.colostate.edu). The increasing desirability of the Rocky Mountain region for settlement has resulted in the progressive loss of intact, high-quality wildlife habitat. The area of land developed in the Rocky Mountain region increased by more than 21% during the 1980s and early 1990s (Theobald et al., 2000). Demographic and economic trends throughout the region indicate that residential development will continue, fueled by annual population growth rates of nearly 3%, and will become the predominant influence on wildlife habitat for the coming decade (Cooperrider et al., 1999). Immigrants to the Rocky Mountains are particularly attracted to the amenities in the high mountain valleys, which are being developed rapidly and include prime riparian habitats. As is typical for many human-dominated landscapes (Turner et al., 1998a; Dale et al., 2000), regional changes in land use are the result of many individual decisions made at local scales.

Principles from conservation biology and landscape ecology form an important part of the approach developed by Duerksen et al. (1997). They begin with the premise that residential development influences wildlife at two fundamentally different scales, the broad landscape scale and the more focused site scale. At the landscape scale, development influences the distribution, survival, and persistence of wildlife populations and communities. At the site scale, development influences the behavior, survival, and reproduction of individual animals (Duerksen et al., 1997). These two key concepts of scale (Table 10.1) influence both the effects of development and the usefulness of actions chosen to modify their impacts. Differentiating between these scales of human effects helps to identify approaches that can be used to manage the influences of development on wildlife, a very practical application of the scale concepts presented in Chapter 2. When considering the continuum between rural and urban land uses, Duerksen et al. (1997) suggest that fine-grained site management will be most effective in urban areas where the landscape has already been fragmented by development. Broad-scale landscape management will be most effective in rural areas where natural areas still exist and the potential to plan for development that minimizes the impact on these habitats is still possible.

Duerksen et al. (1997) proposed biological principles for habitat protection at landscape and site scales, as well as seven operational principles of habitat protection (Table 10.2). The operational principles were developed to enhance the collaborative approach to conservation planning, especially how ecologists, citizens, and planners can work together effectively. The biological principles

Table 10.1.
Comparison of the landscape and site scales characterized by Duerksen et al. (1997) with regard to the effects of residential development on wildlife and the appropriate management actions that might mitigate impacts.

Scale of effects of development	Examples of effects of development	Type of protection	Examples of protection tools
Landscape scale	Conversion of habitat patches to residential development	Landscape management	Zoning Clustering Transportation planning
	Fragmentation of habitat patches by roads		Transfer of development rights Conservation easements
Site scale	Increased predation by domestic pets	Site management	Control of pets Buffer requirements
	Increased disturbance from human activity		Maintenance of native plants in landscaping
	Reduced cover of native vegetation		Sensitive lands overlays

Table 10.2.
Principles for habitat protection at the landscape scale

Biological principles for habitat protection at landscape scales	
Principle 1	Maintain large, intact patches of native vegetation by preventing fragmentation of these patches by development
Principle 2	Establish priorities for species protection and protect habitats that constrain the distribution and abundance of these species
Principle 3	Protect rare landscape elements. Guide development toward areas of landscape containing common features
Principle 4	Maintain connections among wildlife habitats by identifying and protecting corridors for movement

(*continued*)

Table 10.2. (*continued*)

Biological principles for habitat protection at landscape scales	
Principles 5	Maintain significant ecological processes such as fires and floods in protected areas
Principle 6	Contribute to the regional persistence of rare species by protecting some of their habitat locally
Principle 7	Balance the opportunity for recreation by the public with the habitat needs of wildlife
Principles for wildlife conservation at the site scale	
Principle 1	Maintain buffers between areas dominated by human activities and core areas of wildlife habitat
Principle 2	Facilitate wildlife movement across areas dominated by human activities
Principle 3	Minimize human contact with large native predators
Principle 4	Control numbers of mid-size predators, such as some pets and other species associated with human-dominated areas
Principle 5	Mimic features of the natural local landscape in developed areas
Operational principles	
Principle 1	Be willing to use rules of thumb based on scientific findings that may someday prove to be false.
Principle 2	Understand that complex environmental problems do not have a single, scientific solution founded on "truth."
Principle 3	Begin all conservation plans with clearly stated, specific goals for wildlife protection.
Principle 4	Insist that the analysis used for setting conservation priorities can be understood by everyone who is affected by it.
Principle 5	Realize that all models are wrong, but some are useful.
Principle 6	Make plans adaptive by evaluating the consequences of actions. Learn by doing.
Principle 7	Seize opportunities to enhance wildlife habitat by intelligent design of developments.

From Duerksen et al., 1997.

summarized the findings of conservation biology that are most relevant to habitat protection in rapidly developing areas, while acknowledging that such knowledge may be imperfect. These principles are very much in accord with our understanding of the interactions between organisms and landscape pattern (see Chapter 8), addressing issues of patch size, edge effects, habitat connectivity, and the role of natural disturbances in maintaining ecological processes (see Chapter 7).

Implementation of the principles proposed by Duerksen et al. (1997) has been undertaken by the Colorado Natural Diversity Information Source (NDIS). NDIS supports planning by local communities by providing readily accessible information on the impacts of development on wildlife (Cooperrider et al., 1999; Theobald et al., 2000). A pilot project for Summit County, Colorado, a rapidly growing area located in a high mountain valley ~100 km west of Denver, resulted in a decision support system through a collaborative process involving both experts (e.g., programmer, geographer, ecologist, GIS analyst, land-use attorney) and users (landowners, environmental advocates, developers, and planners). Implementation of the system took advantage of existing data (e.g., distribution maps for 30 species of vertebrates) linked to vegetation mapped from satellite imagery and habitat maps for all vertebrate species that occurred in the county. Five indexes were used to assess habitat value: (1) local diversity, (2) neighborhood diversity, (3) user-defined local diversity, (4) corridors, and (5) patch value. Habitat value, in turn, was linked with a stochastic model to forecast the distribution of building units across the county, and the potential impacts of development at a particular location can be assessed (Cooperrider et al., 1999). Through the World Wide Web (see http://ndis.nrel.colostate.edu), users can interactively specify an area to be developed in the future and assess potential impacts on wildlife (Figure 10.1). Many developers are enthusiastic about the system, because it is much to their advantage to make their plans as ecologically sensitive as possible and to identify potentially serious problems before proceeding too far along in the design process (N. T. Hobbs, Colorado Division of Wildlife, personal communication).

FIGURE 10.1.
The home page (a) and an example of the habitat distribution for a selected species (b) from the Colorado Natural Diversity Information Source (NDIS) (http://ndis.nrel.colostate.edu).

(a)

Colorado
Natural Diversity Information Source

What is NDIS? | How to Contact Us? | Related Links | Site Help

Enhancing land-use planning decisions...

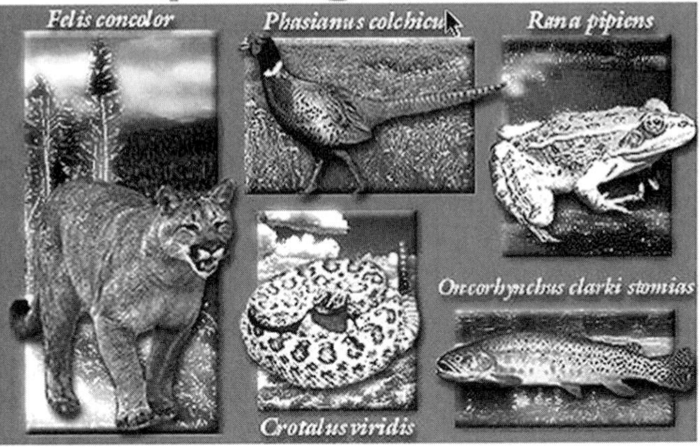

http://ndis.nrel.colostate.edu

Colorado Riparian Mapping Project!

Funding for this site is provided by:
Great Outdoors Colorado and Colorado Division of Wildlife
Updated 04.03.00

(b)

Click Thumbnail to Enlarge

Potentially Suitable Habitat

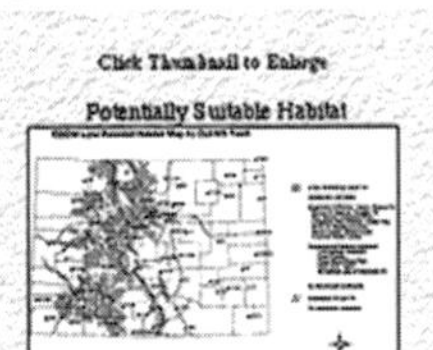

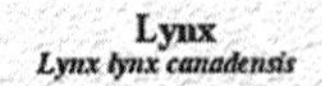

Vegetation Affinities

Aspen, Spruce fir, Spruce-fir clearcut, Lodgepole pine, Lodgepole pine clearcut, Limber pine, Ponderosa Pine, White fir, Bristlecone pine, Mixed conifer, Exposed rock, Prostrate shrub and tundra, Meadow tundra, Subalpine meadow, Bare ground tundra, Mixed tundra

Download Lynx data from the NDIS Ftp Site!

Last updated: 12/22/99

Land-Use Alternatives in the Agricultural Midwest, USA

Further illustration of the value of using spatial models that compare alternative future scenarios of landscape change to help decision makers to visualize and evaluate alternative choices for a particular region is provided in an ongoing study in the agricultural Midwest, USA (Santelmann et al., in press). In contrast to the Rocky Mountain region, the Midwest has been subjected to extensive land conversion, largely to intensive agriculture, since the mid 1800s. A variety of ecological problems have ensued, including severe fragmentation of remaining natural habitat (e.g., Curtis, 1959; Burgess and Sharpe, 1981) and deteriorating water quality (e.g., Kitchell, 1992; Carpenter et al., 1998). Land-use patterns that minimize deleterious effects on both terrestrial and aquatic ecosystems while maintaining agricultural productivity are desired.

Three alternative future scenarios were designed for two different agricultural watersheds in Iowa to represent their potential landscape composition ~25 years into the future (Figure 10.2). In consultation with a range of disciplinary experts, a team of landscape architects led the development of the scenarios (J. Nassauer, University of Michigan, personal communication; http://www.snre.umich.edu/faculty-research/nassauer/Webpage3.html). Scenarios included (1) continuation of present trends in which food production and economic profit receive highest priority, (2) an effort to preserve biodiversity and improve water quality using conventional methods; and (3) incorporation of a greater range of innovative agricultural practices, coupled with effort to preserve and restore native biodiversity and improve water quality. The future landscapes are then linked with a constellation of different modeling approaches to explore the consequences for water quality; aquatic, wetland, and terrestrial biodiversity; and economic impact on farmers (Santelmann et al., in press; see also http://www.iastate.edu/~codi/Watershed/). In addition, farm planning is addressed to incorporate input from local farmers and decision makers and to explore how socioeconomic constraints translate into land-use and land-management decisions. The long-term significance of this approach rests in its ability to inform landowners and policy makers (e.g., those crafting legislation that affects agricultural policy) about the ecological and social effects of land-use and land-management in agricultural landscapes like those in the Western Corn Belt Region.

Forecasting Land-Use Changes in Southern Appalachian Landscapes

In the Southern Appalachian Mountains in the southeast United States, studies during the past 10 years have quantified land-cover changes (e.g., Turner et al.,

FIGURE 10.2.
Alternative future scenarios for two watersheds in Iowa (http://www.snre.umich.edu/faculty-research/nassauer/Webpage3.html). (a) Initial conditions. (b) Continuation of current trends. (c) Effort to preserve biodiversity and improve water quality using conventional methods. (d) Incorporation of more innovative agricultural techniques to restore biodiversity and improve water quality. (Refer to the CD-ROM for a four-color reproduction of this figure.)

1996), related land-cover changes to other ecological response variables (e.g., Harding et al., 1998; Pearson et al., 1998; Wear et al., 1998; Pearson et al., 1999), and explored simulation methods for projecting future landscape patterns (e.g., Wear and Flamm, 1993; Flamm and Turner, 1994; Wear et al., 1996; Wear and Bolstad, 1998) (Figure 10.3). Esthetics, climate, and access to recreation have promoted substantial population growth in the Southern Appalachian Highlands during the past 20 years. As a result, land values have shifted away from agricultural and extractive uses and toward residential development. The consequences of these

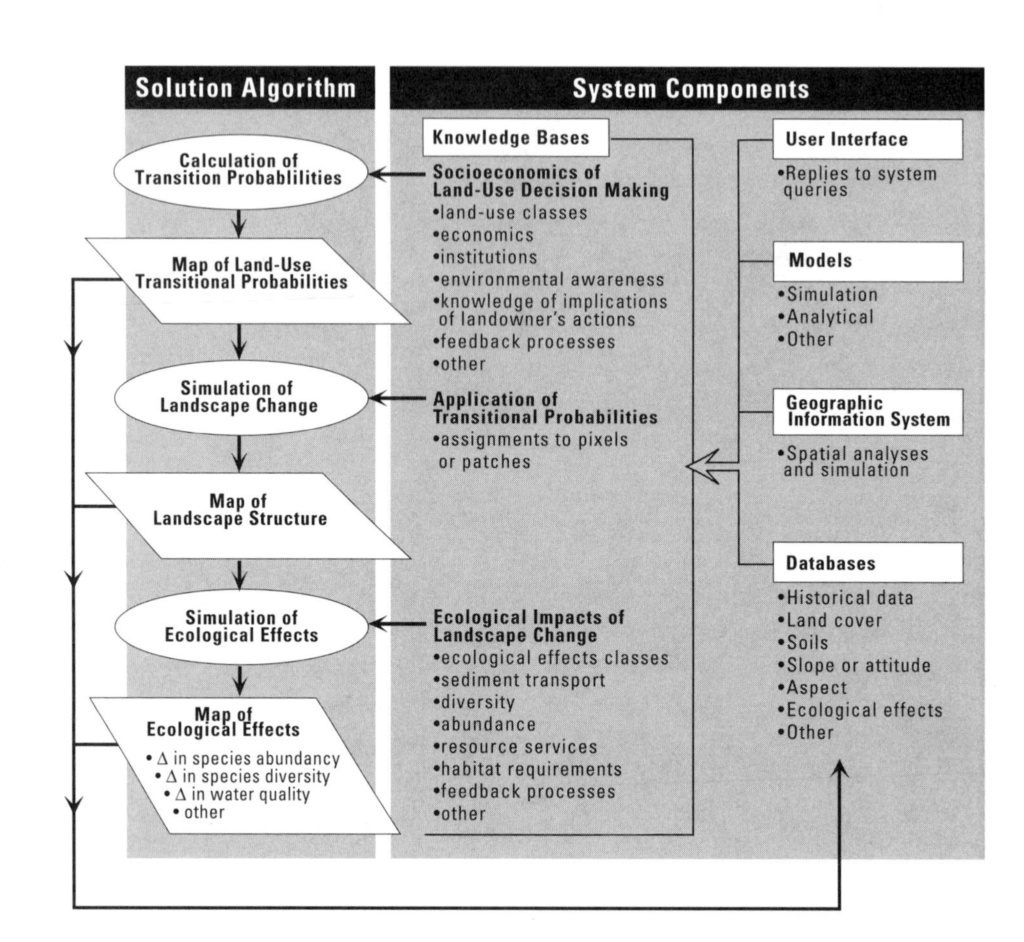

FIGURE 10.3.

Flow chart of steps for modeling land-use changes and their ecological impacts based on probabilities assigned to individual land parcels in a land-use database.

MODIFIED FROM LEE ET AL., 1992.

changes, however, may result in degradation of the very attributes that have drawn people to the region. Spatial models that forecast land use can help planners to evaluate the long-term effects of development patterns on landscape structure and the values derived from them (Wear and Bolstad, 1998).

Wear and Bolstad (1998) developed a spatially explicit model to examine land-use change over a 40-year period (1) to test hypotheses about the effects of various physical and human factors in determining where land uses occur and (2) to construct and evaluate a model for forecasting land uses. One innovation in their approach was the incorporation of building density as an indicator of the spatial diffusion of the human population. The model itself involved linking a negative binomial regression model of building density with a logistic regression model of land cover and fitting the model with spatially referenced data for the region. Models were developed for four study areas that captured the range of development that has occurred in the region. The forecasting performance of the models was evaluated by using a separate validation data set for each study area. The model performed well, explaining 80% to 89% of the variance in land cover, 73% to 76% of broad land-use classes, and 68% to 75% of the fine land-use classes in 1990 based on conditions observed in 1950.

Results of this analysis found that topography and the primary road network, which has been stable since 1950, strongly influenced building density and land-cover change. Land-cover changes occurred more frequently at lower elevations and at locations nearer to roads. These trends, however, also suggest that development may concentrate in riparian areas, which are important functional elements of the landscape (Wear et al., 1998; also see Chapter 9). Water quality and the structure and function of aquatic ecosystems in the region have been strongly influenced by the land-use changes of the 20th century (Harding et al., 1998). This study suggests that the region might benefit more by land-use planning that buffers development in these sensitive riparian zones than by attempting to maintain habitat connectivity in upland areas where land cover has remained relatively stable (Wear and Bolstad, 1998). As with the two prior examples, this approach might be used to drive hazard or risk assessments (discussed later in this chapter) where land-use models are linked to ecological impact models.

Landscape Management in the Australian Wheatbelt

The Australian wheatbelt encompasses an 18,000,000-ha region in western Australia that has undergone extensive conversion to agriculture following European settlement in the mid 1800s. The original vegetation consisted of a mosaic of plant communities, including tall, open woodland, dense shrubland, and low heathland, that reflect the underlying pattern of landforms and soils. Wheat fields and sheep grazing lands dominated the landscape by the early 1960s; only about 7% of the original vegetation remains (Hobbs et al., 1993), and remnant patches of natural vegetation are generally small and isolated. Thirty percent of the native mammals are extinct, half of the bird species are showing declines in population size, and at least 24 plant species are known to have become extinct as a result of this reduction and fragmentation of natural habitats. The wheatbelt region contains some of the highest numbers of rare and endangered species in Australia. In addition, the extensive clearing of native vegetation for agriculture resulted in significantly reduced rates of evapotranspiration and altered patterns of water flow through the soil. As a result, the naturally saline water table has risen, and approximately 10% of the agricultural region has been affected by salinization; wind and water erosion have also become severe.

Ecologists in Australia have been working to restore this landscape by considering the management of individual fragments, the landscape connectivity of natural vegetation, and the introduction of landscape elements (such as shelterbelts) to mitigate undesirable flows of nutrients and materials (Hobbs and Saunders, 1992) (Figure 10.4). Lambeck (1997) developed a novel approach based on the use of *focal* or *umbrella* species, rather than on a single species, to assist in restoration of this landscape and stem the rates of species extinctions. In this approach, a suite of selected species serves as surrogates for the broader assemblage of species. The context for this approach is the desire to prevent further loss of species from landscapes used for productive enterprises such as agriculture, forestry, and grazing by determining the composition, quantity, and configuration of landscape elements required to meet the needs of the species present. Lambeck's (1997) approach builds on the concept of umbrella species whose requirements may be used to represent the needs of other species.

Existing species are considered by a group of experts and grouped based on the kinds of management strategies needed to ensure their protection. Species considered at risk are grouped according to the processes that threaten their persistence, which may include habitat loss, habitat fragmentation, weed invasion, and fire. Within each group, the species most sensitive to the threat is used to define the minimum acceptable level of that threat. For example, the single species with the greatest area requirement becomes a focal species that sets the minimum area for patch size in the landscape design. Similarly, the species with the least mobility becomes a focal species for determining connectivity patterns within the landscape. The premise of this approach is that a landscape designed and managed to meet the needs of the focal species will encompass the requirements of all other species because the most demanding species are used to specify landscape characteristics (Lambeck, 1997). Lambeck (1997) also acknowledges a limitation in this approach: it does not indicate the area over which the solution must be implemented to assure persistence in the face of environmental change or demographic stochasticity, and thus it does not provide a method of achieving a viable landscape over time.

Landscape ecologists have been active in developing practical plans for restoring degraded lands in the Australian wheatbelt (e.g., Lefroy et al., 1991). For example, revegetation might be implemented to reclaim degraded areas, prevent further degradation, beautify the landscape, provide habitat for wildlife, increase farmland productivity, and provide a future renewable source of income, such as

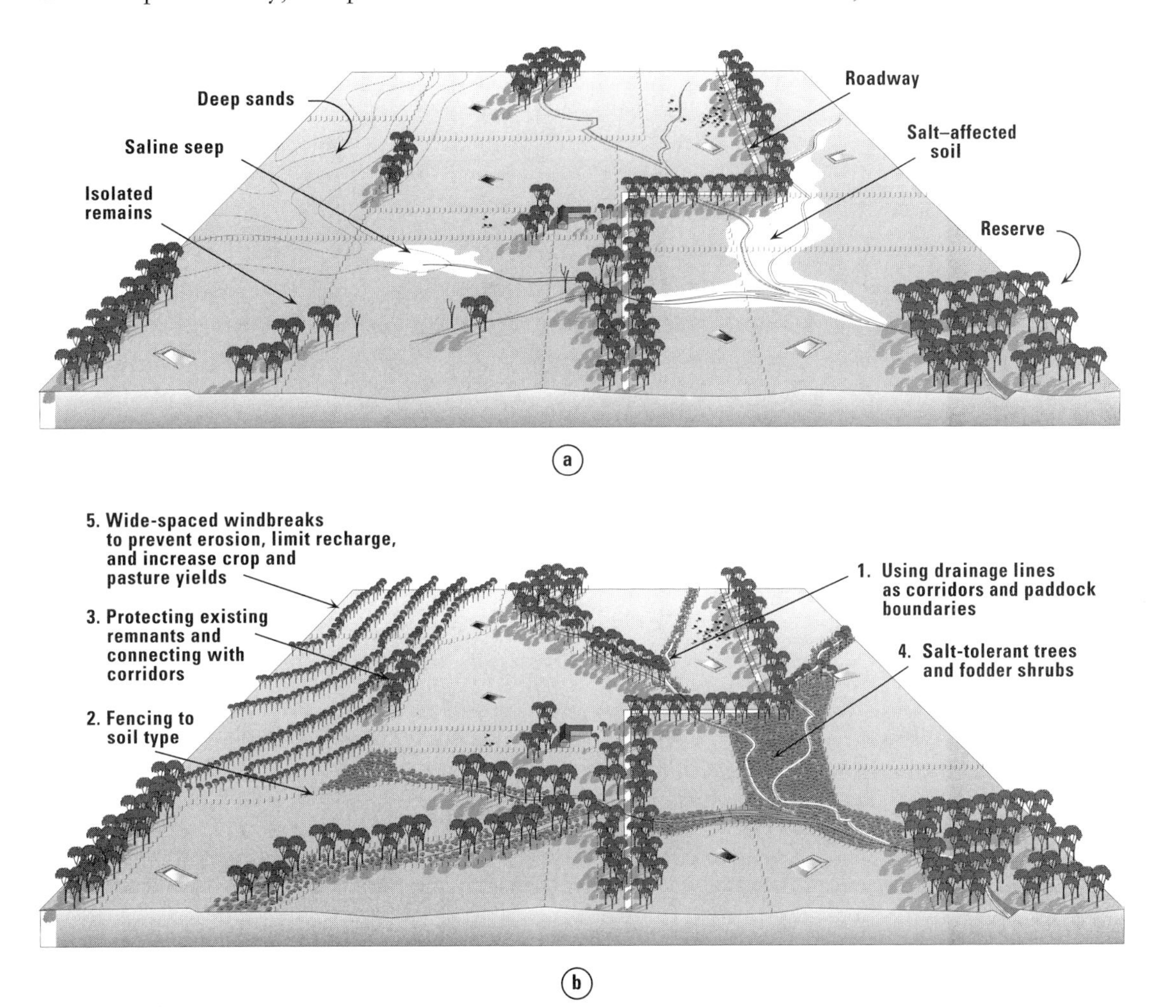

FIGURE 10.4.
Restoration plan for a degraded farm in the Australian wheatbelt. (a) Initial conditions on the farm showing infrastructure and problem areas. (b) Recommended pattern of plantings to reduce salinization and preserve biodiversity.

ADAPTED FROM LEFROY ET AL., 1991.

firewood, honey, or wildflowers (Lefroy et al., 1991). An example of the spatial implementation of such a revegetation plan is shown in Figure 10.4. These types of revegetation plans are developed and implemented at the scale of the farm, but they may eventually produce a connected mosaic across the whole landscape. Landscape restoration can help to mitigate problems of salinization and loss of biodiversity.

Tropical Slash-and-Burn Agriculture

Agricultural land use is of particular concern in the tropics, where soils are often low in organic matter and nutrients (Dixon et al., 1994). The carbon and nutrients are incorporated into the massive tree boles. As a result, a traditional agricultural practice known as slash and burn has been employed (Southworth et al., 1991). The tropical forest is cut and burned, and crops are then planted on the burned site. The burning transfers the nutrients from the trees back into the soil, permitting a few years of relatively high crop production. After the fertilizing effect of the burning is exhausted, the farmer moves to a new plot and repeats the process.

The landscape impacts of these practices were studied in Rondonia, Brazil, an area where development is being encouraged as a means of easing population pressures in the major cities (Southworth et al., 1991). A spatially explicit simulation model (Dale et al., 1993) was developed that included socioeconomic and ecological factors involved in the colonization process. The region was first divided into individual plots available to colonists and each simulated colonist was then allowed to choose a lot based on local site conditions and distance to market. Colonists then decided on the percentage of the lot to clear, crops to plant, and, eventually, duration of occupancy. As colonization continued, the model calculated the carbon lost to the atmosphere as a result of both combustion and decomposition, changes to landscape structure, and the potential impacts on wildlife. In the worst-case scenario, the rate of land clearing was rapid and the natural habitat was quickly fragmented. In the best-case scenario, more sustainable land-use practices were implemented and less land was cleared. The worst-case scenario produced about three times more carbon released to the atmosphere than the best-case scenario (Dale et al., 1994a).

The impacts of slash-and-burn agriculture on biodiversity were assessed by consideration of the habitat requirements for more than 100 individual species (Offerman et al., 1995). In tropical forests, many species, such as primates and tree frogs, never descend from the canopy and are therefore unable to disperse across gaps. Other species, such as bees that pollinate rare orchids, will not cross gaps greater than 100 m. Therefore, fragmentation in these ecosystems would appear to have a drastic impact on the animal community, particularly those species that have large area requirements and low gap-crossing ability (Figure 10.5). For example, under the worst-case scenario of land use, 40% of the area is unusable by the tree frogs within 7 years (Dale et al., 1994b).

The Rondonia project required an interdisciplinary approach to address complex problems in agricultural land-use management. The results showed that individual decisions made by the colonists based on socioeconomic considerations, such as the cost to transport their product to the marketplace, have serious consequences for land-use change in this region. However, collective decisions by the

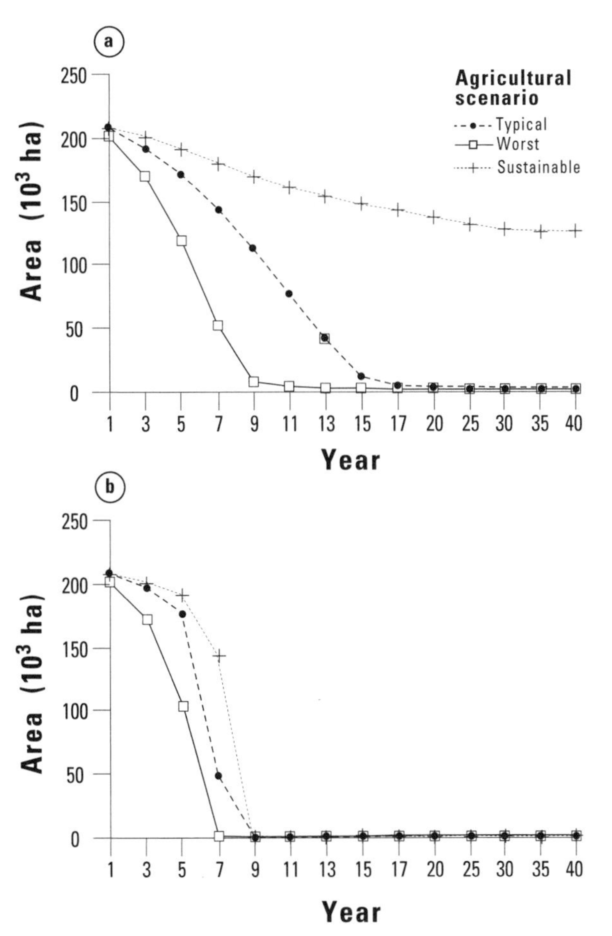

FIGURE 10.5.
Area of suitable habitat available to fauna (a) with area requirements proportional to gap-crossing ability and (b) with large area requirements and low gap-crossing ability under alternative scenarios of deforestation in Rondonia, Brazil. The simulation model predicted patterns of forest clearing under the typical scenario, in which farmers clear ~3 ha per year; the worst-case scenario, based on the most extreme land-use practices of colonists who settled along the Transamazon highway; and the best-case scenario, based on an ecologically innovative farming system in which burning is rare, there is no income from cattle, and a variety of perennial crops are grown.

FROM DALE ET AL., 1994B.

community can either enhance or mitigate these effects; for instance, higher taxes on properties nearer to market result in distant lots being more attractive economically, and colonists then disperse over a wider region. Thus, an integrated landscape model makes it possible to extrapolate the management practices and land-use pattern to determine potential environmental impacts.

Land Use: Synthesis

The question about land use is not whether we should or should not use land, but rather how we can best use the land (Turner et al., 1998a). The answer, however, is complex; there are no "cook book" approaches for identifying the optimal arrangement. Landscape ecology contributes principles and techniques for considering how to arrange human structures spatially and identify potential ecological implications of alternative arrangements. This has been illustrated in the preceding examples, although models that truly integrate a wide range of ecological impacts (e.g., habitat for various species, water quality, biogeochemistry) have been rare, and they reflect an area of active current research.

Santelmann et al. (in press) identified five conditions that must be met for practical application of ecological principles in land-use decision making.

1. Decision makers must understand the need and share the goal.
2. Abstract principles must be translated into specific land-use decisions.
3. Responsibility for associated costs and labor (which tend to occur up front and are specific to place) must be assigned and accepted (made economically feasible).
4. Benefits (which tend to be realized in the longer term and diffuse in space) must be understood and shown to have value.
5. Practices must be culturally acceptable (this includes respect for the rights of property owners).

Land-use models that truly integrate social, economic, and ecological considerations are in their infancy, and no consensus has yet been reached about what approaches are best for this task (Dale et al., 2000). Although recent models of land-use change often seem simplistic, they are at the forefront of landscape integration (e.g., Wilkie and Finn, 1988; Southworth et al., 1991; Baker, 1992; Lee et al., 1992; Dale et al., 1993, 1994a, b; Riebsame et al., 1994; Gilruth et al., 1995; Turner et al., 1996; Wear et al., 1996; Wear and Bolstad, 1998). Although the many diverse factors [including human perceptions, economic systems, market and resource demands, foreign relations (e.g., trade agreements), fluctuations in interest rates, and pressure for environmental conservation and maintenance of ecosystem goods and services] that interact to determine patterns of land use are recognized conceptually, the quantitative relationships among these variables are poorly known. This interface is a key challenge facing the scientific community in the coming decades.

Interdisciplinary studies are often complex (e.g., Naiman, 1999; Pickett et al., 1999; Turner and Carpenter, 1999; Wear, 1999), but there should be strong encouragement to develop the required integration. Riebsame et al. (1994) make several suggestions for improved, integrated land-use, land-cover modeling: (1) improved methods and approaches for integrating sociocultural factors, because social driving forces must be coupled with their ecological effects and feedbacks to society; (2) modeling interactions among multiple resources, not just one or two; (3) modeling cumulative effects, particularly when a threshold response (e.g., sudden disconnection of habitat) may be likely; (4) dealing with surprise, such as unusual conditions, rapid change, and others that may come from the environment or society. It is important to reiterate the sentiments expressed by Riebsame et al. (1994), who wrote, "Our limited ability to simulate realistic land-use patterns is not just a modeling problem, but a reflection of the real world." We concur with this evaluation and suggest that this area of applied research presents a compelling challenge to landscape ecology that will persist for the coming years.

FOREST MANAGEMENT

Forest management practices have an immediate impact on the spatial pattern on the landscape. Natural disturbance regimes create openings that vary in size, shape, and location (see Chapter 7). Harvesting creates clear-cut patches of distinct size and arrangement, and different harvesting strategies produce different landscape patterns (Franklin and Forman, 1987). Landscape ecological research in forest management has focused on (1) establishing the relationship between harvesting strategy and the resultant spatial pattern on the landscape and (2) determining how to minimize adverse effects such as habitat fragmentation and loss of old-growth forest. Spatial models have been important tools in studies evaluating the implications of alternative harvest strategies (e.g., Franklin and Forman, 1987; Li et al., 1993; Gustafson and Crow, 1996; Mladenoff and Baker, 1999). We have discussed many issues of forest management in the preceding chapters. In this section, we highlight two real-world examples in which landscape ecology principles have been applied: (1) forest planning and management in the Pacific Northwest region of the United States over the past decade and (2) sustainable forest management in Ontario, Canada.

FEMAT and the Pacific Northwest, USA

One of the most extensive applications of landscape principles to forest management occurred in the Pacific Northwest region, USA (FEMAT, 1993). Timber cutting and other operations on federal lands in the Pacific Northwest (PNW) had been virtually halted in the early 1990s because of court injunctions based on conservation of threatened species [northern spotted owl (*Strix occidentalis caurina*) and marbled murrelet (*Brachyramphus marmoratus*)], anadromous fish, and late-successional, old-growth forest. President Bill Clinton commissioned the Forest Ecosystem Management Assessment Team (FEMAT) to formulate within three months an array of management options and to assess their consequences with the goal of identifying a solution to the crisis. The team included scientists and technical experts from a variety of disciplines, agencies (Forest Service, Bureau of Land Management, Environmental Protection Agency, National Park Service, and National Marine Fisheries Service), and several universities.

The predicament in the Pacific Northwest had developed over many decades and was extremely contentious. Cutting of the extensive forests that covered much of the landscape began in the 1800s when early non-Indian immigrants began settling the region. However, logging increased dramatically following World War II under the assumption that forests could be cut and regrown at relatively short intervals (40 to 80 yr) without negative effects on other resources. By 1985, nearly two-thirds of the old-growth forest in the Pacific Northwest had been cut, and the remaining old growth was highly fragmented (Figure 10.6). Old-growth timber was a lucrative resource, worth approximately $4000/acre in 1985, and it fueled local and regional economies; 44% of Oregon's and 28% of Washington's economies depended on timber. Research through the 1970s had produced an increased knowledge base regarding the characteristics of old-growth forests and the species that depended on these forests. Studies indicated that many resources in the region, and ultimately even the economies, would not be sustained if the cutting rates and patterns were continued.

The northern spotted owl was the first species for which a very close association with old-growth forests was recognized. Northern spotted owls are forest-dwelling birds that select old-growth conifers. The old-growth stands provide open understories for low flight and hunting, high structural diversity, tree cavities for nesting, abundant prey, and cool temperatures relative to surrounding habitats

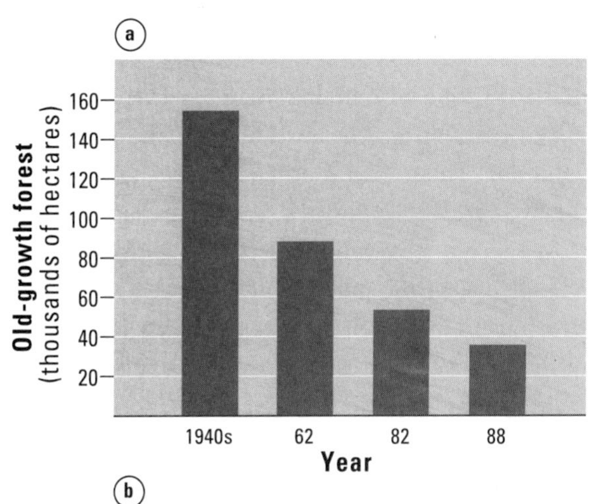

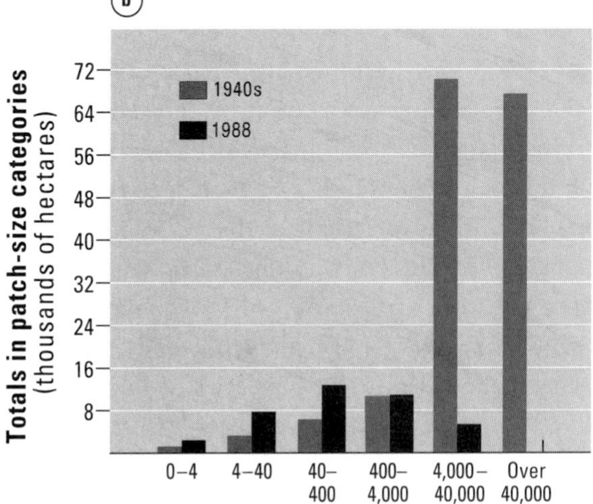

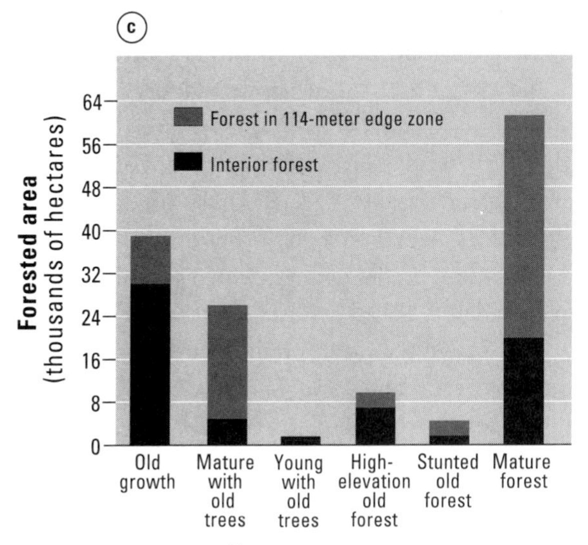

FIGURE 10.6.

Recent and historical changes in the abundance of old-growth forest and its spatial pattern on the Olympic Peninsula, Washington: (a) area of old-growth forest through time; (b) frequency distribution of old-growth by patch size in the 1940s and 1980s showing the dramatic decline in the number of large patches; (c) area of different forest types considered to constitute interior forest, located more than 114 m from the edge of a patch in 1988.

AFTER MORRISON ET AL., 1991.

(the owls have a narrow thermal-neutral zone with makes them susceptible to heat stress). Historically, the owls ranged between southern California and the Sierra Nevada Mountains to British Columbia. Pairs form long-term bonds and occupy large home ranges (>800 ha). The owls are also long-lived, with high annual adult survival rates (0.8 to 0.95) and low reproductive rates (~1 fledgling/yr with variable nesting frequencies). Juveniles fledge in the autumn and disperse over long distance (up to 150 km), even over unsuitable habitat, in their search for territories and mates. The spotted owl population was estimated to have declined by 50% from historical numbers, with metapopulation structure having been produced by human-induced habitat changes. Forest harvest was primarily clear-cutting, which eliminated suitable habitat completely.

Although the owl was the focus of much attention (and legal action), FEMAT was given a broader charge—to develop long-term management alternatives for maintaining biological diversity that met the following objectives (FEMAT, 1993): maintenance and/or restoration of habitat conditions for the northern spotted owl and the marbled murrelet that will provide for viability of each species; maintenance and/or restoration of habitat conditions to support viable populations, well distributed across their current range, of species known to be associated with old-growth forest conditions; maintenance and/or restoration of spawning and rearing habitat on federal lands to support recovery and maintenance of viable populations of anadromous fish species and other at-risk fish species and stocks; and maintenance and/or creation of a connected or interactive old-growth forest ecosystem on the federal lands within the region. In addition, management options for habitat adjacent to streams were developed to conserve aquatic, riparian, and wetland habitats, and stream corridors were to be protected.

In addition to consideration of the ecological effects of management options, FEMAT was charged with addressing a broad range of forest resource outputs and their economic implications. The economic assessments were designed to evaluate resource yields and values, local and regional economic conditions (e.g., employment), National Forest product markets, and other policy considerations. Commodity products considered included timber, special forest products (e.g., mushrooms, boughs, ferns), livestock grazing, commercial fisheries, and minerals. Noncommodity outputs such as recreation, scenic quality, water and air quality, and other public goods were also considered. Social assessment of the options was also conducted to provide policy makers with an understanding of how potential policy options might affect important social values and activities.

FEMAT eventually developed and compared a total of ten options, which varied in four principal respects: the quantity and location of land placed in some form of reserve; the activities permitted within the reserve area; the delineation of the matrix, or the areas outside the reserves; and the activities allowed within the matrix. Reserves were of two types, either old growth or riparian. Buffers of varying widths around permanent and intermittent streams were considered in all options. For timber harvesting in the matrix, guidelines were established to reduce contrast between the reserves and the matrix and facilitate movement of juvenile and spotted owls across the landscape. For example, the 50–11–40 rule called for 50% of the federal forested land within each quarter-township to be in forested condition with trees averaging at least 11 inches in diameter at breast height and with a canopy closure of at least 40%. In addition, guidelines that varied the retention of live or green trees within cut areas were considered.

Landscape and metapopulation concepts both played an important role in the FEMAT assessment. Shaffer (1985) had first suggested the application of metapopulation concepts to owl management. Shaffer (1985) considered the arrangement of habitat, recommending maintenance of the same spatial distribution of average size and spacing of present-day patches and moving away from population viability analysis, which focused only on population size. Lande (1987, 1988) developed the first models, and his results demonstrated that populations of territorial animals would collapse when habitat abundance declines below a threshold (he suggested 20%). These models were not spatially explicit, assuming that every patch or territory was equally accessible to all others. The spatially explicit models developed by Kevin McKelvey, Roland Lamberson, and Barry Noon were first introduced in the Thomas Report (Thomas et al., 1990; Noon and McKelvey, 1996b). Results suggested that a viable metapopulation could be maintained by a network of habitat conservation areas greater than 20 territories and spaced <19.2 km apart. Results also highlighted a time lag between habitat fragmentation and population decline.

The importance of landscape structure and landscape context was evident throughout the FEMAT process and in the development of the owl recovery plan. Analysis of landscape structure, especially the amount and spatial arrangement of old growth, was fundamental to the assessment process, as were the spatial projections of the alternatives. The effects of landscape context were reflected in the 50–11–40 rule for management of the matrix surrounding habitat conservation areas. This acknowledged that landscape surroundings may be more important than narrow habitat corridors in promoting movements and linkages among subpopulations. The plan set aside 7.4 million acres as a reserve for species that depend on old-growth forest and 2.6 million acres as riparian reserves; harvesting is not permitted in these reserves. An additional 1.5 million acres was set aside for experimentation with new harvesting approaches designed to avoid habitat fragmentation and damage to riparian zones. Commercial harvesting would focus on the remaining 4 million acres, although 15% of any harvesting unit was to be left uncut to provide a source area for secondary succession.

There is an interesting contrast between recommendations for conservation of the northern spotted owl in the PNW and in southern California. In contrast to the PNW, the habitat in southern California is largely intact, and metapopulation structure in the owl population results from natural patchiness of suitable habitat in the mountain ranges (Verner et al., 1992). Habitat has declined in quality with selective harvest, but this decline occurred at fine spatial scales and with no evidence of an associated owl population decline. Catastrophic fire, which could wreak havoc with habitat conservation areas, was of concern in California; thus, in the Sierra Nevada a goal of maintaining the current (patchy) distribution of habitat was established.

Basic principles derived from landscape ecology have permeated the land planning and management activities in the PNW. Decisions have been based on broad-scale rather than local analyses. The need to preserve large contiguous areas of habitat and riparian zones has been acknowledged. The current goal is to achieve harvesting patterns that result in fragmentation patterns that are within the natural range of variability (Morgan et al., 1994) experienced by the biota.

Sustainable Forest Management in Ontario, Canada

The Ontario Ministry of Natural Resources (OMNR) is responsible for managing the extensive, publicly owned, or Crown forests, of Ontario, Canada, for the long-term health of the forest. Eighty-eight percent of the land in Ontario is held under Crown ownership. The legal authority for OMNR's responsibilities comes from the Crown Forest Sustainability Act (CFSA), which was passed on April 1, 1995, and which mandates landscape planning. Forest sustainability is identified as the first priority and overriding principle. In addition, there is a requirement that the area of Ontario's Crown forest, on a provincial scale, must not decline and will be expanded whenever feasible, and that all forests must be assessed and evaluated for their contribution to forest diversity and wildlife habitat. OMNR has strict rules for sustaining Crown forests, and these are set out in the Forest Management Planning Manual (Ontario Ministry of Natural Resources, 1996), which has the force of law. Similar to procedures followed in the United States, a Forest Management Plan is prepared for each Crown forest by a local planning team with input from a Local Citizens Committee. The plan covers a 20-yr period and is revised at a 5-yr interval.

What is particularly interesting about Ontario's forest management process is the explicit requirement for landscape assessment that is now required by law. Among the stated principles for sustaining forests is the following: "*Forest practices, including all methods of harvesting, must emulate, within the bounds of silvicultural requirements, natural disturbances and landscape patterns.*" To assess and report on how well Ontario is achieving its objectives for forest sustainability, a series of critical elements, indicators, and indicator standards was identified for reporting at 5-yr intervals. The full set of Criteria and Indicators for Conservation and Sustainable Management of Temperate and Boreal Forests was developed through a set of international conferences begun in Montreal in September 1993. Among the six criteria and 22 indictors are found the following (Canadian Standards Association, 1996):

> *Conservation of biological diversity*: Ecosystem diversity is conserved if the variety and landscape-level patterns of communities and ecosystem that naturally occur on the defined forest area are maintained through time.
>
> *Maintenance and enhancement of forest ecosystem condition and productivity*: Forest health is conserved if biotic (including anthropogenic) and abiotic disturbances and stresses maintain both ecosystem process and ecosystem conditions within the range of natural variability.

These two indicators clearly mandate that the state of the landscape be quantified and reported and that spatial indicators (landscape metrics, such as edge, richness, evenness, core area, patch size, interspersion, and juxtaposition) remain within their natural range of variability. OMNR was directed to prepare an assessment of the forest resources every 5 yr, starting in 1996, and thus the practical applications of these goals, which very much reflect current scientific understanding of forest landscape, are being put to the test.

The challenges posed by implementation of the CFSA were the focus of a symposium at the 1999 International Congress of Landscape Ecology and have been summarized by Perera et al. (2000). One practical challenge is defining naturalness in an operationally useful way (Duinker, 1999). Which of the natural patterns in a landscape are most important to mimic, and how should this be done? Naturalness is appealing conceptually, but difficult to define and mimic. Among the identified information needs is a description of range of natural variability in forces that create change in the forest, such as fire, insects, disease, and natural succession; these can be estimated through historical records. However, over what time scale should the definition of naturalness be applied? The scale dependence of the interactions between disturbance dynamics and landscape patterns has been discussed elsewhere (see Chapter 7); operationally, what scales should be selected for forest management? The use of natural patterns as a model for the patterns and timing of forest harvesting activities has been debated for some years (e.g., Hunter, 1993; and see Chapter 7), and it is clear, for example, that clear-cutting and fire differ in many important ways. In addition, the lack of stationarity in fire regimes has been described for many forested landscapes, revealing no single fire cycle for forest harvesting to mimic.

Second, landscape metrics remain difficult to apply and understand; more empirical data are sorely needed to develop a robust understanding of the quantitative relationships between landscape metrics and ecological processes, including their sensitivities and scale dependencies. Duinker (1999) suggested that (1) edge-based metrics would be most useful, largely because they are among the best understood with regard to their effect on wildlife, and (2) the landscape metrics should be combined with wildlife habitat models, an approach that permeated the FEMAT assessment and is proving useful in the Front Range of Colorado (Duerksen et al., 1997). Furthermore, Duinker (1999) recommended that the objective for forest management should focus on values important to management, such as wildlife, and be measured in simple, understandable ways.

Finally, Baker (1999) raised important conceptual questions that are implicit in this approach. If implemented effectively, the CFSA will produce future patterns in the boreal landscape of Ontario that are similar to current patterns. It is assumed that the ecological processes will also be maintained by sustaining natural landscape patterns. However, Baker (1999) raised the question about transferability of the utility of particular metrics among landscapes; for example, patch size and connectivity are of prime importance in fragmented agricultural and urban landscapes, but are they insightful in the boreal forest? Again, these are very important points for landscape ecologists to consider as the science and its tools are applied.

The experience gained in Ontario should prove extremely useful and informative to landscape ecologists in many other systems. The attempt to redirect forest management goals based on the best current scientific understanding is laudable and progressive, and it will serve as a valuable test bed for broad-scale applications of landscape ecology methods. The challenges posed to the field of landscape ecology should be taken up, discussed, and addressed.

REGIONAL RISK ASSESSMENT

Ecological risk assessment is a key concept in natural resource management and ecological policy making (O'Neill et al., 1982; Bartell et al., 1992; Suter, 1993; Lackey, 1996). The goal of ecological risk assessment is to provide (1) a quantitative basis for balancing and assessing risks associated with environmental hazards and (2) a systematic means of improving the estimation and understanding of these risks (Hunsaker et al., 1990). Risk assessment is widely used in many fields (e.g., the nuclear industry, insurance industry, and disaster management) to estimate the likelihood of occurrence of an adverse event (Molak, 1996). The endpoint for the assessment (e.g., species decline or extinction; see Table 10.3) must

TABLE 10.3.
DEFINITION OF TERMS USED IN REGIONAL RISK ASSESSMENT.

Term	Definition	Examples
Endpoint	Environmental entity of concern and the descriptor or quality of the entity	Extinction of an endangered species; concentration of nitrogen or phosphorus in freshwater ecosystems
Exposure–habitat modification	Intensity of the chemical and physical exposures of an endpoint to a hazard	Amount of habitat, for an endangered species, that is lost; amount of nutrient input to aquatic ecosystem
Hazard	Pollutant or activity and its disruptive influence on the ecosystem containing the endpoint	Forest cutting practices that eliminate critical habitat for an endangered species; land-use practices that result in increased transport and loss of sediment and nutrients
Reference environment	Geographic location and temporal period for the risk assessment	Piedmont of the United States in the next 10 yr; watershed of the Wisconsin River in the next 20 yr
Source terms	Qualitative and quantitative descriptions of the source of the hazard	Forest cutting practices and the laws and economic factors that influence them; agricultural practices (e.g., fertilizer application rates) and zoning or regulations that govern these practices

FROM HUNSAKER ET AL., 1990.

be defined a priori and the probability of observing the endpoint is estimated, given all the stochastic effects and uncertainties involved (O'Neill et al., 1982).

Concepts of landscape ecology have been particularly useful in developing risk assessment approaches that can address questions of long-term management of resources over larger areas (Hunsaker et al., 1990; Graham et al., 1991). Environmental hazards occur over a wide range of spatial and temporal scales. Some hazards affect large regions (e.g., acid deposition, global climatic warming, nitrogen saturation) and some are clearly influenced by landscape configuration (e.g., nonpoint pollution; see Chapter 9). Regional risk assessment was developed for situations in which either the cause or the consequence of the environmental hazard was regional in its extent (Hunsaker et al., 1990). Note that a local phenomenon can have a regional consequence (e.g., single-source pollutants that become widely dispersed or loss of a species' critical source habitat, which results in a broad-scale population decline), or multiple, local factors can combine to create a regional hazard to a population, species, or ecosystem that is widely dispersed.

Regional (and local) risk assessments have two distinct phases. The first phase focuses on a hazard definition that establishes the endpoints, source terms, and reference environment of the problem being studied. In essence, this is the conceptualization of the problem, and much discussion often surrounds this aspect of risk assessment (Lackey, 1996). In the second phase, or problem solution, the exposures and effects are assessed, and the exposure levels are related to the effects levels to determine risk. A variety of specialized techniques is employed to determine the probability of occurrence of the adverse endpoint. In a regional risk assessment, the spatial characteristics of the regional landscape and any spatial characteristics associated with exposure or effects of exposure are included (Graham et al., 1991).

A prototype regional risk assessment was developed by Graham et al. (1991) for the Adirondack Mountains, located in north central New York. The Adirondack Mountains have been studied extensively with regard to acid deposition, and the region has experienced increased atmospheric concentrations of ozone. Elevated ozone concentrations result in physiological stress on trees, particularly conifers (Heck et al., 1986), and stressed trees are more susceptible to bark beetle infestations that, in turn, result in tree-killed patches that range in size from a few trees to >25 ha. The bark beetle attacks reduce interior forest habitat and increase forest-edge habitat. The loss of conifers with their acidic needle litter can also change the pH of small mountain lakes that receive runoff from the affected watershed.

Graham et al. (1991) used regional risk assessment to examine the ozone–beetle complex. They began with a map of land use for a 300,000-ha area in the Adirondacks that contained 66 small lakes. A simulation model was used to impose an ozone stress (low or high) across the study region, which then increased the probability of beetle infestation. The location and size of bark beetle attacks were selected at random, and the resulting change in landscape pattern (e.g., forest edge) was determined. Most bark beetle infestations produced patches that were one grid cell (4 ha) in size. However, the high-ozone concentration resulted in more and larger patches of beetle-induced conifer mortality than did the low-ozone concentration scenario. The model was run 100 times for two scenarios of elevated ozone concentrations. Each altered landscape then was used as input in a second model that related land cover to lake water quality (Hunsaker et al., 1986). To calculate the risk of a significant impact, the percentage of the 100 model runs that showed a significant (≥25%) change in pH was calculated. If there was a 25% change in 50% of the runs, then the risk, or probability, of the ozone–beetle complex causing a significant impact on water acidity was 0.5.

Simulation results revealed that forest edge increased in all cases (risk = 1.00 for both ozone concentration levels) and that significant shifts in lake pH were likely (risk = 0.89). A variety of pattern-sensitive endpoints (e.g., forest-edge lengths, contagion) was computed in the assessment, and most of these were influenced by the ozone scenarios. Thus, this regional risk assessment approach combined existing information on land cover, ozone, and beetle attacks to estimate alterations in the landscape pattern and then to translate this change in pattern into the probability of an ecological impact. Methods employed included landscape pattern analysis and spatial and stochastic modeling, with existing empirical data used for parameterization.

An example of a regional risk assessment is provided by Riitters et al. (1997). Land-cover data for a region around the Chesapeake Bay, located in the mid-Atlantic region of the United States, were used to assess habitat suitability for a variety of organisms. Within a square window of grid cells (for which size was determined based on the home range of the organism) on the land-cover map, simple rule-based algorithms were used to determine whether all habitat requirements for the organism could be met within that window. The window was then moved over the entire region to determine the percentage of the area that provides adequate habitat. Riitters et al. (1997) used home ranges of 5, 45, and 410 ha and asked whether 90% of the window was forested. More complex habitat requirements could be implemented. Risk to different species because of habitat loss under alternative land-cover change scenarios could then be assessed for a variety of species. Although not called regional risk assessment by the authors, White et al. (1997; see Chapter 8) employed this type of approach with alternative land-use scenarios and potential consequences on biodiversity in the Pocono Mountains, Pennsylvania.

A similar approach was also used by Hansen et al. (1993) to evaluate management strategies in the Willamette National Forest in the Cascade Mountains, Oregon. Habitat data, including forest type, territory size, and sensitivity to edges, were collected for 51 species of birds. Using a land-cover map of the area, 140 years of land-cover changes were simulated under a variety of management options, such as intense forestry, no forestry, and a modified multiple-use harvest option. Thus, this study linked spatial and temporal dynamics of the study region. Each management strategy was evaluated in terms of how it changed the percentage of the area suitable for habitat for each of the 51 species, and the resulting change in biodiversity as habitat was lost was estimated. This permitted a direct connection to be made between management strategies and risk to biodiversity.

The Gap Analysis Program of the U.S. Geological Survey, Biological Resources Division (Scott et al., 1996) also utilizes a landscape approach for assessing regional risk to biodiversity. The greatest risk comes from the appropriation and fragmentation of natural cover for human use. At the scale of the ecoregion (a geographic region with a characteristic set of integrated ecosystems, such as tundra or temperate desert; Bailey, 1998), remote imagery, aerial photography, and ground surveys are combined in a geographic information system to produce land-cover maps. Existing land cover is compared to maps of potential vegetation cover to identify subregions in which impacts to ecological communities may be greatest. Maps of ownership are also examined to determine whether current ownership patterns offer appropriate levels of protection to biodiversity or whether protected areas miss some critical subregions.

One example of the application of gap analysis is in the Mojave Desert, California (Thomas and Davis, 1996). A vegetation map recognized 32 natural vegetation types, such as desert grassland and holly scrub. A land management map was superimposed on the vegetation map that distinguished (1) public ownership with well-developed preservation plans, (2) public ownership with a variety of land uses, and (3) private ownership. The overlays permitted each vegetation type to be characterized as adequately represented, poorly represented, or critically underrepresented in protected areas. Thus, gap analysis identifies subregions where ecosystem types are threatened; that is, there are gaps in regional protection. Rather than focusing on lists of individual endangered species, gap analysis recognizes that preservation of the intact ecosystem is of primary importance. Taking a broad landscape approach may be the most practical approach to identifying threats and prioritizing policy decisions at the regional scale (Franklin, 1993).

CONTINENTAL-SCALE MONITORING

An important application of landscape ecology is in the area of broad-scale monitoring. Satellite imagery of Earth's surface is now collected routinely and is widely available. Classification of repeated imagery can be analyzed over time to quantify changes in land cover and landscape pattern. Landscape ecology provides a key step in the process, making it possible to relate changes in surface patterns to ecological impacts, such as the loss of wildlife habitat and changes in water quality.

The Environmental Protection Agency's Landscape Ecology Program is designed to establish the feasibility of this approach to monitoring and assessment (O'Neill et al., 1994). The program combines available data sets, such as land cover, major roads, topography, and population density, to quantify landscape changes through time. These changes are then related to ecological variables of interest to determine the potential impact or risk. An example of this broad-scale monitoring is provided for watersheds in the Chesapeake Bay region (Riitters and Wickham, 1995). Streams that run through agricultural fields receive nutrients and silt from erosion. The effect is intensified if slopes are steep, since gravity accelerates the erosion process. Information on land cover and topography are then combined to assess the risk to water quality. Figure 10.7 shows the watersheds in the region that are at the highest risk of impact on water quality when farming is adjacent to streams on slopes of >3%.

The concept of continental-scale monitoring has been expanded to the entire United States (O'Neill et al., 1997). In one analysis, the potential vegetation, determined by climate and topography (Kuchler, 1964), was compared to actual vegetation, determined by satellite imagery (Loveland et al., 1991). The comparisons were summarized by ecoregion (Omernick, 1987) to identify the areas with the greatest risk to biodiversity that might occur as a consequence of habitat loss. In another analysis, land cover in major water resource regions (Seaber et al., 1984) was used to estimate nitrogen concentrations in surface waters using empirical relationships established by Omernick (1977). As time series of higher-resolution imagery become available, it will be possible to monitor the risk to dozens of environmental variables across the continent on an annual basis. However, the abil-

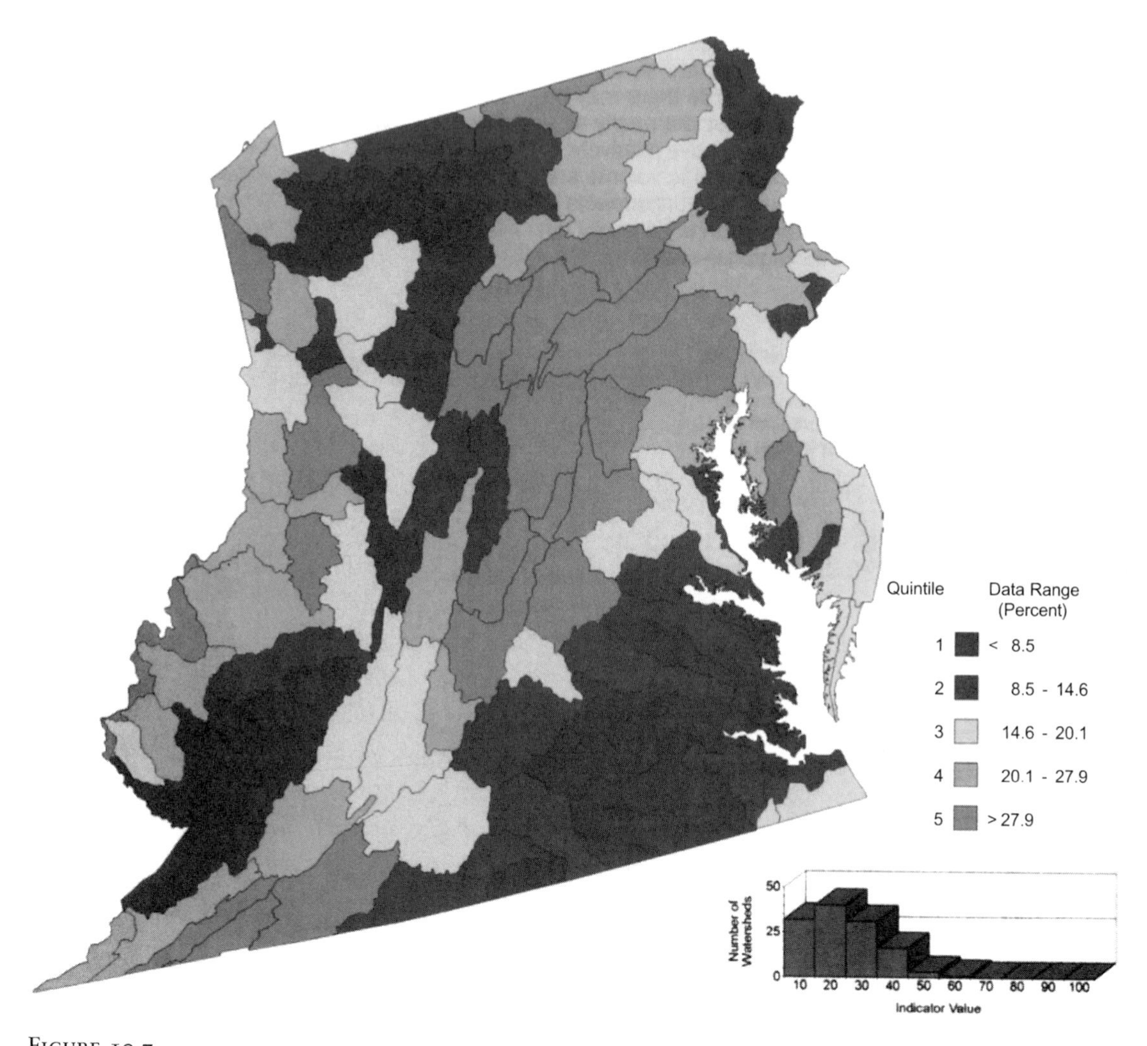

FIGURE 10.7.

Proportion of total stream length with adjacent agricultural land cover in the mid-Atlantic region of the United States. The index was determined for each watershed by overlaying land cover and stream maps. The index is the total length of stream with agricultural land cover divided by the total length of all streams in the watershed. (Refer to the CD-ROM for a four-color reproduction of this figure.)

FROM JONES ET AL., 1997.

ity of scientists to quantify spatial patterns in numerous ways (see Chapter 5) still exceeds our ability to relate these spatial patterns to ecological processes; further development of empirical data demonstrating linkages between landscape pattern and ecological indicators remains a critical need. We suggest that the potential for landscape monitoring to be truly effective depends on development of a much stronger and more extensive library of case studies that relate the broad-scale landscape metrics to measured ecological processes.

SUMMARY

Landscape ecology contributes to many different aspects of applied ecology. Resource managers have shifted their emphasis from management of separate resources to management focused on the integrity of entire ecosystems and landscapes. Challenges facing resource managers often involve spatial interdependencies among landscape components, and demand for the scientific underpinnings of managing landscapes is high. Despite the acknowledged importance of a landscape perspective by both scientists and resources managers, determining how to implement management at broader scales is very much a work in progress (Turner et al., in press). Thus far, landscape ecology has contributed to a variety of applications, including land use, forest management, risk assessment, and broad-scale monitoring. The demand for input from landscape ecologists to resource management decisions at all scales is very high.

Can lessons be extracted thus far from applications of landscape ecology? We think so. First, determining what is optimal for any given landscape is not straightforward. Optimal is a value judgment, and the development of a consensus opinion is often an arduous process. It is difficult to balance the multiple criteria by which a landscape may be considered—economic output, persistence of biodiversity, maintenance of water quality, and adequate land for human development. Even when naturalness is identified as a goal for landscape condition, implementing such an objective is not straightforward. The landscape ecologist can provide a rigorous analysis of alternative landscapes, including the potential implications for various ecological processes. However, the science still stumbles when an optimal target condition must be unambiguously identified, in part because such a decision requires more than science.

Second, applied landscape ecology often requires a team approach; indeed, none of the examples that we have highlighted in this chapter have resulted from single-investigator science! This means that landscape ecologists must be attentive to the requirements of building effective teams; Likens (1998) nicely summarized many of the factors that so strongly influence the effectiveness and productivity of collaborative (and often interdisciplinary) research (Table 10.4). We strongly recommend that collaborative research be encouraged, and rewarded, at all levels and that graduate training include opportunities for team research. Furthermore, practical applications of landscape ecology involve communication between scientists and decision makers. Thus, teams addressing applied problems will often be broader in composition and interests than those dealing with topics of concern only to basic science.

TABLE 10.4.
CHARACTERISTICS NEEDED FOR BUILDING AN EFFECTIVE TEAM TO CONDUCT EFFECTIVE RESEARCH IN ECOSYSTEM SCIENCE.

1. Brightness
2. Trusting and trustworthy
3. Abundant common (or good) sense
4. Creativity and willingness to share
5. Appropriately trained
6. Collective ability to make up deficiencies
   - Shared experiences
7. Willing to give team time
8. Personality
   - Ability and willingness to listen
   - Enjoy working with other people
   - Curiosity and interested
   - Openness of mind
9. Keeping eyes open (serendipity reigns)
10. Liking each other
    [Luck helps!]

FROM LIKENS, 1998.
APPLICATIONS OF LANDSCAPE ECOLOGY ALSO USUALLY REQUIRE A TEAM APPROACH, AND THE ABOVE FACTORS SHOULD BE HELPFUL WHEN BUILDING AN EFFECTIVE TEAM.

Third, there are still basic research needs that have direct implications for applications of landscape ecology. For example, the insufficient development of clear principles for development and analysis of spatial data and models to estimate both the causes and consequences of landscape pattern and change remain a serious limiting factor. The effectiveness of broad-scale landscape monitoring is still limited by uncertainty in what the changes in landscape metrics really mean for ecological processes. Progress in these arenas is clearly needed, and results will transfer quickly to applied landscape ecology.

Fourth, there is an urgent need for training in advanced analytical methods, including modeling, spatial statistics, and remote sensing. Modeling offers a means for exploring the causes and consequences of alternative spatial patterns. But the integration of models and their evaluation is a technologically challenging process. Greater exposure to the development, parameterization, and interpretation of models should be required in the training of landscape ecologists. Landscape ecologists must have more than a passing familiarity with model development and interpretation, as well as with techniques for quantitative spatial analysis.

Fifth, there will never be enough data upon which to base management recommendations. As GIS databases become more and more common in land and resource management, some of the tedious work required to build databases for landscape analysis and applications has been reduced. However, there is still never enough data—for species occurrence, demography, or dispersal; water quality; productivity or nutrient dynamics; and the like. As is true for other areas of ecology, landscape ecologists should recognize that decisions must and will be made in the face of uncertainty, and that the data represent only one of the inputs to the decision-making process. The best current science and data must be brought to bear on applied questions, with full recognition that our understanding may well change in the future.

Finally, applications of landscape ecology require integration and synthesis. Throughout this book, and as students are typically taught, concepts are separated into seemingly discrete units, for example, species responses to landscape patterns, disturbance dynamics, or spatial patterns of nutrient cycling. In landscape ecology applications, however, these separate concepts must be considered synthetically. Therefore, thinking across traditional boundaries must be encouraged. We are optimistic that applications of landscape ecology will continue to develop and that applied problems will continue to stimulate progress in our basic understanding of the relationship between pattern and process at multiple scales.

~ DISCUSSION QUESTIONS

1. Disturbance regimes can alter landscape structure and, in turn, influence the persistence and abundance of organisms across a landscape. For each of the three "species" listed below, (a) evaluate its ability to persist under the disturbance regime and the expected trend of its population during the coming 200 yr, and (b) describe how you might manage the landscape under the specified disturbance regime to maximize and minimize the species' population.

   Species:

   i. The *odd-looking purple bird* is an old-growth forest obligate that requires large patches of suitable habitat.
   ii. The *small-eared hoary vole* requires forested habitat (not meadows) and disperses only along forested corridors.
   iii. The *dominant-tree root rot fungus* invades ~100 m into the edges of the forest and can also be transported downhill by water.

   Disturbance regimes:

   i. Forest fire affects 30% of a landscape at approximately 100-yr intervals; vegetation takes 125 yr to regrow following disturbance.
   ii. Road length in a forested landscape increases by 20% during each decade for 100 yr.
   iii. Each year, about 2% of a forested landscape is affected by small disturbances; vegetation regrows within 75 yr.

2. You have been named the director of the newly created Agency for Monitoring Landscape Change (AMLC). Your charge is to develop a set of metrics that can be used to detect change through time and differences between regions in the southeastern United States (or a regional landscape of your choice). Describe a set of at least five landscape metrics that will constitute your initial monitoring plan for the AMLC. For each index, describe (a) what attributes of the pattern and process of the landscape the index measures and (b) the advantage and limitations affecting its use. Explain why the set that you have selected will best meet the needs of the AMLC; that is, justify your choices based on what the set will accomplish in toto.

3. Describe the procedures that you would use to incorporate landscape ecological principles and techniques in development of a 20-yr land-use plan for a county of your choice.

4. Compare and contrast *basic* and *applied* research in landscape ecology. Do you think the distinction is useful? Why or why not?

RECOMMENDED READINGS

Dale, V. H., S. M. Pearson, H. L. Offerman, and R. V. O'Neill. 1994. Relating patterns of land-use change to faunal biodiversity in the Central Amazon. *Conservation Biology* 8:1027–1036.

Duerksen, C. J., D. L. Elliott, N. T. Hobbs, E. Johnson, and J. R. Miller. 1997. *Habitat Protection Planning: Where the Wild Things Are.* American Planning Association, Planning Advisory Service, Report 470/471.

Franklin, J. F., and R. T. T. Forman. 1987. Creating landscape pattern by forest cutting: ecological consequences and principles. *Landscape Ecology* 1:5–18.

O'Neill, R. V., C. T. Hunsaker, K. B. Jones, K. H. Riitters, J. D. Wickham, P. M. Schwartz, I. A. Goodman, B. L. Jackson, and W. S. Baillargeon. 1997. Monitoring environmental quality at the landscape scale. *BioScience* 47:513–519.

Wear, D. N., and P. Bolstad. 1998. Land-use changes in southern Appalachian landscapes: spatial analysis and forecast evaluation. *Ecosystems* 1:575–594.

CHAPTER 11

Conclusions and Future Directions

The exponential growth of human populations has resulted in dramatic changes in Earth's ecosystems (Walker et al., 1999). The accelerating pace of landscape conversion has caused significant deterioration of the Earth's natural resources and extensive fragmentation of undisturbed ecosystems. During the past few decades, the consequences of these changes have been studied by a multitude of scientists representing a diversity of disciplines. Motivated in part by issues associated with broad-scale environmental change, landscape ecology has emerged as a synthetic discipline that has generated new concepts, theory, and methods to reveal the importance of spatial patterning on the dynamics of interacting ecosystems. Landscape ecology is now a well-recognized subdiscipline within ecology, as well as an interdisciplinary area of research and application that extends well beyond ecology. This text has emphasized the current ecological understanding of the causes of spatial pattern, the effect of spatial pattern on ecological process, and how these patterns and processes change through time. Although the implications for management, conservation, and design have been discussed, the primary emphasis has been on the ecological component of landscape ecology. Here we highlight some of what has been learned through landscape ecology during the past two decades and identify directions for future research.

WHAT HAVE WE LEARNED?

It is difficult to distill from an entire book, which itself is already a distillation of many contributions from the primary literature, a few summary statements about what we have learned from landscape ecology. Recognizing the potential pitfall of superficiality, we nonetheless suggest a few general insights that are directly related to the growth and development of landscape ecology.

First and foremost, landscape ecology has clearly demonstrated that spatial heterogeneity can be an important influence on a wide range of ecological patterns and processes. Ecologists now routinely consider the implications of spatial pattern as a potential factor influencing ecological responses, regardless of whether they study individual organisms, populations, communities, or entire ecosystems. In addition, the pervasive influence of scale is appreciated, despite the inherent challenges that it presents. The necessity of understanding ecological patterns and processes over areas of broad spatial extent is acknowledged both in research and applied ecology. Landscape ecology has contributed to all these developments.

In an earlier essay (Turner et al., 1995b), we summarized insights relating to ecological dynamics at broad scales with a particular emphasis on biodiversity. Here we extend this list to consider more broadly the contributions of landscape ecology.

- The complex patterns observed on today's landscape result from many causes, including variability in the abiotic template, biotic interactions, natural disturbances, and both past and present patterns of human settlement and land use.
- There is no right scale for landscape ecological studies, but scale effects must be considered carefully; concepts such as equilibrium and species persistence are scale dependent.
- Many metrics are available for quantifying landscape patterns; one metric is insufficient for characterizing a landscape, yet there is no standard recipe for determining how many and which ones should be used.
- Organisms are influenced by spatial pattern, but pattern–process interactions involving organisms are scale dependent and require an organism-based view.
- Disturbances both create and respond to landscape heterogeneity, and thus landscapes may be strongly influenced by shifts in disturbance regimes.
- Because pattern matters, natural disturbance can be very important for biodiversity and for ecosystem function.
- Populations or guilds can produce important feedbacks to ecosystem processes and landscape patterns.

- Ecosystem function in terrestrial and aquatic ecosystems can be influenced by landscape position.
- Elements of the landscape may serve as sources or sinks for nutrients that move across the terrestrial landscape or from the land to aquatic ecosystems.
- Human influences (e.g., land-use change) may be dominant factors controlling ecological dynamics at broad scales.

RESEARCH DIRECTIONS

Current trends in research and land management suggest that a landscape focus in ecology will remain prominent for some time to come. What are the frontiers likely to drive research and lead to new insights during the coming decade? Although it is always risky to make predictions about future research directions, we identify several of what we consider are among the most pressing and exciting research directions and challenges for landscape ecologists.

Spatial Heterogeneity and Ecosystem Processes

Landscape ecology has made tremendous progress in quantifying spatial patterns and how they change (see Chapters 4 through 7) and in understanding the consequences of landscape patterns, especially habitat fragmentation, for populations (see Chapter 8). However, our understanding of ecosystem processes in the landscape—how rates vary over space and at different spatial scales and what controls this variation—is in its infancy (Carpenter and Turner, 1998; see Chapter 9). Process-based studies are costly and often require sophisticated laboratory equipment and methods, and it may be very difficult to collect an adequate number of samples (across both space and time). However, this represents an important challenge to landscape and ecosystem ecologists, and we suggest that the integration of these two areas of ecology should be a high priority.

Relating Landscape Metrics to Ecological Processes

The development of landscape pattern analysis has been rapid (see Chapter 5), but there are three major areas in which further understanding is sorely needed: (1) the statistical properties and behavior of metrics need to be better understood, (2) the relative sensitivity of different metrics to detecting changes in the landscape is not known, and (3) the empirical relationships between landscape patterns and ecological processes of interest must be better documented and the underlying mechanisms understood. Collectively, progress in these three research areas will help ecologists determine what is worth measuring and why and when a change in a metric is significant both statistically and ecologically. It remains a critical research task to determine what constitutes a significant change, both statistically and ecologically, in spatial metrics and to relate such changes to ecologically relevant responses. Applications of landscape ecology depend heavily on such understanding (see Chapter 10).

Thresholds, Nonlinearities, and Rules for Scaling

Critical thresholds in landscape pattern (see Chapter 6) provide an example of a nonlinearity with important implications for understanding the relationship between pattern and process. Identifying and understanding the nonlinearities associated with changes in spatial and temporal scale provide exciting opportunities for research and very practical applied challenges. The effects of scale are now well recognized, but the need for improved quantitative understanding remains critical. Ecologists still struggle with identifying the right scale(s) for studying and understanding particular patterns and processes and extrapolating the knowledge gained at one scale to other scales. The rules of thumb that have been suggested for scaling (see Chapters 2 and 5) need to be tested more widely, and the qualitative differences associated with changes in spatial pattern (e.g., critical thresholds; see Chapter 6) need to be considered in actual landscapes. Many of the ideas developing in the realm of complex adaptive systems (e.g., Milne, 1998), in which emphasis is frequently placed on slow and fast variables, may also shed light on issues associated with spatial scaling. We see nonlinear dynamics and scaling as topics that will continue to motivate a considerable volume of basic and applied research in landscape ecology.

Feedbacks between Organisms and Ecosystems in Space

Identifying the interactions and feedbacks between organisms and ecosystem processes in a spatial context remains an important goal. A large body of literature describes the effects of spatial pattern on the presence or abundance of populations, but there is relatively little research on the relationship between species dynamics and ecosystem processes in a spatial context. The potential feedbacks of populations to ecosystem and landscape dynamics remain poorly understood.

Causes and Consequences of Land-use Change

The size of the human population on Earth reached 6 billion during the fall of 1999, and it will continue to increase. The human population is already transforming much of the surface of Earth and co-opting much of the world's resources (Vitousek et al., 1997b). Increases in the extent and intensity of human land uses are primary drivers of landscape change worldwide. Land-use patterns and changes are spatial phenomena, and landscape ecologists have an opportunity to contribute toward understanding and predicting these patterns and their ecological consequences (see Chapter 10). This area should receive increasing attention from landscape ecologists. In addition, greater emphasis is needed in understanding land-use legacies, that is, the types, extents, and durations of persistent effects of prior land use on ecological patterns and processes.

Sampling

Landscape ecology is certainly not constrained to address questions over large areas, but we must recognize that many landscape ecological studies do so. The problems inherent in sampling across large regions in a way that permits inference of the effects of spatial heterogeneity remain challenging. We need to develop improved ways of sampling over large areas, using appropriate (and possibly new) statistical methods for data analysis and creative combinations of the assortment of available methods, including field sampling, experimentation, remote sensing, and modeling.

Collectively, these research directions will continue to contribute toward furthering our knowledge of what Wiens (1999) summarized as the four central themes of landscape ecology: spatial variation, scaling, boundaries, and flows. We expect that landscape ecology research will continue to complement the understanding obtained from research at other levels in ecology. We also hope that a spatially explicit view of ecological systems will enhance integration across traditional boundaries within ecology. For example, there is an opportunity to improve the widely noted boundary between population dynamics and ecosystem processes (Schulze and Mooney, 1993; Jones and Lawton, 1995) when the patterns and controls on biodiversity are considered in a landscape context. Similarly, metapopulation dynamics have been emphasized in population ecology, landscape ecology, and conservation biology, but there is much to be gained by enhanced communication and collaborative studies.

CONCLUSION

Landscapes are characterized by complexity, change, and scale dependencies. We remain excited about the basic questions and applied challenges that face landscape ecologists and optimistic that significant progress will continue in the coming decades. Simple cause–effect relationships are not likely to be established in landscape systems, and this argues for a pluralism of approaches that embraces this complexity (Wiens, 1999). Landscape ecologists must continue to learn about the causes of spatial heterogeneity in landscapes and how these patterns and their dynamics influence ecological processes. Instruction of the next generation of ecologists, resource managers, and landscape architects requires that landscape ecological principles be clearly defined and articulated. The potential contribution of landscape ecology (and landscape ecologists) to address the serious consequences of landscape change is enormous. We trust that this text will be a stimulus for critical and productive discussion about how spatial pattern and ecological processes interact, and we hope it provides a foundation on which new ideas, approaches, and applications of landscape ecology can continue to build.

DISCUSSION QUESTIONS

1. Has landscape ecology made unique contributions to ecology? What are these? Should landscape ecology be considered separately within ecology, or should it be folded into the broader discipline? Explain your rationale.
2. You are the program officer in charge of $10 million of new funds targeted for cutting-edge research in landscape ecology. Write the one-page Request for Proposals (RFP) that identifies the areas for which you would want to see proposals submitted.
3. Wiens (1999) characterized landscape ecology as a scientifically immature discipline, suggesting that landscape ecology lacks the conceptual unity expected of a mature science. Do you agree or disagree with this assessment? Explain your thinking.

RECOMMENDED READINGS

Hobbs, R. J. 1996. Future landscapes and the future of landscape ecology. *Landscape and Urban Planning* 37:1–9.

Risser, P. G. 1999. Landscape ecology: does the science only need to change at the margin? In J. M. Klopatek and R. H. Gardner, eds. *Landscape Ecological Analysis: Issues and Applications,* pp. 3–10. Springer-Verlag, New York.

Wiens, J. A. 1992. What is landscape ecology, really? *Landscape Ecology* 7:149–150.

Wiens, J. A. 1999. The science and practice of landscape ecology. In J. M. Klopatek and R. H. Gardner, eds. *Landscape Ecological Analysis: Issues and Applications,* pp. 372–383. Springer-Verlag, New York.

References

Acevedo, M. F., D. L. Urban, and H. H. Shugart. 1996. Models of forest dynamics based on roles of tree species. *Ecological Modeling* 87:267–284.

Addicott, J. F., J. M. Aho, M. F. Antolin, D. K. Padilla, J. S. Richardson, and D. A. Soluk. 1987. Ecological neighborhoods: scaling environmental patterns. *Oikos* 49:340–346.

Agee, J. K., and D. R. Johnson, editors. 1988. *Ecosystem Management for Parks and Wilderness.* University of Washington Press, Seattle, Washington, USA.

Ahern, J. 1999. Spatial concepts, planning strategies, and future scenarios: a framework method for integrating landscape ecology and landscape planning. *In* J. M. Klopatek and R. H. Gardner, eds. *Landscape Ecological Analysis: Issues and Applications,* pp. 175–201. Springer-Verlag, New York, New York, USA.

Aho, J. 1978. Freshwater snail populations and the equilibrium theory of island biogeography. *Annales Zoologici Fennici* 15:146–176.

Aide, T. M., and J. Cavelier. 1994. Barriers to tropical lowland forest restoration in the Sierra Nevada de Santa Marta, Colombia. *Restoration Ecology* 2:219–229.

Allen, T. F. H. 1998. The landscape "level" is dead: persuading the family to take it off the respirator. *In* D. L. Peterson and V. L. Parker, eds. *Ecological Scale,* pp. 35–54. Columbia University Press, New York, New York, USA.

Allen, T. F. H., and T. W. Hoekstra. 1992. *Toward a Unified Ecology.* Columbia University Press, New York, New York, USA.

Allen, T. F. H., and T. B. Starr. 1982. *Hierarchy: Perspectives for Ecological Complexity.* University of Chicago Press, Chicago, Illinois, USA.

Anderson, D. J. 1971. Spatial pattern in some Australian dryland plant communities. *In* G. P. Patil, E. C. Pielou, and W. E. Waters, eds. *Statistical Ecology*, Volume 1, pp. 272–285. *Spatial Patterns and Statistical Distributions.* Pennsylvania State University Press, University Park, Pennsylvania, USA.

Anderson, J. R., E. E. Hardy, J. T. Roach, and R. E. Witmer. 1976. *A land use and land cover classification system for use with remote sensor data.* U.S. Geological Survey Professional Paper 964, 28 p.

Andersson, M. 1981. Central place foraging in the whinchat, *Saxicolor rubetra. Ecology* 62:538–544.

Ando, A., J. Camm, S. Polasky, and A. Solow. 1998. Species distributions, land values, and efficient conservation. *Science* 279:2126–2128.

Andow, D. A., P. M. Kareiva, S. A. Levin, and A. Okubo. 1990. Spread of invading organisms. *Landscape Ecology* 4:177–188.

Andren, H. 1992. Corvid density and nest predation in relation to forest fragmentation: a landscape perspective. *Ecology* 73:794–804.

Andren, H. 1994. Effects of habitat fragmentation on birds and mammals in landscapes with different proportions of suitable habitat. *Oikos* 71:355–366.

Andrewartha, H. G., and L. C. Birch. 1954. *The Distribution and Abundance of Animals.* University of Chicago Press, Chicago, Illinois, USA.

Aronson, R. B., and T. J. Givnish. 1983. Optimal central place foragers: a comparison with null hypothesis. *Ecology* 64:395–399.

Attiwill, P. M. 1994. The disturbance of forest ecosystems: the ecological basis for conservative management. *Forest Ecology and Management* 63:247–300.

Bailey, R. G. 1996. *Ecosystem Geography.* Springer-Verlag, New York, New York, USA.

Bailey, R. G. 1998. *Ecoregions: The Ecosystem Geography of the Oceans and Continents.* Springer-Verlag, New York, New York, USA.

Baines, S. B., K. E. Webster, T. K. Kratz, S. R. Carpenter, and J. J. Magnuson. 2000. Synchronous behavior of temperature, calcium and chlorophyll in lakes of northern Wisconsin. *Ecology* 81:815–825.

Bak, P., C. Tang, and K. Wiesenfeld. 1988. Self-organized criticality. *Physical Review* 38:364–374.

Baker, J. A. 1999. Fragmentation concerns and forest wildlife habitat management: have we adopted the wrong prescriptions for the right reasons? *5th World Congress, International Association for Landscape Ecology, Abstracts Volume 1*: 9.

Baker, W. L. 1989a. Landscape ecology and nature reserve design in the Boundary Waters Canoe Area, Minnesota. *Ecology* 70:23–35.

Baker, W. L. 1989b. A review of models of landscape change. *Landscape Ecology* 2:111–133.

Baker, W. L. 1989c. Effect of scale and spatial heterogeneity on fire-interval distributions. *Canadian Journal of Forest Research* 19:700–706.

Baker, W. L. 1992. Effects of settlement and fire suppression on landscape structure. *Ecology* 73:1879–1887.

Baker, W. L. 1995. Long-term response of disturbance landscapes to human intervention and global change. *Landscape Ecology* 10:143–159.

Baker, W. L., and Y. Cai. 1992. The r.le programs for multiscale analysis of landscape structure using the GRASS geographic information system. *Landscape Ecology* 7:291–302.

Baker, W. L., and G. M. Walford. 1995. Multiple stable states and models of riparian vegetation succession on the Animas River, Colorado. *Annals of the Association of American Geographers* 85:320–338.

Baker, W. L., S. L. Egbert, and G. F. Frazier. 1991. A spatial model for studying the effects of climatic change on the structure of landscapes subject to large disturbances. *Ecological Modeling* 56:109–125.

Ball, S. T., D. J. Mulla, and C. F. Konzak. 1993. Spatial heterogeneity affects variety trial interpretation. *Crop Science* 33:931–935.

Band, L. E., D. L. Peterson, S. W. Running, J. Coughlan, R. Lammers, J. Dungan, and R. Nemani. 1991. Forest ecosystem processes at the watershed scale: basis for distributed simulation. *Ecological Modeling* 56:171–196.

Barnosky, C. W., P. M. Anderson, and P. J. Bartlein. 1987. The northwestern U.S. during deglaciation: vegetational history and paleoclimatic implications. *In* W. F. Ruddiman and H. E. Wright, Jr., eds. *North America and Adjacent Oceans During the Last Deglaciation*, volume K-3, pp. 289–321. Geological Society of North America, Boulder, Colorado, USA.

Barnsley, M. F., R. L. Devaney, B. Mandelbrot, H. O. Peitgen, D. Saupe, and R. F. Vos. 1988. *The Science of Fractal Images*. Springer-Verlag, New York, New York, USA.

Bartell, S. M., and A. L. Brenkert. 1991. A spatial–temporal model of nitrogen dynamics in a deciduous forest watershed. *In* M. G. Turner and R. H. Gardner, eds. *Quantitative Methods in Landscape Ecology*, pp. 379–398. Springer-Verlag, New York, New York, USA.

Bartell, S. M., R. H. Gardner, and R. V. O'Neill. 1992. *Ecological Risk Estimation*. Lewis Publishers, Chelsea, Michigan, USA.

Bartlein, P. J., C. Whitlock, and S. L. Shafer. 1997. Future climate in the Yellowstone National Park Region and its potential impact on vegetation. *Conservation Biology* 11:782–792.

Bartuska, A. M. 1999. Cross-boundary issues to manage for healthy forest ecosystems. *In* J. M. Klopatek and R. H. Gardner, eds. *Landscape Ecological Analysis: Issues and Applications*, pp. 24–34. Springer-Verlag, New York, New York, USA.

Bascompte, J., and R. V. Sole. 1996. Habitat fragmentation and extinction thresholds in spatially explicit models. *Journal of Animal Ecology* 65:465–473.

Baskent, E. Z. 1997. Assessment of structural dynamics in forest landscape management. *Canadian Journal of Forest Research* 27:1675–1684.

Baskent, E. Z., and G. A. Jordan. 1995. Characterizing spatial structure of forest landscapes. *Canadian Journal of Forest Research* 25:1830–1849.

Basta, D. J., and B. T. Bower, editors. 1982. *Analyzing Natural Systems*. Johns Hopkins University, Baltimore, Maryland, USA.

Beier, P., and R. Noss. 1998. Do habitat corridors provide connectivity? *Conservation Biology* 12:1242–1252.

Bell, G., M. J. Lechowicz, A. Appenzeller, M. Chandler, E. DeBlois, L. Jackson, B. Mackenzie, R. Preziosi, M. Schallenberg, and N. Tinker. 1993. The spatial structure of the physical environment. *Oecologia* 96:114–121.

Bell, W. J. 1991. *Searching Behavior.* Chapman and Hall, London, UK.

Bender, D. J., T. A. Contreras, and L. Fahrig. 1998. Habitat loss and population decline: a metaanalysis of the patch size effect. *Ecology* 79:517–533.

Bennett, E. M., T. Reed-Andersen, J. N. Houser, J. R. Gabriel, and S. R. Carpenter. 1999. A phosphorus budget for the Lake Mendota watershed. *Ecosystems* 2:69–75.

Benning, T. L., and T. R. Seastedt. 1995. Landscape-level interactions between topoedaphic features and nitrogen limitation in tallgrass prairie. *Landscape Ecology* 10:337–348.

Benson, B. J., and M. D. MacKenzie. 1995. Effects of spatial sensor resolution on landscape structure parameters. *Landscape Ecology* 10:13–20.

Bergeron, Y. 1991. The influence of island and mainland lakeshore landscapes on boreal forest fire regimes. *Ecology* 72:1980–1992.

Bergeron, Y., and P. R. Dansereau. 1993. Predicting the composition of Canadian southern boreal forest in different fire cycles. *Journal of Vegetation Science* 4:827–832.

Bergin, T. M. 1992. Habitat selection by the western kingbird in western Nebraska: a hierarchical analysis. *Condor* 94:903–911.

Berry, J. K., and J. K. Sailor. 1987. Use of a geographic information system for storm runoff prediction from small urban watersheds. *Environmental Management* 11:21–27.

Bessie, W. C., and E. A. Johnson. 1995. The relative importance of fuels and weather on fire behavior in subalpine forests. *Ecology* 76:747–762.

Bevers, M., and C. H. Flather. 1999. Numerically exploring habitat fragmentation effects on populations using cell-based coupled map lattices. *Theoretical Population Biology* 55:61–76.

Binkley, D., D. Richter, M. B. David, and B. Caldwell. 1992. Soil chemistry in a loblolly/longleaf pine forest with interval burning. *Ecological Applications* 2:157–164.

Birge, E. A., and C. Juday. 1911. The inland lakes of Wisconsin: the dissolved gases of the water and their biological significance. *Wisconsin Geological and Natural History Bulletin No.* 22.

Bissonette, J. 1997. Scale-sensitive ecological properties: historical context, current meaning. *In* J. Bisonnette, ed. *Wildlife and Landscape Ecology*, pp. 3–31. Springer-Verlag, New York, New York, USA.

Boerner, R. E. J., and J. G. Kooser. 1989. Leaf litter redistribution among forest patches within an Allegheny Plateau watershed. *Landscape Ecology* 2:81–92.

Bolger, D. T., A. C. Alberts, R. M. Sauvajot, P. Potenza, C. McCalvin, D. Tran, S. Mazzoni, and M. E. Soule. 1997a. Response of rodents to habitat fragmentation in coastal southern California. *Ecological Applications* 7:552–563.

Bolger, D. T., T. A. Scott, and J. T. Rotenberry. 1997b. Breeding bird abundance in an urbanizing landscape in coastal southern California. *Conservation Biology* 11:406–421.

Bolin, B. 1977. Changes of land biota and their importance for the carbon cycle. *Science* 196:613–615.

Bolstad, P. V., and W. T. Swank. 1997. Cumulative impacts of land use on water quality in a southern Appalachian watershed. *Journal of the American Water Resources Association* 33:519–533.

Bonan, G. B., and H. H. Shugart. 1989. Environmental factors and ecological processes in boreal forests. *Annual Review of Ecology and Systematics* 20:1–28.

Bonham-Carter, G. F. 1994. *Geographic Information Systems for Geoscientists: Modeling with GIS*. Pergamon, Tarrytown, New York, USA.

Boose, E. R., D. R. Foster, and M. Fluet. 1994. Hurricane impacts to tropical and temperate forest landscapes. *Ecological Monographs* 64:369–400.

Bormann, F. H., and G. E. Likens. 1979. *Pattern and Process in a Forested Ecosystem*. Springer-Verlag, New York, New York, USA.

Botkin, D. B., and M. J. Sobel. 1975. Stability in time-varying ecosystems. *American Naturalist* 109:625–646.

Botkin, D. B., J. F. Janak, and J. R. Wallis. 1972. Rationale, limitations and assumptions of a northeast forest growth simulation. *IBM Journal of Research and Development* 16:10.

Botkin, D. B., J. M. Melillo, and L. S. Y. Wu. 1981. How ecosystem processes are linked to large mammal population dynamics. *In* C. F. Fowler and T. D. Smith, eds. *Dynamics of Large Mammal Populations*, pp. 373–387. John Wiley & Sons, New York, New York, USA.

Box, E. O. 1978. Geographical dimensions of terrestrial net and gross primary productivity. *Radiation and Environmental Biophysics* 15:305–322.

Boychuk, D., and A. H. Perera. 1997. Modeling temporal variability of boreal landscape age-classes under different fire disturbance regimes and spatial scales. *Canadian Journal of Forest Research* 27:1083–1094.

Boychuk, D., A. H. Perera, M. T. Ter-Mikaelian, D. L. Martell, and C. Li. 1997. Modeling the effect of spatial scale and correlated fire disturbances on forest age distribution. *Ecological Modeling* 95:145–164.

Boynton, W. R., J. H. Garber, R. Summers, and W. M. Kemp. 1995. Inputs, transformations, and transport of nitrogen and phosphorus in Chesapeake Bay and selected tributaries. *Estuaries* 18:285–314.

Bradford, E., and J. R. Philip. 1970. Stability of steady state distributions of asocial populations dispersing in one dimension. *Journal of Theoretical Biology* 29:13–26.

Braun-Blanquet, J., translated by G. D. Fuller and H. S. Conard. 1932. *Plant Sociology, The Study of Plant Communities*. McGraw-Hill Book Company, Inc., New York, New York, USA.

Breckling, B., and F. Muller. 1994. Current trends in ecological modeling and the 8th ISEM conference on the state-of-the-art. *Ecological Modeling* 75:667–675.

Brock, T. D. 1985. *A Eutrophic Lake: Lake Mendota, Wisconsin*. Springer-Verlag, New York, New York, USA.

Brokaw, N. V. L. 1982. Treefalls: frequency, timing, and consequences. *In* E. G. Leigh, A. S. Rand, and M. M. Windsor, eds. *The Ecology of a Tropical Forest: Seasonal Rhythms and Long-term Changes*, pp. 101–108. Smithsonian Institution Press, Washington, DC, USA.

Brokaw, N. V. L. 1985. Treefalls, regrowth, and community structure in tropical forests. *In* S. T. A. Pickett and P. S. White, eds. *The Ecology of Natural Disturbance and Patch Dynamics*, pp. 53–59. Academic Press, New York, New York, USA.

Brokaw, N. V. L. 1987. Gap-phase regeneration of three pioneer tree species in a tropical forest. *Journal of Ecology* 75:9–19.

Browder, J. A., L. N. May, A. Rosen, J. G. Gosslink, and R. H. Baumann. 1989. Modeling future trends in wetland loss and brown shrimp production in Louisiana using thematic mapper imagery. *Remote Sensing of the Environment* 28:45–59.

Brown, J. H., and A. Kodric-Brown. 1977. Turnover rates in insular biogeography: effects of immigration on extinction. *Ecology* 58:445–449.

Brunet, R. C., and K. B. Astin. 1997. Spatio-temporal variations in sediment nutrient levels: the River Adour. *Landscape Ecology* 12:171–184.

Brush, G. S. 1984. Patterns of recent sediment accumulation in Chesapeake Bay (Virginia–Maryland, USA) tributaries. *Chemical Geology*, 44: 227–242.

Brush, G. S. 1986. Geology and paleoecology of the Chesapeake Bay: a long-term monitoring tool for management. *Journal of the Washington Academy of Sciences*, 76(3):146–160.

Brush, G. S. 1989. Rates and patterns of estuarine sediment accumulation. *Limnology and Oceanography*, 34:1235–1246.

Brush, G. S. 1997. History and impact of human activities on the Chesapeake Bay. *In* R. D. Simpson and N. L. Christensen, eds, pp. 125–145. *Ecosystem Function and Human Activities*. Chapman and Hall, New York, New York, USA.

Burgan, R. E., and R. A. Hartford. 1988. Computer mapping of fire danger and fire locations in the continental United States. *Journal of Forestry* 86:25–30.

Burgess, R. L., and D. M. Sharpe, editors. 1981. *Forest Island Dynamics in Man-dominated Landscapes*. Springer-Verlag, New York, New York, USA.

Burke, I. C., D. S. Schimel, W. J. Parton, C. M. Yonker, L. A. Joyce, and W. K. Lauenroth. 1990. Regional modeling of grassland biogeochemistry using GIS. *Landscape Ecology* 4:45–54.

Burke, I. C., E. T. Elliott, and C. V. Cole. 1995. Influence of macroclimate, landscape position, and management on soil organic matter in agroecosystems. *Ecological Applications* 5:124–131.

Burkey, T. V. 1989. Extinction in nature reserves: the effect of fragmentation and the importance of migration between reserve fragments. *Oikos* 55:75–81.

Burrough, P. A. 1981. Fractal dimension of landscapes and other environmental data. *Nature* 294:241–243.

Burrough, P. A. 1986. *Principles of Geographic Information Systems for Land Resources Assessment*. Oxford University Press, Oxford, UK.

Burrough, P. A., and R. A. McDonnell. 1998. *Principles of Geographic Information Systems*. Oxford University Press, New York, New York, USA.

Butcher, J. B. 1999. Forecasting future land use for watershed assessment. *Journal of the American Water Resources Association* 35:555–565.

Cain, D. H., K. Riitters, and K. Orvis. 1997. A multi-scale analysis of landscape statistics. *Landscape Ecology* 12:199–212.

Cain, M. L. 1985. Random search by herbivorous insects: a simulation model. *Ecology* 66:876–888.

Caldarelli, G., P. G. Higgs, and A. J. McKane. 1998. Modeling coevolution in multispecies communities. *Journal of Theoretical Biology* 193:345–358.

Campbell, I. D., and J. H. McAndrews. 1993. Forest disequilibrium caused by rapid Little Ice Age cooling. *Nature* 334:233–235.

Canadian Standards Association. 1996. *A Sustainable Forest Management System: Guidance Document.* Canadian Standards Association, Etobicoke. National Standard of Canada CAN/CSA-Z808-96.

Cardille, J. A. 1998. *Wildfires in the Northern Great Lakes Region: Assessment at Multiple Scales of Factors Influencing Wildfires in Minnesota, Wisconsin and Michigan.* Masters thesis, University of Wisconsin, Madison, Wisconsin, USA.

Cardille, J. A., D. W. Bolgrien, R. H. Wynne, and J. W. Chipman. 1996. Variation in landscape metrics derived from multiple independent classifications. *In Proceedings, Eco-Informa '96: Global Networks for Environmental Information*, pp. 749–754. Environmental Research Institute of Michigan, USA.

Cardille, J. A., S. J. Ventura, and M. G. Turner. 2001. Environmental and social factors influencing wildfires in the Upper Midwest, USA. *Ecological Applications*, 11:111–127.

Carlile, D. W., J. R. Skalski, J. I. Batker, J. M. Thomas, and V. I. Cullinan. 1989. Determination of ecological scale. *Landscape Ecology* 2:203–214.

Carlquist, S. 1974. *Island Biology*. Columbia University Press, New York, New York, USA.

Carpenter, S., T. Frost, L. Persson, M. Power, and D. Soto. 1995. Freshwater ecosystems: linkages of complexity and processes. *In* H. A. Mooney, ed. *Functional Roles of Biodiversity: A Global Perspective*, Chapter 12. John Wiley & Sons, New York, New York, USA.

Carpenter, S. R. 1998. Ecosystem Ecology. *In* S. I. Dodson et al. *Ecology*, pp. 123–162. Oxford University Press, New York, New York, USA.

Carpenter, S. R., and J. F. Kitchell. 1987. Plankton community structure and limnetic primary production. *American Naturalist* 124:159–172.

Carpenter, S. R., and M. G. Turner. 1998. At last: a journal devoted to ecosystem science. *Ecosystems* 1:1–5.

Carpenter, S. R., N. F. Caraco, D. L. Correll, R. W. Howarth, A. N. Sharpley, and V. H. Smith. 1998. Nonpoint pollution of surface waters with phosphorus and nitrogen. *Ecological Applications* 8:559–568.

Castello, J. D., D. J. Leopold, and P. J. Smallidge. 1995. Pathogens, patterns, and processes in forest ecosystems. *BioScience* 45:16–24.

Caswell, H. 1976. Community structure: a neutral model analysis. *Ecological Monographs* 46:327–354.

Caswell, H., and J. E. Cohen. 1995. Red, white and blue—environmental variance spectra and coexistence in metapopulations. *Journal of Theoretical Biology* 176:301–316.

Caswell, H., and M. C. Trevisan. 1994. Sensitivity analysis of periodic matrix models. *Ecology* 75:1299–1303.

Ceva, H. 1998. On the asymptotic behavior of an earthquake model. *Physics Letters* A 245:413–418.

Chayes, J. T., L. Chayes, and R. Durret. 1988. Connectivity properties of Mandelbrot's percolation process. *Probability Theory and Related Fields* 77:307–324.

Chen, J., J. F. Franklin, and T. A. Spies. 1992. Vegetation responses to edge environments in old-growth Douglas-fir forests. *Ecological Applications* 2:387–396.

Chen, J., J. F. Franklin, and T. A. Spies. 1995. Growing-season microclimatic gradients from clearcut edges into old-growth Douglas-fir forests. *Ecological Applications* 5:74–86.

Chen, J., J. F. Franklin, and J. S. Lowe. 1996. Comparison of abiotic and structurally defined patch patterns in a hypothetical forest landscape. *Conservation Biology* 10:854–862.

Cherrill, A., and C. McClean. 1995. An investigation of uncertainty in field habitat mapping and the implications for detecting land-cover change. *Landscape Ecology* 10:5–21.

Chou, Y. H., R. A. Minnich, and R. A. Chase. 1993. Mapping probability of fire occurrence in San Jacinto Mountains, California, USA. *Environmental Management* 17:129–140.

Christensen, N. L., J. K. Agee, P. F. Brussard, J. Hughes, D. H. Knight, G. W. Minshall, J. M. Peek, S. J. Pyne, F. J. Swanson, J. W. Thomas, S. Wells, S. E. Williams, and H. A. Wright. 1989. Interpreting the Yellowstone fires of 1988. *BioScience* 39:678–685.

Christensen, N. L., A. M. Bartuska, J. H. Brown, S. R. Carpenter, C. D'Antonio, R. Francis, J. F. Franklin, J. A. MacMahon, R. F. Noss, D. J. Parsons, C. H. Peterson, M. G. Turner, and R. G. Woodmansee. 1996. The scientific basis for ecosystem management. *Ecological Applications* 6:665–691.

Cirmo, C. P., and J .J. McDonnell. 1997. Linking the hydrologic and biogeochemical controls of nitrogen transport in near-stream zones of temperate forested catchments: a review. *Journal of Hydrology* 199:88–120.

Clark, J. S. 1988. Effect of climate change on fire regimes in northwestern Minnesota. *Nature* 334:233–235.

Clark, J. S. 1989. Ecological disturbance as a renewal process: theory and application to fire history. *Oikos* 56:17–30.

Clark, J. S. 1990. Fire and climate change during the last 750 years in northwestern Minnesota. *Ecological Monographs* 60:135–159.

Clark, J. S., and P. R. Royall. 1994. Pre-industrial particulate emissions and carbon sequestration from biomass burning in North America. *Biogeochemistry* 23:35–51.

Clark, W. C. 1980. *Witches, Floods, and Wonder Drugs.* Institute of Resource Ecology, R-22. University of British Columbia, Vancouver, British Columbia, Canada.

Clark, W. C., D. D. Jones, and C. S. Holling. 1978. Patches, movements, and population dynamics in ecological systems: a terrestrial perspective. *In* J. S. Steele, ed. *Spatial Pattern in Plankton Communities*, pp. 105–126. Plenum Press, New York, New York, USA.

Clark, W. C., D. D. Jones, and C. S. Holling. 1979. Lessons for ecological policy design: a case study of ecosystem management. *Ecological Modeling* 7:1–53.

Clements, F. E. 1915. *Plant Succession: An Analysis of the Development of Vegetation.* Carnegie Institute Publication No. 242, Washington, D.C. 512 p.

Cluis, D. A., D. Couillarg, and L. Potvin. 1979. A square grid transport model relating land use exports to nutrient loads in rivers. *Water Resources Research* 15:630–636.

Cody, M. L. 1983. The land birds. *In* T. J. Case and M. L. Cody, eds. *Island Biogeography in the Sea of Cortez*, pp. 210–245. University of California Press, Berkeley, California, USA.

Cohen, A. E., A. Gonzalez, J. H. Lawton, O. L. Petchey, D. Wildman, and J. E. Cohen. 1998. A novel experimental apparatus to study the impact of white noise and 1/f noise on animal populations. *Proceedings of the Royal Society of London Series B—Biological Sciences* 265:11–15.

Cohen, D., and S. A. Levin. 1991. Dispersal in patchy environments: the effects of temporal and spatial structure. *Theoretical Population Biology* 39:63–99.

Cole, J. J., C. L. Peierls, N. F. Caraco, and M. L. Pace. 1993. Nitrogen loading of rivers as a human-driven process. *In* M. J. McDonnell and S. T. A. Pickett, eds. *Humans as Components of Ecosystems*, pp. 141–157. Springer-Verlag, New York, New York, USA.

Cole, L. C. 1951. Population cycles and random oscillations. *Journal of Wildlife Management* 15:233–252.

Cole, L. C. 1954. Some features of random cycles. *Journal of Wildlife Management* 18:107–109.

Collins, S. L., A. K. Knapp, J. M. Briggs, J. M. Blair, and E. M. Steinhauer. 1998. Modulation of diversity by grazing and mowing in native tallgrass prairie. *Science* 280:745–747.

Comeleo, R. L., J. F. Paul, P. V. August, J. Copeland, C. Baker, S. S. Hale, and R. W. Latimer. 1996. Relationships between watershed stressors and sediment contamination in Chesapeake Bay estuaries. *Landscape Ecology* 11:307–319.

Comins, H. N., and D. W. E. Blatt. 1974. Pre-predation models in spatially heterogeneous environments. *Journal of Theoretical Biology* 48:75–83.

Connell, J. H., and R. O. Slatyer. 1977. Mechanisms of succession in natural communities and their role in community stability and organization. *American Naturalist* 111:1119–1144.

Connell, J. H., and W. P. Sousa. 1983. On the evidence needed to judge ecological stability or persistence. *American Naturalist* 121:789–824.

Cooper, J. R., J. W. Gilliam, R. B. Daniels, and W. P. Robarge. 1987. Riparian areas as filters for agricultural sediment. *Soil Science Society of America Journal* 51:416–420.

Cooper, S. R., and G. S. Brush. 1991. Long-term history of Chesapeake Bay anoxia. *Science*, 254:992–996.

Cooper, W. S. 1913. The climax forest of Isle Royale, Lake Superior, and its development. I. *Botanical Gazette* 55:1–44.

Cooperrider, A., L. R. Garrett, and N. T. Hobbs. 1999. Data collection, management, and inventory. *In* N. C. Johnson, A. J. Malk, W. T. Sexton, and R. Szaro, eds. *Ecological Stewardship: A Common Reference for Ecosystem Management*, pp. 604–627. Elsevier Science Limited, Oxford, UK.

Correll, D. L. 1997. Buffer zones and water quality protection: general principles. *In* N. E. Haycock, T. P. Burt, K. W. T. Goulding, and G. Pinay, eds. *Buffer Zones: Their Processes and Potential in Water Protection,* pp. 7–20. Quest Environmental, Harpenden, UK.

Correll, D. L., T. E. Jordan, and D. E. Weller. 1992. Nutrient flux in a landscape: effects of coastal land use and terrestrial community mosaic on nutrient transport to coastal waters. *Estuaries* 15:431–442.

Costa, J .E. 1975. Effects of agriculture on erosion and sedimentation in the Piedmont province, Maryland. *Geological Society of America Bulletin* 86: 1281–1286.

Costanza, R., and T. Maxwell. 1994. Resolution and predictability: an approach to the scaling problem. *Landscape Ecology* 9:47–57.

Costanza, R., F. H. Sklar, and M. L. White. 1990. Modeling coastal landscape dynamics:

process-based dynamic spatial ecosystem simulation can examine long-term natural changes and human impacts. *Bioscience* 40:91–107.

Costanza, R., D. Duplisea, and U. Kautsky. 1998. Ecological modeling and economic systems with STELLA—Introduction. *Ecological Modeling* 110:1–4.

Coughenour, M. B. 1991. Spatial components of plant–herbivore interactions in pastoral, ranching, and native ungulate ecosystems. *Journal of Range Management* 44:530–542.

Coulson, R. N., C. N. Lovelady, R. O. Flamm, S. L. Spradling, and M. C. Saunders. 1991. *In*: M. G. Turner and R. H. Gardner, eds. *Quantitative Methods in Landscape Ecology*, pp. 153–172. Springer-Verlag, New York.

Cressie, N. A. C. 1991. *Statistics for Spatial Data*. John Wiley & Sons, New York, New York, USA.

Criminale, W. O., and D. F. Winter. 1974. The stability of steady-state depth distributions of marine phytoplankton. *American Naturalist* 108:679–687.

Crist, T. O., and J. A. Wiens. 1994. Scale effects of vegetation on forager movement and seed harvesting by ants. *Oikos* 69:37–46.

Cronin, T. M., and C. E. Schneider. 1990. Climatic influences on species: evidence from the fossil record. *Trends in Ecology and Evolution* 5:275–279.

Crowley, T. J., and K. Y. Kim. 1994. Milankovitch forcing of the last interglacial sea level. *Science* 265:1566–1568.

Crutzen, P. J., and J. G. Goldhammer, editors. 1993. *Fire in the Environment: The Ecological, Atmospheric, and Climatic Importance of Vegetation Fires*. John Wiley & Sons, New York, New York, USA.

Cullinan, V. I., and J. M. Thomas. 1992. A comparison of quantitative methods for examining landscape pattern and scale. *Landscape Ecology* 7:211–227.

Culver, D. C. 1970. Analysis of simple cave communities. I. Caves as islands. *Evolution* 29:463–474.

Curtis, J. T. 1956. The modification of mid-latitude grasslands and forests by man. *In* W. L. Thomas, ed. *Man's Role in Changing the Face of the Earth*, pp. 721–736. University of Chicago Press, Chicago, Illinois, USA.

Curtis, J. T. 1959. *The Vegetation of Wisconsin*. University of Wisconsin Press, Madison, Wisconsin, USA.

Cwynar, L.C. 1987. Fire and the forest history of the North Cascade Range. *Ecology* 68:791–802.

da Silva, J. M., C. Uhl, and G. Murray. 1996. Plant succession, landscape management, and the ecology of frugivorous birds in abandoned Amazonian pastures. *Conservation Biology* 10:491–503.

Dale, V. H. 1991. The debris avalanche at Mount St. Helens: vegetation establishment in the ten years since the eruption. *National Geographic Research and Exploration* 7(3):328–341.

Dale, V. H., F. Southworth, R. V. O'Neill, A. Rosen, and R. Frohn. 1993. Simulating spatial patterns of land-use change in Rondonia, Brazil. *Lectures on Mathematics in the Life Sciences* 23:29–55.

Dale, V. H., R. V. O'Neill, F. Southworth, and M. Pedlowski. 1994a. Modeling effects of land management in the Brazilian Amazonian settlement of Rondonia. *Conservation Biology* 8:196–206.

Dale, V. H., S. M. Pearson, H. L. Offerman, and R. V. O'Neill. 1994b. Relating patterns of land-use change to faunal biodiversity in the Central Amazon. *Conservation Biology* 8:1027–1036.

Dale, V. H., A. E. Lugo, J. A. MacMahon, and S. T. A. Pickett. 1998. Ecosystem management in the context of large, infrequent disturbances. *Ecosystems* 1:546–557.

Dale, V. H., S. Brown, R. A. Haeuber, N. T. Hobbs, N. Huntly, R. J. Naiman, W. E. Riebsame, M. G. Turner, and T. J. Valone. 2000. Ecological principles and guidelines for managing the use of land: a report from the Ecological Society of America. *Ecological Applications* 10:639–670.

Danielson, B. J. 1991. Communities on a landscape: the influence of habitat heterogeneity on the interactions between species. *American Naturalist* 138:1105–1120.

Danielson, B. J. 1992. Habitat selection, interspecific interactions and landscape composition. *Evolutionary Ecology* 6:399–411.

Dannell, K., L. Edenius, and P. Lundberg. 1991. Herbivory and tree stand composition: moose patch use in winter. *Ecology* 72:1350–1357.

Darwen, P. J., and D. G. Green. 1996. Viability of populations in a landscape. *Ecological Modeling* 85:165–171.

Davis, M. B. 1983. Quaternary history of deciduous forests of eastern North America and Europe. *Annals Missouri Botanical Garden* 70:550–563.

Davis, S. M., and J. C. Ogden, editors. 1994. *Everglades: The Ecosystem and Its Restoration*. St. Lucie Press, Delray Beach, Florida, USA.

Day, T. A., and J. K. Detling. 1990. Grassland patch dynamics and herbivore grazing preference following urine deposition. *Ecology* 71:180–188.

DeAngelis, D. L., and J. C. Waterhouse. 1987. Equilibrium and nonequilibrium concepts in ecological models. *Ecological Monographs* 57:1–21.

DeAngelis, D. L., L. J. Gross, M. A. Huston, W. F. Wolff, D. M. Fleming, E. J. Comiskey, and S. M. Sylvester. 1998. Landscape modeling for Everglades ecosystem restoration. *Ecosystems* 1:64–75.

Delcourt, H. R. 1987. The impact of prehistoric agriculture and land occupation on natural vegetation. *Trends in Ecology and Evolution* 2:39–44.

Delcourt, H. R., and P. A. Delcourt. 1987. *Long-term Forest Dynamics of the Temperate Zone*. Springer-Verlag, New York, New York, USA.

Delcourt, H. R., and P. A. Delcourt. 1991. *Quaternary Ecology: A Paleoecological Perspective*. Chapman and Hall, New York, New York, USA.

Delcourt, H. R., and P. A. Delcourt. 1996. Presettlement landscape heterogeneity: evaluating grain of resolution using General Land Office Survey data. *Landscape Ecology* 11:363–381.

Delcourt, H. R., and W. F. Harris. 1980. Carbon budget of the southeastern US biota: analysis of historical change in trend from source to sink. *Science* 210:321–323.

Delcourt, H. R., P. A. Delcourt, and T. Webb. 1983. Dynamic plant ecology: the spectrum of vegetational change in space and time. *Quaternary Science Review* 1:153–175.

Del Moral, R., and L. C. Bliss. 1993. Mechanisms of primary succession: insights resulting from the eruption of Mount St. Helens. *Advances in Ecological Research* 24:1–66.

Delong, S. C., and D. Tanner. 1996. Managing the pattern of forest harvest: lessons from wildfire. *Biodiversity and Conservation* 5:1191–1205.

Denevan, W. M. 1992. The pristine myth: the landscape of the Americas in 1492. *Annals of the Association of American Geographers* 82:369–385.

Denslow, J. D. 1980a. Gap partitioning among tropical trees. *Biotropica* 12 (Supplement):47–55.

Denslow, J. D. 1980b. Patterns of plant species diversity during succession under different disturbance regimes. *Oecologia* 46:18–21.

Denslow, J. S. 1987. Tropical rain forest gaps and tree species diversity. *Annual Review of Ecology and Systematics* 18:431–451.

Detenbeck, N., C. A. Johnston, and G. Niemi. 1993. Wetland effects on lake water quality in the Minneapolis/St. Paul metropolitan area. *Landscape Ecology* 8:39–61.

Deutsch, C. V., and A. G. Journel. 1992. *GSLIB: Geostatistical Software Library and User's Guide.* Oxford University Press, New York, New York, USA.

Diamond, J. M. 1972. Biogeographic kinetics: estimation of relaxation times for avifauna of southwestern Pacific Islands. *Proceedings National Academy of Sciences USA* 69: 3199–3203.

Diamond, J. M., and R. M. May. 1976. Island biogeography and the design of natural reserves. *In* J. Diamond and T. J. Case, eds. *Theoretical Ecology: Principles and Applications*, pp. 163–186. Harper & Row, New York, New York, USA.

Diamond., J. M., and E. Mayr. 1976. Species–area relations for birds of the Solomon archipelago. *Proceedings of the National Academy of Sciences, USA* 73:262–266.

Dias, P. C. 1996. Sources and sinks in population biology. *Tree* 11:326–329.

Dixon, R. K., S. Brown, R. A. Houghton, A. M. Solomon, M. C. Trexler, and J. Wisniewski. 1994. Carbon pools and flux of global forest ecosystems. *Science* 263: 185–190.

Dobzhansky, T., and S. Wright. 1947. Genetics of natural populations. XV. Rate of diffusion of a mutant gene through a population of *Drosophila pseudoobscura. Genetics* 31:303–324.

Dubois, D. M. 1975. A model of patchiness for prey–predator plankton populations. *Ecological Modeling* 1:67–80.

Duerksen, C. J., D. L. Elliott, N. T. Hobbs, E. Johnson, and J. R. Miller. 1997. *Habitat Protection Planning: Where the Wild Things Are.* American Planning Association, Planning Advisory Service, Report number 470/471.

Duinker, P. N. 1999. Landscape ecology in forest management and policy: are reality and expectations diverging? *5th World Congress, International Association for Landscape Ecology, Abstracts Volume* 1:38–39.

Dunn, C. P., D. M. Sharpe, G. R. Guntenspergen, F. Stearns, and Z. Yang. 1991. Methods for analyzing temporal changes in landscape pattern. *In* M. G. Turner and R. H. Gardner, eds. *Quantitative Methods in Landscape Ecology*, pp. 173–198. Springer-Verlag, New York, New York, USA.

Dunning, J. B., B. J. Danielson, and H. R. Pulliam. 1992. Ecological processes that affect populations in complex landscapes. *Oikos* 65:169–175.

Dunning, J. B., D. J. Steware, B. J. Danielson, B. R. Noon, T. L. Root, R. H. Lamberson, and E. E. Stevens. 1995. Spatially explicit population models: current forms and future uses. *Ecological Applications* 5:3–11.

Dunwiddie, P.W. 1986. A 6000-year record of forest history on Mount Rainier, Washington. *Ecology* 67:58–68.

Durrett, R., and S. A. Levin. 1994. Stochastic spatial models: a user's guide to ecological applications. *Philosophical Transactions of the Royal Society of London B* 343:329–350.

Dyer, J. M., and P. R. Baird. 1997. Wind disturbance in remnant forest stands along the prairie–forest ecotone, Minnesota, USA. *Plant Ecology* 129:121–134.

Dyer, M. I., D. L. DeAngelis, and W. M. Post. 1986. A model of herbivore feedback on plant productivity. *Mathematical Biosciences* 79:171–184.

Edelstein-Keshet, L. 1986. *Mathematical Models in Biology.* Random House, New York, New York, USA.

Egerton, F. N. 1973. Changing concepts of the balance of nature. *Quarterly Review of Biology* 48:322–350.

Ehrlich, P. R., D. D. Murphy, M. C. Singer, C. B. Sherwood, R. R. White, and I. L. Brown. 1980. Extinction, reduction, stability and increase: the responses of the checkerspot butterfly (Euphydryas) populations to the California drought. *Oecologia* 46:101–105.

Elliott, N. C., R. W. Kieckhefer, J. H. Lee, and B. W. French. 1999. Influence of within-field and landscape factors on aphid predator populations in wheat. *Landscape Ecology* 14:239–252.

Emanuel, W. R. 1996. Modeling carbon cycling on disturbed landscapes. *Ecological Modelling* 89:1–12.

Emanuel, W. R., G. G. Killough, W. M. Post, and H. H. Shugart. 1984. Modeling terrestrial ecosystems in the global carbon cycle with shifts in carbon storage capacity by land-use change. *Ecology* 65:970–983.

Epperson, B. K. 1994. Spatial and space–time correlations in systems of subpopulations with stochastic migration. *Theoretical Population Biology* 46:160–197.

Epstein, H. E., I. C. Burke, and A. R. Mosier. 1998. Plant effects on spatial and temporal patterns of nitrogen cycling in shortgrass steppe. *Ecosystems* 1:374–385.

Estades, C. F., and S. A. Temple. 1999. Deciduous-forest bird communities in a fragmented landscape dominated by exotic pine plantations. *Ecological Applications* 9:573–585.

Fabos, J. G. 1985. *Land-use Planning: From Global to Local Challenge.* Chapman and Hall, New York, New York, USA.

Fagan, W. E., R. S. Cantrell, and C. Cosner. 1999. How habitat edges change species interactions. *American Naturalist* 153:165–182.

Fahrig, L. 1997. Relative effects of habitat loss and fragmentation on population extinction. *Journal of Wildlife Management* 61:603–610.

Fahrig, L., and G. Merriam. 1994. Conservation of fragmented populations. *Conservation Biology* 8:50–59.

Fahse, L., C. Wissel, and V. Grimm. 1998. Reconciling classical and individual-based approaches in theoretical population ecology: a protocol for extracting population parameters from individual-based models. *American Naturalist* 152:838–852.

Fall, A., and J. Fall. 1996. *A Hierarchical Organization of Landscape Models.* Simon Fraser University CSS/LCCR Technical Report TR96–01.

Farina, A. 1998. *Principles and Methods in Landscape Ecology.* Chapman and Hall, London, UK.

Fastie, C. L. 1995. Causes and ecosystem consequences of multiple pathways of primary succession at Glacier Bay, Alaska. *Ecology* 76:1899–1916.

Feder, J. 1988. *Fractals*. Plenum Press, New York, New York, USA.

FEMAT (Forest Ecosystem Management Assessment Team). 1993. *Forest Ecosystem Management: An Ecological, Economic, and Social Assessment*. Report of the Forest Ecosystem Management Assessment Team, July 1993. U. S. Government Printing Office: Washington, DC.

Finegan, B. 1984. Forest succession. *Nature* 312:109–114.

Fisher, R. A. 1935. *The Design of Experiments*. Oliver and Boyd, London, UK.

Fisher, R. A. 1937. The wave of advance of advantageous genes. *Annals of Eugenics (London)* 7:355–369.

Fitz, H. C., E. B. DeBellevue, R. Costanza, R. Boumans, T. Maxwell, L. Wainger, and F. H. Sklar. 1996. Development of a general ecosystem model for a range of scales and ecosystems. *Ecological Modeling* 88:263–295.

Flamm, R. O., and M. G. Turner. 1994. Alternative model formulations of a stochastic model of landscape change. *Landscape Ecology* 9:37–46.

Flather, C. H., and J. R. Sauer. 1996. Using landscape ecology to test hypotheses about large-scale abundance patterns in migratory birds. *Ecology* 77:28–35.

Fleming, D. M., W. F. Wolff, and D. L. DeAngelis. 1994. Importance of landscape heterogeneity to wood storks in Florida Everglades. *Environmental Management* 18:743–757.

Fonseca, M. S., S. S. Bell, M. O. Hall, and D. L. Meyer. 1996. *Evidence for Threshold Effects of Seagrass Landscape Structure on Associated Fauna*. 24th Annual Benthic Ecology Meeting, Columbia, South Carolina, USA.

Forbes, S. A. 1887. The lake as a microcosm. *Bulletin Science Association of Peoria, Illinois*, 1887, pp. 77–87.

Forel, F. A. 1892. *Lac Leman: Monographie Limnologique*. Lausanne, Rouge, France.

Forester, J. W. 1969. *Urban Dynamics*. Massachusetts Institute of Technology Press, Cambridge, Massachusetts, USA.

Forman, R. T. T. 1983. An ecology of the landscape. *BioScience* 33:535.

Forman, R. T. T. 1990. The beginnings of landscape ecology in America. *In* I. S. Zonneveld and R. T. T. Forman, eds. *Changing Landscapes: An Ecological Perspective*, pp. 35–41. Springer-Verlag, New York, New York, USA.

Forman, R. T. T. 1995. *Land Mosaics*. Cambridge University Press, Cambridge, UK.

Forman, R. T. T., and J. Baudry. 1984. Hedgerows and hedgerow networks in landscape ecology. *Environmental Management* 8:495–510.

Forman, R. T. T., and M. Godron. 1981. Patches and structural components for a landscape ecology. *Bioscience* 31:733–740.

Forman, R. T. T., and M. Godron. 1986. *Landscape Ecology*. John Wiley & Sons, New York, New York, USA.

Foster, D. R. 1983. Fire history and landscape patterns in southeastern Labrador. *Canadian Journal of Botany* 61:2459–2471.

Foster, D. R. 1988a. Disturbance history, community organization and vegetation dynamics of the old-growth Pisgah Forest, southwestern New Hampshire, USA. *Journal of Ecology* 76:105–134.

Foster, D. R. 1988b. Species and stand response to catastrophic wind in central New England, USA. *Journal of Ecology* 76:135–151.

Foster, D. R. 1992. Land-use history (1730–1990) and vegetation dynamics in central New England, USA. *Journal of Ecology* 80:753–772.

Foster, D. R., and E. R. Boose. 1992. Patterns of forest damage resulting from catastrophic wind in central New England, USA. *Journal of Ecology* 80:79–98.

Foster, D. R., and G. A. King. 1986. Vegetation pattern and diversity in S. E. Labrador, Canada: *Betula papyrifera* (birch) forest development in relation to fire history and physiography. *Journal of Ecology* 74:465–483.

Foster, D. R., D. H. Knight, and J. F. Franklin. 1998. Landscape patterns and legacies resulting from large infrequent forest disturbances. *Ecosystems* 1:497–510.

Foster, J., and M. S. Gaines. 1991. The effects of a successional habitat mosaic on a small mammal community. *Ecology* 72:1358–1373.

Fotheringham, S., and P. Rogerson. 1994. *Spatial Analysis and GIS*. Taylor and Francis, Bristol, Pennsylvania, USA.

Franklin, J. F. 1993. Preserving biodiversity: species, ecosystems, or landscapes? *Ecological Applications* 3:202–205.

Franklin, J. F,. and R. T. T. Forman. 1987. Creating landscape patterns by forest cutting: ecological consequences and principles. *Landscape Ecology* 1:5–18.

Freeman, W., and J. Fox. 1995. ALAWAT: A spatially allocated watershed model for approximating stream, sediment and pollutant flows in Hawaii, USA. *Environmental Management* 19:567–577.

Freemark, K. E., and H. G. Merriam. 1986. Importance of area and habitat heterogeneity to bird assemblages in temperate forest fragments. *Biological Conservation* 36:115–141.

Frelich, L. E., and C. G. Lorimer. 1991. Natural disturbance regimes in hemlock–hardwood forests of the Upper Great Lakes region. *Ecological Monographs* 61:145–164.

Fries, C., M. Carlsson, B. Dahlin, T. Lamas, and O. Sallnas. 1998. A review of conceptual landscape planning models for multiobjective forestry in Sweden. *Canadian Journal of Forest Research* 28:159–167.

Fryer, G. I., and E. A. Johnson. 1988. Reconstructing fire behavior and effects in a subalpine forest. *Journal of Applied Ecology* 25:1063–1072.

Galipeau, C., D. Kneeshaw, and Y. Bergeron. 1997. White spruce and balsam fir colonization of a site in the southeastern boreal forest as observed 68 years after fire. *Canadian Journal of Forest Research* 27:139–147.

Gardner, R. H. 1999. RULE: A program for the generation of random maps and the analysis of spatial patterns. *In* J. M. Klopatek and R. H. Gardner, eds. *Landscape Ecological Analysis: Issues and Applications*, pp. 280–303. Springer-Verlag, New York, New York, USA.

Gardner, R. H. 1998. Pattern, process, and the analysis of spatial scales. *In* E. L. Peterson and V. T. Parker, eds. *Ecological Scale, Theory and Applications*, pp. 17–34. Columbia University Press, New York, New York, USA.

Gardner, R. H., and R. V. O'Neill. 1991. Pattern, process and predictability: the use of neutral models for landscape analysis. *In* M. G. Turner and R. H. Gardner, eds. *Quan-*

titative Methods in Landscape Ecology, pp. 289–308. Springer-Verlag, New York, New York, USA.

Gardner, R. H., W. G. Cale, and R. V. O'Neill. 1982. Robust analysis of aggregation error. *Ecology* 63:1771–1779.

Gardner, R. H., R. V. O'Neill , J. B. Mankin, and J. H. Carney. 1981. A comparison of sensitivity analysis and error analysis based on a stream ecosystem model. *Ecological Modeling* 12:177–194.

Gardner, R. H., B. T. Milne, M. G. Turner, and R. V. O'Neill. 1987. Neutral models for the analysis of broad-scale landscape patterns. *Landscape Ecology* 1:19–28.

Gardner, R. H., and J. R. Trabalka. 1985. *Methods of Uncertainty Analysis for a Global Carbon Dioxide Model.* Department of Energy Technical Report. DOE/OR/21400-4. Washington, DC.

Gardner, R. H., R. V. O'Neill, M. G. Turner, and V. H. Dale. 1989. Quantifying scale-dependent effects of animal movement with simple percolation models. *Landscape Ecology* 3:217–227.

Gardner, R. H., V. H. Dale, and R. V. O'Neill. 1990. Error propagation and uncertainty in process modeling. *In* R. K. Dixon, R. S. Meldahl, G. A. Ruark, and W. G. Warren, eds. *Process Modeling of Forest Growth Responses to Environmental Stress,* pp. 208–219. Timber Press, Portland, Oregon, USA.

Gardner, R. H., V. H. Dale, R. V. O'Neill, and M. G. Turner. 1992a. A percolation model of ecological flows. *In* A. J. Hansen and F. di Castri, eds. *Landscape Boundaries: Consequences for Biotic Diversity and Ecological Flows*, pp. 259–269. Springer-Verlag, New York, New York, USA.

Gardner, R. H., M. G. Turner, R. V. O'Neill, and S. Lavorel. 1992b. Simulation of the scale-dependent effects of landscape boundaries on species persistence and dispersal. *In* M. M. Holland, P. G. Risser, and R. J. Naiman, eds. *The Role of Landscape Boundaries in the Management and Restoration of Changing Environments*, pp. 76–89. Chapman and Hall, New York, New York, USA.

Gardner, R. H., R. V. O'Neill and M. G. Turner. 1993a. Ecological implications of landscape fragmentation. *In* S. T. A. Pickett and M. J. McDonnell, eds. *Humans as Components of Ecosystems*, pp. 208–226. Springer-Verlag, New York, New York, USA.

Gardner, R. H., A. W. King, and V. H. Dale. 1993b. Interactions between forest harvesting, landscape heterogeneity, and species persistence. *In* D. C. LeMaster and R. A. Sedjo, eds. *Modeling Sustainable Forest Ecosystems,* American Forests, Washington, DC, pp. 65–75.

Gardner, R. H., W. W. Hargrove, M. G. Turner, and W. H. Romme. 1996. Climate change, disturbances and landscape dynamics. *In* B. Walker and W. Steffen, eds. *Global Change and Terrestrial Ecosystems*, pp. 149–172. Cambridge University Press, Cambridge, UK.

Garrabou, J., J. Riera, and M. Zabala. 1998. Landscape pattern indices applied to Mediterranean subtidal rocky benthic communities. *Landscape Ecology* 13:225–247.

Garten, C. T., Jr., M. A. Huston, and C. A. Thoms. 1994. Topographic variation of soil nitrogen dynamics at Walker Branch watershed, Tennessee. *Forest Science* 40:497–512.

Gause, G. F. 1934. *The Struggle for Existence.* Williams and Wilkin, Baltimore, Maryland, USA.

Geier, T. W., J. A. Perry, and L. Queen. 1994. Improving lake riparian source area management using surface and subsurface runoff indices. *Environmental Management* 18:569–586.

Gergel, S. E., and M. G. Turner, editors. 2000. *Learning Landscape Ecology*. Springer-Verlag, New York, New York, USA.

Gergel, S. E., M. G. Turner, and T. K. Kratz. 1999. Scale-dependent landscape effects on north temperate lakes and rivers. *Ecological Applications* 9:1377–1390.

Gilbert, F. S. 1980. The equilibrium theory of island biogeography: fact or fiction? *Journal of Biogeography* 7:209–235.

Gillman, M., and R. Hails. 1997. *An Introduction to Ecological Modeling. Putting Practice into Theory*. Blackwell Scientific, Oxford, UK.

Gilliland, M. W., and W. Baxter-Potter. 1987. A geographic information system to predict nonpoint source pollution potential. *Water Resources Bulletin* 23:281–291.

Gilruth, P. T., S. E. Marsh, and R. Itami. 1995. A dynamic spatial model of shifting cultivation in the highlands of Guinea West Africa. *Ecological Modelling* 79:179–197.

Givnish, T. J. 1981. Serotiny, geography, and fire in the Pine Barrens of New Jersey. *Evolution* 35:101–123.

Glenn, S. M., and S. L. Collins. 1992. Disturbances in tallgrass prairie—local and regional effects on community heterogeneity. *Landscape Ecology* 7:243–251.

Glenn, S. M., and S. L. Collins. 1993. Experimental-analysis of patch dynamics in tallgrass prairie plant-communities. *Journal of Vegetation Science* 4:157–162.

Glenn-Lewin, D. C., and E. van der Maarl. 1992. Patterns and processes of vegetation dynamics. *In* D. C. Glenn-Lewin, R. K. Peet, and T. T. Veblen, eds. *Plant Succession*, pp. 11–59. Chapman and Hall, New York, New York, USA.

Glenn-Lewin, D. C., R. K. Peet, and T. T. Veblen, editors. 1992. *Plant Succession*. Chapman and Hall, New York, New York, USA.

Glitzenstein, J. S., W. J. Platt, and D. R. Strong. 1995. The effects of fire regime and habitat on tree dynamics in North Florida longleaf pine savannas. *Ecological Monographs* 65:441–476.

Gluck, M. J., and R. S. Rempel. 1996. Structural characteristics of post-wildfire and clearcut landscapes. *Environmental Monitoring and Assessment* 39:435–450.

Golley, F. B. 1993. *A History of the Ecosystem Concept in Ecology: More Than the Sum of Its Parts*. Yale University Press, New Haven, Connecticut, USA.

Goodman, D. 1987. The demography of chance extinction. *In* M. E. Soule, ed. *Viable Populations for Conservation*, pp. 11–34. Cambridge University Press, Cambridge, UK.

Goward, S. N., C. J. Tucker, and D. G. Dye. 1985. North American vegetation patterns observed with the NOAA-7 advanced very high resolution radiometer. *Vegetatio* 64:3014.

Goward, S. N., A. Kerber, D. G. Dye, and V. Kalb. 1987. Comparison of North and South American biomes from AVHRR observations. *Geocarto* 2:27–40.

Graham, R. L., M. G. Turner, and V. H. Dale. 1990. How increasing CO_2 and climate change affect forests. *BioScience* 40:575–587.

Graham, R. L., C. T. Hunsaker, R. V. O'Neill, and B. L. Jackson. 1991. Ecological risk assessment at the regional scale. *Ecological Applications* 1:196–206.

Grant, W. E., E. K. Pedersen, and S. L. Marin. 1997. *Ecology and Natural Resource Management. Systems Analysis and Simulation.* John Wiley & Sons, Inc., New York, New York, USA.

Green, D. G. 1982. Fire and stability in the postglacial forests of southwest Nova Scotia. *Journal of Biogeography* 9:29–40.

Green, D. G. 1989. Simulated effects of fire, dispersal and spatial pattern on competition within forest mosaics. *Vegetatio* 82:139–153.

Green, D. G. 1994. Simulation studies of connectivity and complexity in landscapes and ecosystems. *Pacific Conservation Biology* 3:194–200.

Greig-Smith, P. 1952. The use of random and contiguous quadrants in the study of the structure of plant communities. *Annals of Botany* 16:293–316.

Greig-Smith, P. 1979. Pattern in vegetation. *Journal of Ecology* 67:755–779.

Greig-Smith, P. 1983. *Quantitative Plant Ecology*. University of California Press, Berkeley, California, USA.

Grimm, N. B. 1995. Why link species and ecosystems? *In* C. G. Jones and J. H. Lawton, eds. *Linking Species and Ecosystems*, pp. 5–15. Chapman and Hall, New York, New York, USA.

Groffman, P. M., and J. M. Tiedje. 1989. Denitrification in north temperate forest soils: spatial and temporal patterns at the landscape and seasonal scales. *Soil Biology and Biochemistry* 21:613–620.

Groffman, P. M., J. M. Tiedje, D. L. Mokma, and S. Simkins. 1992. Regional scale analysis of denitrification in north temperate forest soils. *Landscape Ecology* 7:45–53.

Groffman, P. M., D. R. Zak, S. Christensen, A. Mosier, and J. M. Tiedje. 1993. Early spring nitrogen dynamics in a temperate forest landscape. *Ecology* 74:1579–1585.

Gross, K. L., K. S. Pregitzer, and A. J. Burton. 1995. Spatial variation in nitrogen availability in three successional plant communities. *Journal of Ecology* 83:357–367.

Grumbacher, S. K., K. M. McEwen, D. A. Halverson, D. T. Jacobs, and J. Lindner. 1993. Self-organized criticality—an experiment with sandpiles. *American Journal of Physics* 61:329–335.

Grumbine, R. E. 1994. What is ecosystem management? *Conservation Biology* 8:27–38.

Gulve, P. S. 1994. Distribution and extinction patterns within a northern metapopulation of the pond frog, *Rana lessonae*. *Ecology* 75:1357–1367.

Gustafson, E. J. 1998. Quantifying landscape spatial pattern: What is the state of the art? *Ecosystems* 1:143–156.

Gustafson, E. J., and T. R. Crow. 1994. Modeling the effects of forest harvesting on landscape structure and the spatial distribution of cowbird brood parasitism. *Landscape Ecology* 9:237–248.

Gustafson, E. J., and T. R. Crow. 1996. Simulating the effects of alternative forest management strategies on landscape structure. *Journal of Environmental Management* 46:77–94.

Gustafson, E. J., and R. H. Gardner. 1996. The effect of landscape heterogeneity on the probability of patch colonization. *Ecology* 77:94–107.

Gustafson, E. J., and G. R. Parker. 1992. Relationships between landcover proportion and indices of landscape spatial pattern. *Landscape Ecology*, 7:101–110.

Gutzwiller, K. J., and S. H. Anderson. 1992. Interception of moving organisms: influences of patch shape, size, and orientation on community structure. *Landscape Ecology* 6:293–303.

Habeck, J. R., and R. W. Mutch. 1973. Fire-dependent forests in the northern Rockies. *Quaternary Research* 3:408–424.

Haber, W. 1990. Using landscape ecology in planning and management. *In* I. S. Zonneveld and R. T. T. Forman, eds. *Changing Landscapes: An Ecological Perspective*, pp. 217–232. Springer-Verlag, New York, New York, USA.

Haefner, J. W. 1988a. Niche shifts in Greater Antillean *Anolis* communities: effects of niche metric and biological resolution on null model tests. *Oecologia* 77:107–117.

Haefner, J. W. 1988b. Assembly rules for Greater Antillean *Anolis* lizards: competition and random models compared. *Oecologia* 74:551–565.

Hagget, P. 1963. Scale components in geographical problems. *In* R. J. Chorley and P. Haggett, eds. *Frontiers in Geographical Teaching*, pp. 164–185. Methuen & Company Limited, London, UK.

Haines-Young, R., and M. Chopping. 1996. Quantifying landscape structure: a review of landscape indices and their application to forested landscapes. *Progress in Physical Geography* 20:418–445.

Haines-Young, R., D. R. Green, and S. Cousins, editors. 1993. *Landscape Ecology and Geographic Information Systems*. Taylor & Francis, Bristol, Pennsylvania, USA.

Halpern, C. B. 1988. Early successional pathways and the resistance and resilience of forest communities. *Ecology* 69:1703–1715.

Halpern, C. B. 1989. Early successional patterns of forest species: interactions of life history traits and disturbance. *Ecology* 70:704–720.

Halpin, P. N. 1997. Global climate change and natural-area protection: management responses and research directions. *Ecological Applications* 7:828–843.

Hamilton, H. R., S. E. Goldstone, J. W. Milliman, A. L. Pugh, E. B. Roberts, and Z. Zeller. 1969. *Systems Simulation for Regional Analysis: An Application to River-basin Planning*. Massachusetts Institute of Technology Press, Cambridge, Massachusetts, USA.

Hamilton, W. D., and R. M. May. 1977. Dispersal in stable habitats. *Nature* 269:578–581.

Hansen, A. J., S. L. Garman, B. Marks, and D. L. Urban. 1993. An approach for managing vertebrate diversity across multiple-use landscapes. *Ecological Applications* 3:481–496.

Hansen, A. J., W. C. McComb, R. Vega, M. G. Raphael, and M. Hunter. 1995. Bird habitat relationships in natural and managed forests in the West Cascades of Oregon. *Ecological Applications* 5:555–569.

Hanski, I. 1981. Coexistence of competitors in patchy environments with and without predation. *Oikos* 37:306–312.

Hanski, I. 1983. Coexistence of competitors in patchy environments. *Ecology* 64:493–500.

Hanski, I. 1989. Metapopulation dynamics: does it help to have more of the same? *Trends in Ecology and Evolution* 4:113–114.

Hanski, I. 1998. Metapopulation dynamics. *Nature* 396:41–49.

Hanski, I. A., and M. E. Gilpin, editors. 1997. *Metapopulation Biology*. Academic Press, New York, New York, USA.

Hanski, I., and D. Simberloff. 1997. The metapopulation approach, its history, conceptual domain, and application to conservation. *In* I. A. Hanski and M. E. Gilpin, eds. *Metapopulation Biology*, pp. 5–26. Academic Press, New York, New York, USA.

Hansson, L., L. Fahrig, and G. Merriam, editors. 1995. *Mosaic Landscapes and Ecological Processes*. Chapman and Hall, London, UK.

Harding, J. S., E. F. Benfield, P. V. Bolstad, G. S. Helfman, and E. B. D. Jones, III. 1998. Stream biodiversity: the ghost of land use past. *Proceedings of the National Academy of Sciences* 95:14843–14847.

Hardt, R. A., and R. T. T. Forman. 1989. Boundary form effects on woody colonization of reclaimed surface mines. *Ecology* 70:1252–1260.

Hargrove, W. W., and J. Pickering. 1992. Pseudoreplication: a sine qua non for regional ecology. *Landscape Ecology*, 6: 251–258.

Harkness, R. D., and N. G. Maroudas. 1985. Central place foraging by an ant (*Cataglyphis bicolor* Fab.): a model of searching. *Animal Behavior* 33:916–928.

Harrington, R. A., J. H. Fownes, P. G. Scowcroft, and C. S. Vann. 1997. Impact of Hurricane Iniki on native Hawaiian Acacia koa forests: damage and two-year recovery. *Journal of Tropical Ecology* 13:539–558.

Harris, G. P. 1998. Predictive models in spatially and temporally variable freshwater systems. *Australian Journal of Ecology* 23:80–94.

Harris, L. D. 1984. *The Fragmented Forest*. University of Chicago Press, Chicago, Illinois, USA.

Harrison, G. W. 1979. Stability under environmental stress: resistance, resilience, persistence and variability. *American Naturalist* 113:659–669.

Harrison, S. 1994. Metapopulations and conservation. *In* P. J. Edwards, R. M. May, and N. R. Webb, eds. *Large-scale Ecology and Conservation Biology*, pp. 111–128. Blackwell Scientific Publications, Oxford, UK.

Harton, J. C., and N. O. E. Smart. 1984. The effect of long-term exclusion of large herbivores on soil nutrient status in Murchison Falls National Park, Uganda. *African Journal of Ecology* 22:20–23.

Hartshorn, G. S. 1980. Neotropical forest dynamics. *Biotropica* 12 (Supplement): 23–30.

Hartway, C., M. Ruckelshaus, and P. Kareiva. 1998. The challenge of applying spatially explicit models to a world of sparse and messy data. *In* J. Bascompte and R. V. Sole, eds. *Modeling Spatiotemporal Dynamics in Ecology*, pp. 215–223. Springer-Verlag, Berlin, Germany.

Hastings, A. 1977. Spatial heterogeneity and the stability of predator–prey systems. *Theoretical Population Biology* 12:37–48.

Hastings, A. 1996a. Models of spatial spread: is the theory complete? *Ecology* 77: 1675–1679.

Hastings, A. 1996b. *Population Biology: Concepts and Models*. Springer-Verlag, New York, New York, USA.

Hastings, H. M., and G. Sugihara. 1993. *Fractals: A User's Guide for the Natural Sciences*. Oxford University Press, New York, New York, USA.

Hayden, I. J., J. Faaborg, and R. L. Clawson. 1985. Estimates of minimal area requirements for Missouri forest birds. *Transactions of the Missouri Academy of Science* 19:11–22.

He, H. S., and D. J. Mladenoff. 1999. Spatially explicit and stochastic simulation of forest-landscape fire disturbance and succession. *Ecology* 80:81–99.

He, H. S., D. J. Mladenoff, and T. R. Crow. 1999. Linking an ecosystem model and a landscape model to study forest species response to climate warming. *Ecological Modeling* 114:213–233.

Heck, W. W., A. S. Heagle, and D. S. Shriner. 1986. Effects on vegetation: native, crops, forest. *In* A. Stern, ed. *Air Pollution*, Volume 6, pp. 247–350. Academic Press, New York, New York, USA.

Heinselman, M. L. 1973. Fire in the virgin forests of the Boundary Waters Canoe Area, Minnesota. *Quaternary Research* 3:329–382.

Heinselman, M. L. 1996. *The Boundary Waters Wilderness Ecosystem*. University of Minnesota Press, Minneapolis, Minnesota, USA.

Helliwell, D. R. 1976. The effects of size and isolation on the conservation value of wooded sites in Britain. *Journal of Biogeography* 3:407–416.

Hemens, G. C. 1970. Analysis and simulation of urban activity patterns. *Socio-economic Planning Science* 4:53–66.

Hemstrom, M. A., and J. F. Franklin. 1982. Fire and other disturbances of the forests in Mount Rainier National Park. *Quaternary Research* 18:32–51.

Henderson, M. T., G. Merriam, and J. Wegner. 1985. Patchy environments and species survival: chipmunks in an agricultural mosaic. *Biological Conservation* 31:95–105.

Henderson-Sellers, A., M. F. Wilson, and G. Thomas. 1985. The effect of spatial resolution on archives of land cover type. *Climate Change* 7:391–402.

Henebry, G. M. 1995. Spatial model error analysis using autocorrelation indices. *Ecological Modeling* 82:75–91.

Herbert, J., and B. J. Stevens. 1960. A model for the distribution of residential activities in urban areas. *Journal of Regional Science* 2:21–36.

Hett, J. M. 1971. *Land Use Changes in East Tennessee and a Simulation Model Which Describes These Changes for Three Counties*. ORNL/IBP-71/8. Oak Ridge National Laboratory, Oak Ridge, Tennessee, USA.

Heuvelink, G. B. M. 1998. Uncertainty analysis in environmental modeling under a change of spatial scale. *Nutrient Cycling in Agroecosystems* 50:255–264.

Higgins, S. I., and D. M. Richardson. 1999. Predicting plant migration rates in a changing world: the role of long-distance dispersal. *American Naturalist* 153:464–475.

Hilborn, R. 1975. The effect of spatial heterogeneity on the persistence of predator–prey interactions. *Theoretical Population Biology* 8:346–355.

Hobbs, R. J. 1992. The role of corridors in conservation: solution or bandwagon? *Trends in Ecology and Evolution* 7:389–392.

Hobbs, R. J. 1993. Effects of landscape fragmentation on ecosystem processes in the western-Australian wheat-belt. *Biological Conservation* 64:193–201.

Hobbs, R. J., and D. A. Saunders, eds. 1992. *Reintegrating Fragmented Landscapes*. Springer-Verlag, New York, New York, USA.

Hobbs, R. J., D. A. Saunders, L. A. Lobry de Bruyn, and A. R. Main. 1993. Changes in biota. *In* R. J. Hobbs and D. A. Saunders, eds. *Reintegrating Fragmented Landscapes*, pp. 65–106. Springer-Verlag, New York, New York, USA.

Hogeweg, P. 1988. Cellular automata as a paradigm for ecological modeling. *Applications in Mathematics and Computation* 27:81–100.

Hohn, M. E., A. M. Liebhold, and L. S. Gribko. 1993. Geostatistical model for forecasting spatial dynamics of defoliation caused by the gypsy moth (Lepidoptera: Lymantriidae). *Environmental Entomology* 22:1066–1075.

Holland, E. A., W. J. Parton, J. K. Detling, and D. L. Coppock. 1992. Physiological responses of plant populations to herbivory and other consequences of ecosystem nutrient flow. *American Naturalist* 140:685–706.

Holling, C. S. 1978. *Adaptive Environmental Assessment and Management*. John Wiley & Sons, New York, New York, USA.

Holling, C. S. 1992. Cross-scale morphology, geometry and dynamics of ecosystems. *Ecological Monographs* 62:447–502.

Holmes, E. E., M. A. Lewis, J. E. Banks, and R. R. Veit. 1994. Partial differential equations in ecology: spatial interactions and population dynamics. *Ecology* 75:17–29.

Holmes, R. M., J. B. Jones, Jr., S. G. Fisher, and N. B. Grimm. 1996. Denitrification in a nitrogen-limited stream ecosystem. *Biogeochemistry* 33:125–146.

Host, G. E., K. S. Pregitzer, C. W. Ramm, J. B. Hart, and D. T. Cleland. 1987. Landform-mediated differences in successional pathways among upland forest ecosystems in northwestern Lower Michigan. *Forest Science* 33:445–457.

Host, G. E., K. S. Pregitzer, C. W. Ramm, D. P. Lusch, and D. T. Cleland. 1988. Variation in overstory biomass among glacial landforms and ecological land units in northwestern Lower Michigan. *Canadian Journal of Forest Research* 18:659–668.

Houghton, R. A. 1995. Land-use change and the carbon cycle. *Global Change Biology* 1:275–287.

Huffaker, C. B. 1958. Experimental studies on predation: dispersion factors and predator–prey oscillations. *Hilgardia* 27:343–383.

Huffaker, C. B., K. P. Shea, and S. G. Herman. 1963. Experimental studies on predation: complex dispersion and levels of food in an acarine predator–prey interaction. *Hilgardia* 34:305–330.

Hunsaker, C. T., and D. A. Levine. 1995. Hierarchical approaches to the study of water quality in rivers. *Bioscience* 45:193–203.

Hunsaker, C. T., S. W. Christensen, J. J. Beauchamp, R. J. Olson, R. S. Turner, and J. L. Malenchuk. 1986. *Empirical Relationships Between Watershed Attributes and Headwater Chemistry in the Adirondack Region*. ORNL/TM-9838. Oak Ridge National Laboratory, Oak Ridge, Tennessee, USA.

Hunsaker, C. T., R. L. Graham, L. W. Barnthouse, R. H. Gardner, R. V. O'Neill, and G. W. Suter III. 1990. Assessing ecological risk on a regional scale. *Environmental Management* 14:325–332.

Hunsaker, C. T., D. A. Levine, S. P. Timmons, B. L. Jackson, and R. V. O'Neill. 1992. Landscape characterization for assessing regional water quality. *In* D. H. McKenzie, D. E. Hyatt, and V. J. McDonald, eds. *Ecological Indicators*, Volume 2, pp. 997–1006. Elsevier Applied Science, New York, New York, USA.

Hunt, E. R., Jr., F. C. Martin, and S. W. Running. 1991. Simulating the effects of climatic variation on stem carbon accumulation of a ponderosa pine stand: comparison with annual growth increment data. *Tree Physiology* 9:161–171.

Hunter, M. L., Jr. 1991. Coping with ignorance: the coarse-filter strategy for maintaining biodiversity. *In* K. A. Kohm, ed. *Balancing on the Brink of Extinction*, pp. 266–281. Island Press, Washington, DC, USA.

Hunter, M. L., Jr. 1993. Natural fire regimes as spatial models for managing boreal forests. *Biological Conservation* 65:115–120.

Hurlbert, S. H. 1984. Pseudoreplication and the design of ecological field experiments. *Ecological Monographs* 54:187–211.

Hurley, J. P., J. M. Benoit, C. L. Babiarz, M. M. Shafer, A. W. Andre, J. R. Sullivan, R. Hammond, and D. A. Webb. 1995. Influence of watershed characteristics on mercury levels in Wisconsin rivers. *Environmental Science and Technology* 29:1867–1875.

Hyman, J. B., J. B. McAninch, and D. L. DeAngelis. 1991. An individual-based simulation model of herbivory in a heterogeneous landscape. *In* M. G. Turner and R. H. Gardner, eds. *Quantitative Methods in Landscape Ecology*, pp. 443–475. Springer-Verlag, New York, New York, USA.

Hynes, H. B. N. 1975. The stream and its valley. *Verhandlungren der Internationalen Vereiningung fuer Theoretische und Angewandte Limnologie* 19:1–15.

Imes, R. A., J. Rolstad, and P. Wegge. 1993. Predicting space use responses to habitat fragmentation: can voles *Microtus oeconomus* serve as an experimental model system (EMS) for capercaille grouse in boreal forest? *Biological Conservation* 63:261–268.

Isard, W. 1960. *Methods of Regional Analysis: An Introduction to Regional Science*. Massachusetts Institute of Technology Press, Cambridge, Massachusetts, USA.

Isard, W. 1972. *Ecologic–economic Analysis for Regional Development*. The Free Press, New York, New York, USA.

Isard, W. 1975. *Introduction to Regional Science*. Prentice Hall, Upper Saddle River, New Jersey, USA.

ISI. 1999. *The Web of Science* (http//webofscience.com), the Web interface providing access to the Science Citation Index Expanded (TM), Institute for Scientific Information, 3501 Market Street, Philadelphia, Pennsylvania, 19104, USA.

Istock, C. A., and S. M. Scheiner. 1987. Affinities and high-order diversity within landscape mosaics. *Evolutionary Ecology* 1:11–29.

Istok, J. D., J. D. Smythe, and A. L. Flint. 1993. Multivariate geostatistical analysis of groundwater contamination: a case history. *Ground Water* 31: 63–74.

Iverson, L. R. 1988. Land-use changes in Illinois, USA: the influence of landscape attributes on current and historic land use. *Landscape Ecology* 2:45–62.

Iverson, L. R. 1991. Forest resources of Illinois: What do we have and what are they doing for us? *Illinois Natural History Survey Bulletin* 34:361–374.

Ives, A. R., M. G. Turner, and S. M. Pearson. 1998. Local explanations of landscape patterns: can analytical approaches approximate simulation models of spatial processes? *Ecosystems* 1:35–51.

Jackson, D. A., K. M. Somers, and H. H. Harvey. 1992. Null models and fish communities: evidence of nonrandom patterns. *American Naturalist* 139:930–951.

Jacobson, R. B., and D. J. Coleman. 1986. Stratigraphy and recent evolution of Maryland Piedmont flood plains. *American Journal of Science*, 286:617–637.

Jeffries, R. L., D. R. Klein, and G. R. Shaver. 1994. Vertebrate herbivores and northern plant communities: reciprocal influences and responses. *Oikos* 71:193–206.

Jensen, J. R. 1996. *Introductory Digital Image Processing: A Remote Sensing Perspective.* Prentice Hall, Upper Saddle River, New Jersey, USA.

Jewett, K., D. Daugharty, H. H. Krause, and P. A. Arp. 1995. Watershed responses to clear-cutting: effects on soil solutions and stream water discharge in central New Brunswick. *Canadian Journal of Soil Science* 75:475–490.

Jin, K. R., and Y. G. Wu. 1997. Boundary-fitted grid in landscape modeling. *Landscape Ecology* 12:19–26.

Johnes, P., B. Moss, and G. Phillips. 1996. The determination of total nitrogen and total phosphorus concentrations in freshwaters from land use, stock headage and population data: testing of a model for use in conservation and water quality management. *Freshwater Biology* 36:451–473.

Johnson, A. R., B. T. Milne, and J. A. Wiens. 1992a. Diffusion in fractal landscapes—simulations and experimental studies of tenebrionid beetle movements. *Ecology* 73:1968–1983.

Johnson, A. R., J. A. Wiens, B. T. Milne, and T. O. Crist. 1992b. Animal movements and population dynamics in heterogeneous landscapes. *Landscape Ecology* 7:63–75.

Johnson, D. W., and R. I. Van Hook, editors. 1989. *Analysis of Biogeochemical Cycling Processes in Walker Branch Watershed.* Springer-Verlag, New York, New York, USA.

Johnson, E. A. 1979. Fire recurrence in the subarctic and its implications for vegetation composition. *Canadian Journal of Botany* 57:1374–1379.

Johnson, E. A. 1992. *Fire and Vegetation Dynamics: Studies from the North American Boreal Forest.* Cambridge University Press, Cambridge, UK.

Johnson, E. A., and S. L. Gutsell. 1994. Fire frequency models, methods, and interpretations. *Advances in Ecological Research* 25:239–287.

Johnson, E. A., and C. E. Van Wagner. 1985. The theory and use of two fire history models. *Canadian Journal of Forest Research* 15:214–220.

Johnson, L. B., C. Richards, G. Host, and J. W. Arthur. 1997. Landscape influences on water chemistry in Midwestern streams. *Freshwater Biology* 37:193–208.

Johnson, W. C., D. M. Sharpe, D. L. DeAngelis, D. E. Fields, and R. J. Olson. 1981. Modeling seed dispersal and forest island dynamics. *In* R. Burgess and D. M. Sharpe, eds. *Forest Island Dynamics in Man Dominated Landscapes*, pp. 215–239. Springer-Verlag, New York, New York, USA.

Johnston, C., and R. J. Naiman. 1990a. Aquatic patch creation in relation to beaver population trends. *Ecology* 71:1617–1621.

Johnston, C. A., and R. J. Naiman. 1990b. The use of a geographic information system to analyze landscape alteration by beaver. *Landscape Ecology* 4:5–19.

Jones, C. G., and J. H. Lawton, editors. 1995. *Linking Species and Ecosystems.* Chapman and Hall, New York, New York, USA.

Jones, D. D. 1975. *Stability Implications of Dispersal Linked Ecological Models.* International Institute for Applied Systems Analysis Report 75–44.

Jones, K. B., K. H. Riitters, J. D. Wickham, R. D. Tankersley, Jr., R. V. O'Neill, D. J. Chaloud, E. R. Smith, and A. C. Neale. 1997. *An Ecological Assessment of the United States Mid-Atlantic Region: A Landscape Atlas.* EPA/600/R-97/130. U.S. Environmental Protection Agency, Office of Research and Development, Washington, DC, USA.

Jordan, T. E., D. L. Correll, and D. E. Weller. 1997a. Nonpoint source discharges of nutrients from Piedmont watersheds of Chesapeake Bay. *Journal of the American Water Resources Association* 33(3):631–645.

Jordan, T. E., D. L. Correll, and D. E. Weller. 1997b. Relating nutrient discharges from watersheds to land use and streamflow variability. *Water Resources Research* 33(11):2579–2590.

Jorgensen, S. E. 1994. *Fundamentals of Ecological Modeling* (2nd edition). Elsevier, Amsterdam, The Netherlands.

Justice, C., J. Townshend, B. Holben, and C. Tucker. 1985. Analysis of the phenology of global vegetation using meteorological satellite data. *International Journal of Remote Sensing* 6:1271–1318.

Kadmon, R., and A. Schmida. 1990. Spatiotemporal demographic processes in plant populations: an approach and a case study. *American Naturalist* 135:382–397.

Kareiva, P. 1994. Space: the final frontier for ecological theory. *Ecology* 75:1.

Kareiva, P. M. 1983. Influence of vegetation texture on herbivore populations: resource concentration and herbivore movements. *In* R. F. Denno and M. McClure, eds. *Variable Plants and Herbivores in Natural and Managed Systems*, pp. 259–289. Academic Press, New York, New York, USA.

Kareiva, P. M. 1990. Population dynamics in spatially complex environments: theory and data. *Philosophical Transactions of the Royal Society of London B* 330:175–190.

Katori, M., S. Kizaki, Y. Terui, and T. Kubo. 1998. Forest dynamics with canopy gap expansion and stochastic Ising model. *Fractals—An Interdisciplinary Journal on the Complex Geometry of Nature* 6:81–86.

Keane, R. E., P. Morgan and S. W. Running. 1996a. *FIRE-BGC—A Mechanistic Ecological Process Model for Simulating Fire Succession on Coniferous Forest Landscapes of the Northern Rocky Mountains.* USDA Forest Service Research Paper INT-RP-484.

Keane, R. E., K. C. Ryan, and S. W. Running. 1996b. Simulating effects of fire on northern Rocky Mountain landscapes with the ecological process model FIRE-BGC. *Tree Physiology* 16:319–331.

Keitt, T. H. 2000. Spectral representation of neutral landscapes. *Landscape Ecology* 15: 479–493

Kelsey, K. A., and S. D. West. 1998. Riparian wildlife. *In* R. J. Naiman and R. E. Bilby, eds. *River Ecology and Management*, pp. 235–258. Springer-Verlag, New York, New York, USA.

Kencairn, B. 1996. *Peril on Common Ground: The Applegate Experiment.* Rowan and Littlefield, Lanham, Maryland, USA.

Kennedy, A. D. 1997. Bridging the gap between general circulation model (GCM) output and biological microenvironments. *International Journal of Biometeorology* 40:119–122.

Kesner, B. T., and B. Meentemeyer. 1989. A regional analysis of total nitrogen in an agricultural landscape. *Landscape Ecology* 2:151–164.

Kienast, F. 1993. Analysis of historic landscape patterns with a Geographic Information System: a methodological outline. *Landscape Ecology* 8:103–118.

Kierstead, H., and L. B. Slobodkin. 1953. The size of water masses containing plankton blooms. *Journal of Marine Research* 12:141–147.

King, A. W. 1991. Translating models across scales in the landscape. *In* M. G. Turner and R. H. Gardner, eds. *Quantitative Methods in Landscape Ecology*, pp. 479–517. Springer-Verlag, New York, New York, USA.

Kitchell, J. F., editor. 1992. *Food Web Management: A Case Study of Lake Mendota, Wisconsin*. Springer-Verlag, New York, New York, USA.

Kitching, R. L. 1983. *Systems Ecology*. University of Queensland Press, St. Lucia, Queensland, Australia.

Klopatek, J. M., and R. H. Gardner, editors. 1999. *Landscape Ecological Analysis*. Springer-Verlag, New York, New York, USA.

Klopatek, J. M., J. R. Krummel, J. B. Mankin, and R. V. O'Neill. 1983. A theoretical approach to regional environmental conflicts. *Journal of Environmental Management* 16:1–15.

Knick, S. T., and D. L. Dyer. 1997. Distribution of black-tailed jackrabbit habitat determined by GIS in southwestern Idaho. *Journal of Wildlife Management* 61:75–85.

Knight, D. H., and L. L. Wallace. 1989. The Yellowstone fires: issues in landscape ecology. *BioScience* 39:700–706.

Knight, D. H., T. J. Fahey, and S. W. Running. 1985. Water and nutrient outflow from contrasting lodgepole pine forests in Wyoming. *Ecological Monographs* 55:29–48.

Koestler, A. 1967. *The Ghost in the Machine*. Macmillan, New York, New York, USA.

Kolasa, J. 1989. Ecological systems in hierarchicial perspective: breaks in community structure and other consequences. *Ecology* 70:36–47.

Kotliar, N. B., and J. A. Wiens. 1990. Multiple scales of patchiness and patch structure: a hierarchical framework for the study of heterogeneity. *Oikos* 59:253–260.

Kratz, T. K., B. J. Benson, E. R. Blood, G. L. Cunningham, and R. A. Dahlgren. 1991. The influence of landscape position on temporal variability in four North American ecosystems. *American Naturalist* 138:355–378.

Kratz, T. K., K. E. Webster, C. J. Bowser, J. J. Magnuson, and B. J. Benson. 1997. The influence of landscape position on lakes in northern Wisconsin. *Freshwater Biology* 37:209–217.

Krummel, J. R., C. C. Gilmore, and R. V. O'Neill. 1984. Locating vegetation at risk to air pollution: an exploration of a regional approach. *Journal of Environmental Management* 18:279–290.

Krummel, J. R., R. V. O'Neill, and J. B. Mankin. 1986. Regional environmental simulation of African cattle herding societies. *Human Ecology* 14:117–130.

Krummel, J. R., R. H. Gardner, G. Sugihara , R. V. O'Neill, and P. R. Coleman. 1987. Landscape patterns in a disturbed environment. *Oikos* 48:321–324.

Kuchler, A. W. 1964. *Potential Natural Vegetation of the Conterminous United States*. Special publication 36, American Geographical Society, New York, New York, USA.

Kuuluvainen, T., S. Kimmo, and R. Kalliola. 1998. Structure of a pristine *Picea abies* forest in northeastern Europe. *Journal of Vegetation Science* 9:563–574.

Lack, D. 1942. Ecological features of the bird fauna of British small islands. *Journal of Animal Ecology* 11:9–36.

Lackey, R. T. 1996. Challenges to using ecological risk assessment to implement ecosystem management. *Water Resources Update* 103:46–49.

LaGro, J. A., and S. D. DeGloria. 1992. Land-use dynamics within an urbanizing nonmetropolitan county in New York State (USA). *Landscape Ecology* 7:275–289.

Lambeck, R. J. 1997. Focal species: a multi-species umbrella for nature conservation. *Conservation Biology* 11:849–856.

Lambin, E. F. 1997. Modeling and monitoring land-cover change processes in tropical regions. *Progress in Physical Geography* 21:375–393.

Lande, R. 1987. Extinction thresholds in demographic models of territorial populations. *American Naturalist* 130:624–635.

Lande, R. 1988. Demographic models of the northern spotted owl (*Strix occidentalis caurina*). *Oecologia* 75:601–607.

Lavorel, S., R. H. Gardner, and R. V. O'Neill. 1993. Analysis of patterns in hierarchically structured landscapes. *Oikos* 67:521–528.

Lavorel, S., J. Lepart, M. Debussche, J. D. Lebreton, and J. L. Beff. 1994a. Small scale disturbances and the maintenance of species diversity in Mediterranean old fields. *Oikos* 70:455–473.

Lavorel, S., R. V. O'Neill, and R. H. Gardner. 1994b. Spatio-temporal dispersal strategies and annual plant species coexistence in a structured landscape. *Oikos* 71:75–88.

Lavorel, S., R. H. Gardner, and R. V. O'Neill. 1995. Dispersal of annual plants in hierarchically structured landscapes. *Landscape Ecology* 10:277–289.

Lee, R. G., R. O. Flamm, M. G. Turner, C. Bledsoe, P. Chandler, C. DeFerrari, R. Gottfried, R. J. Naiman, N. Schumaker, and D. Wear. 1992. Integrating sustainable development and environmental vitality. *In* R. J. Naiman, ed. *New Perspectives in Watershed Management*, pp. 499–521. Springer-Verlag, New York, New York, USA.

Leenhouts, W. P. 1998. Assessment of biomass burning in the conterminous United States. *Conservation Ecology* 2.

Lefkovitch, L. P., and L. Fahrig. 1985. Spatial characteristics of habitat patches and population survival. *Ecological Modeling* 30:297–308.

Lefroy, E. C., R. J. Hobbs, and L. J. Atkins. 1991. *Revegetation Guide to the Central Wheatbelt.* Department of Agriculture, Western Australia, Bulletin 4231.

Legendre, L., and P. Legendre. 1983. *Numerical Ecology.* Elsevier Scientific Publishing Company, Amsterdam, The Netherlands.

Legendre, P. 1993. Spatial autocorrelation: trouble or new paradigm. *Ecology* 74:1659–1673.

Legendre, P., and M. J. Fortin. 1989. Spatial pattern and ecological analysis. *Vegetatio* 80:107–138.

Leith, H., and R. H. Whittaker, editors. 1975. *Primary Productivity of the Biosphere.* Springer-Verlag, New York, New York, USA.

Leopold, A. S. 1933. *Game Management.* Scribner's, New York, New York, USA.

Lertzman, K., and J. Fall. 1998. From forest stands to landscapes: Spatial scales and the roles of disturbances. *In* D. L. Peterson and V. T. Parker, eds. *Ecological Scale: Theory and Applications*, pp. 339–367. Columbia University Press, New York, New York, USA.

Lertzman, K. P. 1992. Patterns of gap-phase replacement in a subalpine, old-growth forest. *Ecology* 73:657–669.

Levin, S. A. 1976a. Spatial patterning and the structure of ecological communities. *Lectures on Mathematics in the Life Science* 8:1–35.

Levin, S. A. 1976b. Population dynamics models in heterogeneous environments. *Annual Review of Ecology and Systematics* 7:287–310.

Levin, S. A. 1978. Population models and community structure in heterogeneous environments. *In* S. A. Levin, ed. *Studies in Mathematical Biology Series—Studies in Mathematics (Mathematical Association of America) Volume II. Populations and Communities*, pp. 439–476. Mathematical Association of America, Washington, DC, USA.

Levin, S. A. 1992. The problem of pattern and scale in ecology. *Ecology* 73:1943–1983.

Levin, S. A., and R. T. Paine. 1974a. The role of disturbance in models of community structure. *In* S. A Levin, ed. *Ecosystem Analysis and Prediction*, pp. 55–63. Society for Industrial and Applied Mathematics, Philadelphia, Pennsylvania, USA.

Levin, S. A., and R. T. Paine. 1974b. Disturbance, patch formation and community structure. *Proceedings of the National Academy of Science* 71:2744–2747.

Levin, S. A., T. M. Powell, and J. H. Steele, editors. 1993. *Patch Dynamics. Lecture Notes in Biomathematics.* Springer-Verlag, New York, New York, USA.

Levins, R. 1969. Some demographic and genetic consequences of environmental heterogeneity for biological control. *Bulletin of the Entomological Society of America* 15:237–240.

Levins, R. 1970. Extinctions. *In Some Mathematical Questions in Biology. Lectures on Mathematics in the Life Sciences*, pp. 77–107. American Mathematical Society, Providence, Rhode Island, USA.

Levins, R., and D. Culver. 1971. Regional coexistence of species and competition between rare species. *Proceedings of the National Academy of Science* 68:1246–1248.

Lewis, M. A. 1997. Variability, patchiness, and jump dispersal in the spread of an invading population. *In* D. Tilman and P. Kareiva, eds. *Spatial Ecology: The Role of Space in Population Dynamics and Interspecific Interactions*, pp. 46–69. Princeton University Press, Princeton, New Jersey, USA.

Li, B. L. 1995. Stability analysis of a nonhomogeneous Markovian landscape model. *Ecological Modeling* 82:247–256.

Li, H., and J. F. Reynolds. 1993. A new contagion index to quantify spatial patterns of landscapes. *Landscape Ecology* 8:155–162.

Li, H., and J. F. Reynolds. 1994. A simulation experiment to quantify spatial heterogeneity in categorical maps. *Ecology* 75:2446–2455.

Li, H., and J. F. Reynolds. 1995. On definition and quantification of heterogeneity. *Oikos* 73:280–284.

Li, H., J. F. Franklin, F. J. Swanson, and T. A. Spies. 1993. Developing alternative forest cutting patterns: a simulation approach. *Landscape Ecology* 8:63–75.

Lidicker, W. Z., and W. D. Koenig. 1996. Responses of terrestrial vertebrates to habitat edges and corridors. *In* D. R. McCullough, ed. *Metapopulations and Wildlife Conservation*, pp. 85–109. Island Press, Washington, DC, USA.

Liebold, A. M., R. E. Rossi, and W. P. Kemp. 1993. Geostatistics and geographic information systems in applied insect ecology. *Annual Review of Entomology* 38:303–327.

Likens, G. E., editor. 1985. *An Ecosystem Approach to Aquatic Ecology: Mirror Lake and Its Environment*. Springer-Verlag, New York, New York, USA.

Likens, G. E. 1995. *The Ecosystem Approach: Its Use and Abuse. Ecology* Institute, Oldendorf/Luhe, Germany.

Likens, G. E. 1998. Limitations to intellectual progress in ecosystem science. *In* M. L. Pace and P. M. Groffman, eds. *Successes, Limitations and Frontiers in Ecosystem Science*, pp. 247–271. Springer-Verlag, New York, New York, USA.

Likens, G. E., and F. H. Bormann. 1995. *Biogeochemistry of a Forested Ecosystem*. Springer-Verlag, New York, New York, USA.

Lillesand, T. M., and R. W. Kiefer. 1994. *Remote Sensing and Image Interpretation* (third edition). John Wiley & Sons, New York, New York, USA.

Lindenmayer, D. B., and H. A. Nix. 1993. Ecological principles for the design of wildlife corridors. *Conservation Biology* 7:627–630.

Linder, P., B. Elfving, and O. Zackrisson. 1997. Stand structure and successional trends in virgin boreal forest reserves in Sweden. *Forest Ecology and Management* 98:17–33.

Liu, J. 1993. ECOLECON: An ECOLogical–ECONomic model for species conservation in complex forest landscapes. *Ecological Modeling* 70:63–87.

Liu, J., J. B. Dunning, Jr., and H. R. Pulliam. 1995. Potential effects of a forest-management plan on Bachman's sparrow (*Aimophila aestivalis*): linking a spatially explicit model with GIS. *Conservation Biology* 9:62–75.

Liu, J. G., and P. S. Ashton. 1998. FORMOSAIC: An individual-based spatially explicit model for simulating forest dynamics in landscape mosaics. *Ecological Modeling* 106:177–200.

Loehle, C., and B. L. Li. 1996. Statistical properties of ecological and geologic fractals. *Ecological Modeling* 85:271–284.

Lorimer, C. G. 1977. The presettlement forest and natural disturbance cycle of northeastern Maine, USA. *Ecology* 58:139–148.

Los, S. O., C. O. Justace, and C. J. Tucker. 1994. A global 1-degrees-by-1-degrees NDVI data set for climate studies derived from the GIMMS continental NDVI data. *International Journal of Remote Sensing* 15:3493–3518.

Loucks, O. L. 1970. Evolution of diversity, efficiency, and community stability. *American Zoologist* 10:17–25.

Lovejoy, S. 1985. Fractal properties of rain, and a fractal model. *Tellus*, 37A:209–232.

Lovejoy, S., D. Schertzer, and A. A. Tsonis. 1987. Functional box counting and multiple elliptical dimensions in rain. *Science*, 235:1036–1038.

Loveland, T. R., J. W. Merchant, D. P. Ohlen, and J. F. Brown. 1991. Development of a land-cover characteristics database for the conterminous U.S. *Photogrammetric Engineering and Remote Sensing* 57:1453–1463.

Lowrance, R. 1998. Riparian forest ecosystems as filters for nonpoint-source pollution. *In* M. L. Pace and P. M. Groffman, eds. *Successes, Limitations and Frontiers in Ecosystem Science*, pp. 113–141. Springer-Verlag, New York, New York, USA.

Lowrance, R., L. S. Altier, J. D. Newbold, R. R. Schnabel, P. M. Groffman, J. M. Denver, D. L. Correll, J. W. Gilliam, J. L. Robinson, R. B. Brinsfield, K. W. Staver, W. Lucas, and A. H. Todd. 1997. Water quality functions of riparian forest buffer systems in the Chesapeake Bay watershed. *Environmental Management* 21:687–712.

Lowry, I. S. 1967. *Seven Models of Urban Development*. Highway Research Board, National Research Council, Washington, DC, USA.

Lubchenco, J., A. M. Olson, L. B. Brubaker, S. R. Carpenter, M. M. Holland. S. P. Hubbell, S. A. Levin, J. A. McMahon, P. A. Matson, J. M. Mellillo, H. A. Mooney, C. H. Peterson, H. R. Pulliam, L. A. Real, P. J. Regal, and P. G. Risser. 1991. The sustainable biosphere initiative: an ecological research agenda. *Ecology* 72:371–412.

Lynn, B. H., D. Rind, and R. Avissar. 1995. The importance of mesoscale circulations generated by subgrid-scale landscape heterogeneities in general-circulation models. *Journal of Climate* 8:191–205.

MacArthur, R. H. 1972. *Geographical Ecology*. Harper & Row, New York, New York, USA.

MacArthur, R. H., and E. O. Wilson. 1963. An equilibrium theory of insular zoogeography. *Evolution* 17:373–387.

MacArthur, R. H., and E. O. Wilson. 1967. *The Theory of Island Biogeography*. Princeton University Press, Princeton, New Jersey, USA.

Macilwain, C. 1996. Biosphere 2 begins fight for credibility. *Nature* 380:275.

Magnuson, J. J. 1990. Long-term ecological research and the invisible present. *BioScience* 40:495–501.

Magnuson, J. J., and T. K. Kratz, editors. In press. *Lakes in the Landscape*. Oxford University Press, Oxford, UK.

Magnuson, J. J., B. J. Benson, and T. K. Kratz. 1990. Temporal coherence in the limnology of a suite of lakes in Wisconsin, USA. *Freshwater Biology* 23:145–159.

Magnuson, J. J., T. K. Kratz, T. M. Frost, C. J. Bowser, B. J. Benson, and R. Nero. 1991. Expanding the temporal and spatial scales of ecological research and comparison of divergent ecosystems: roles for LTER in the United States. *In* P. G. Risser, ed. *Long-term Ecological Research*, pp. 45–70. John Wiley & Sons, New York, New York, USA.

Magnuson, J. J., W. M. Tonn, A. Banerjee, J. Toivonen, O. Sanchez, and M. Rask. 1998. Isolation vs. extinction in the assembly of fishes in small northern lakes. *Ecology* 79:2941–2956.

Malamud, B. D., G. Morein, and D. L. Turcotte. 1998. Forest fires: an example of self-organized critical behavior. *Science* 281:1840–1842.

Malanson, G. P. 1984. Intensity as a third factor of disturbance regime and its effect on species diversity. *Oikos* 43:411–413.

Malanson, G. P., and M. P. Armstrong. 1996. Dispersal probability and forest diversity in a fragmented landscape. *Ecological Modeling* 87:91–102.

Mandelbrot, B. 1967. How long is the coast of Britain? Statistical self-similarity and fractional dimension. *Science* 156:636–638.

Mandelbrot, B. B. 1983. *The Fractal Geometry of Nature*. W. H. Freeman, New York, New York, USA.

Mandelbrot, B. B. 1985. Self-affine fractals and fractal dimension. *Physical Scripts* 32:257–260.

Mankin, J. B., R. V. O'Neill, H. H. Shugart, and B. W. Rust. 1975. The importance of validation in ecosystem analysis. *New Directions in the Analysis of Ecological Systems*, Part 1, 5:63–71, LaJolla, California, USA.

Mankin, J. B., J. M. Klopatek, R. V. O'Neill, and J. R. Krummel. 1981. A regional modeling approach to an energy–environment conflict. *In* W. J. Mitsch, R. W. Bosserman, and J. M. Klopatek, eds. *Energy and Ecological Modeling*, pp. 535–541. Elsevier, New York, New York, USA.

Marsh, G. P. 1864. *Man and Nature; or, Physical Geography as Modified by Human Action.* Reprinted 1965, Belknap Press of Harvard University Press, Cambridge, Massachusetts, USA.

May, R. M. 1973. *Stability and Complexity in Model Ecosystems.* Princeton University Press, Princeton, New Jersey, USA.

McCarty, H. H., J. C. Hook, and D. S. Knos. 1956. *The Measurement of Association in Industrial Geography.* University of Iowa, Department of Geography Report 1:1–143. Ames, Iowa, USA.

McClanahan, T. R. 1986. The effect of seed source on primary succession in a forest ecosystem. *Vegetatio* 65:175–178.

McCollin, D. 1993. Avian distribution patterns in a fragmented wooded landscape (North Humerside, UK): the role of between patch and within-patch structure. *Global Ecology and Biogeography Letters* 3:48–62.

McGarigal, K., and B. J. Marks. 1995. *FRAGSTATS. Spatial Analysis Program for Quantifying Landscape Structure.* USDA Forest Service General Technical Report PNW-GTR-351.

McGarigal, K., and W. C. McComb. 1995. Relationship between landscape structure and breeding birds in the Oregon Coast Range. *Ecological Monographs* 65:235–260.

McHarg, I. L. 1969. *Design with Nature.* Natural History Press, Garden City, New York, New York, USA.

McInnes, P. F., R. J. Naiman, J. Pastor, and Y. Cohen. 1992. Effects of moose browsing on vegetation and litter of the boreal forest, Isle Royale, Michigan, USA. *Ecology* 73:2059–2075.

McIntosh, R. P. 1962. The forest cover of the Catskill Mountain region, New York, as indicated by land survey records. *American Midland Naturalist* 68:409–423.

McKechnie, J. L. 1979. *Webster's New Universal Unabridged Dictionary.* Simon and Schuster, New York, New York, USA.

McKenzie, D., D. L. Peterson, and E. Alvarado. 1996. Extrapolation problems in modeling fire effects at large spatial scales. *International Journal of Wildland Fire* 6:165–176.

McLaughlin, J. F., and J. Roughgarden. 1991. Pattern and stability in predator–prey communities: how diffusion in spatially variable environments affects the Lotka–Volterra model. *Theoretical Population Biology* 40:148–172.

McNaughton, S. J., R. W. Reuss, and S. W. Seagle. 1988. Large mammals and process dynamics in African ecosystems. *BioScience* 38:794–800.

Meade, R. H. 1982. Sources, sinks, and storage of river sediment in the Atlantic drainage of the United States. *Journal of Geology* 90(3):235–252.

Medley, K. E., B. W. Okey, G. W. Barrett, M. F. Lucas, and W. H. Renwick. 1995. Landscape change with agricultural intensification in a rural watershed, southwestern Ohio, USA. *Landscape Ecology* 10:161–176.

Meentemeyer, V. 1984. The geography of organic decomposition rates. *Annals of the Association of American Geographers* 74:551–560.

Meentemeyer, V., and E. O. Box. 1987. Scale effects in landscape studies. *In* M. G. Turner, ed. *Landscape Heterogeneity and Disturbance*, pp. 15–34. Springer-Verlag, New York, New York, USA.

Meffe, G. K., and C. R. Carroll. 1994. *Principles of Conservation Biology*. Sinauer Associates, Inc., Sunderland, Massachusetts, USA.

Meisel, J. E., and M. G. Turner. 1998. Scale detection in real and artificial landscapes using semivariance analysis. *Landscape Ecology* 13:347–362.

Merriam, G. 1991. Corridors and connectivity: animal populations in heterogeneous environments. *In* D. A. Saunders, R. J. Hobbs, and P. R. Ehrlich, eds. *Nature Conservation 3: Reconstruction of Fragmented Ecosystems*, pp. 133–142. Surrey Beatty & Sons, Chipping Norton, NSW, Australia.

Merriam, G., K. Henein, and K. Stuart-Smith. 1991. Landscape dynamic models. *In* M. G. Turner and R. H. Gardner, eds. *Quantitative Methods in Landscape Ecology*, pp. 399–416. Springer-Verlag, New York, New York, USA.

Metzger, J. N., R. A. Fjeld, J. S. Hammonds, and F. O. Hoffman. 1998. Evaluation of software for propagating uncertainty through risk assessment models. *Human and Ecological Risk Assessment* 4:263–290.

Meybeck, M. 1982. Carbon, nitrogen, and phosphorus transport by world rivers. *American Journal of Science*, 282:401–450.

Meyer, W. B. 1995. Past and present land use and land cover in the USA. *Consequences*, Spring 1995:25–33.

Miller, C., and D. L. Urban. 1999. Forest pattern, fire, and climatic change in the Sierra Nevada. *Ecosystems* 2:76–87.

Miller, J. N., R. P. Brooks, and M. J. Croonquist. 1997. Effects of landscape patterns on biotic communities. *Landscape Ecology* 12:137–153.

Miller, T. E. 1982. Community diversity and interactions between the size and frequency of disturbance. *American Naturalist* 120:533–536.

Milne, B. T. 1988. Measuring the fractal geometry of landscapes. *Applications in Mathematics and Computation* 27:67–79.

Milne, B. T. 1991a. The utility of fractal geometry in landscape design. *Landscape and Urban Planning* 21:81–90.

Milne, B. T. 1991b. Lessons from applying fractal models to landscape patterns. *In* M. G. Turner and R. H. Gardner, eds. *Quantitative Methods in Landscape Ecology*, pp. 199–235. Springer-Verlag, New York, New York, USA.

Milne, B. T. 1992. Spatial aggregation and neutral models in fractal landscapes. *American Naturalist* 139:32–57.

Milne, B. T. 1998. Motivation and benefits of complex systems approaches in ecology. *Ecosystems* 1:449–456.

Milne, B. T., and A. R. Johnson. 1993. Renormalization relations for scale transformation in ecology. *In* R. H. Gardner, ed. *Some Mathematical Questions in Biology: Theoretical Approaches for Predicting Spatial Effects in Ecological Systems*, pp. 109–128. American Mathematical Society, Providence, Rhode Island, USA.

Milne, B. T., K. M. Johnston, and R. T. T. Forman. 1989. Scale-dependent proximity of wildlife habitat in a spatially-neutral Bayesian model. *Landscape Ecology* 2:101–110.

Milne, B. T., A. R. Johnson, T. H. Keitt, C. A. Hatfield, J. David, and P. T. Hraber. 1996. Detection of critical densities associated with pinon–juniper woodland ecotones. *Ecology* 77:805–821.

Minnich, R. A. 1983. Fire mosaics in southern California and northern Baja California. *Science* 219:1287–1294.

Mladenoff, D. J., and W. L. Baker, editors. 1999. *Spatial Modeling of Forest Landscape Change: Approaches and Applications.* Cambridge University Press, Cambridge, UK.

Mladenoff, D. J., and H. S. He. 1999. Design and behavior of LANDIS, an object oriented model of forest landscape disturbance and succession. *In* D. J. Mladenoff and W. L. Baker, eds. *Spatial Modeling of Forest Landscape Change: Approaches and Applications*, pp. 125–162. Cambridge University Press, Cambridge, UK.

Mladenoff, D. J., and T. A. Sickley. 1998. Assessing potential gray wolf restoration in the northeastern United States: a spatial prediction of favorable habitat and potential population levels. *Journal of Wildlife Management* 62:1–10.

Mladenoff, D. J., and F. Stearns. 1993. Eastern hemlock regeneration and deer browsing in the northern Great Lakes region—a reexamination and model simulation. *Conservation Biology* 7:889–900.

Mladenoff, D. J., M. A. White, J. Pastor, and T. R. Crow. 1993. Comparing spatial pattern in unaltered old-growth and disturbed forest landscapes. *Ecological Applications* 3:294–306.

Mladenoff, D. J., M. A. White, T. R. Crow, and J. Pastor. 1994. Applying principles of landscape design and management to integrate old-growth forest enhancement and commodity use. *Conservation Biology* 8:752–762.

Mladenoff, D. J., T. A. Sickley, R. G. Haight, and A. P. Wydeven. 1995. A regional landscape analysis and prediction of favorable gray wolf habitat in the northern Great Lakes region. *Conservation Biology* 9:279–294.

Mladenoff, D. J., G. E. Host, J. Boeder, and T. R. Crow. 1996. LANDIS: a spatial model of forest landscape disturbance, succession, and management. *In* M. F. Goodchild, L. T. Steyaert, and B. O. Parks, eds. *GIS and Environmental Modeling*, pp. 175–180. National Center for Geographic Information and Analysis, Santa Barbara, California.

Mladenoff, D. J., G. J. Niemi, and M. A. White, 1997. Effects of changing landscape pattern and USGS land cover data variability on ecoregion discrimination across a forest-agriculture gradient. *Landscape Ecology* 12:379–396.

Moen, R., Y. Cohen, and J. Pastor. 1997. A spatially explicit model of moose foraging and energetics. *Ecology* 78:505–521.

Moen, R., J. Pastor, and Y. Cohen. 1998. Linking moose population and plant growth models with a moose energetics model. *Ecosystems* 1:52–63.

Molak, V., editor. 1996. *Fundamentals of Risk Analysis and Risk Management.* CRC/Lewis Publishers, New York, New York, USA.

Molofsky, J. 1994. Population dynamics and pattern formation in theoretical population. *Ecology* 75:30–39.

Moody, A., and C. E. Woodcock. 1995. The influence of scale and the spatial characteristics of landscapes on land-cover mapping using remote sensing. *Landscape Ecology* 10:363–379.

Mooney, H. A., and M. Godron, editors. 1983. *Disturbance and Ecosystems.* Springer-Verlag, New York, New York, USA.

Mooney, H. A., P. M. Vitousek, and P. A. Matson. 1987. Exchange of materials between terrestrial ecosystems and the atmosphere. *Science* 238:926–932.

Morgan, P., G. H. Aplet, J. B. Haufler, H. C. Humphries, M. M. Moore, and W. D. Wilson. 1994. Historical range of variability: a useful tool for evaluating ecosystem change. *Journal of Sustainable Forestry* 2:87–111.

Morris, S. J., and R. E. J. Boerner. 1998. Landscape patterns of nitrogen mineralization and nitrification in southern Ohio hardwood forests. *Landscape Ecology* 13:215–224.

Morrison, P. H., D. Kloepfer, D. A. Leversee, C. M. Socha, and D. L. Ferber. 1991. *Ancient Forests in the Pacific Northwest: Analysis and Maps of Twelve National Forests.* Wilderness Society, Washington, DC, USA.

Mueller, D. K., and D. R. Helsel. 1996. *Nutrients in the Nation's Waters: Too Much of a Good Thing?* U.S. Geological Survey Circular 1136.

Murphy, D. D., K. E. Freas, and S. B. Weiss. 1990. An environment–metapopulation approach to population viability for a threatened invertebrate. *Conservation Biology* 4:41–51.

Murray, J. D. 1989. *Mathematical Biology.* Springer-Verlag, New York, New York, USA.

Myers, J. H. 1976. Distribution and dispersal in populations capable of resource depletion. *Oecologia* 23:255–269.

Naiman, R. J. 1996. Water, society and landscape ecology. *Landscape Ecology* 11:193–196.

Naiman, R. J. 1999. A perspective on interdisciplinary science. *Ecosystems* 2:292–295.

Naiman, R. J., and H. Decamps. 1997. The ecology of interfaces: riparian zones. *Annual Review of Ecology and Systematics* 28:621–658.

Naiman, R. J., J. M. Melillo, J. E. Hobbie. 1986. Ecosystem alteration of boreal streams by beaver (*Castor canadensis*). *Ecology* 67:1254–1269.

Naiman, R. J., J. J. Magnuson, D. M. McKnight, and J. A. Stanford, editors. 1995. *The Freshwater Imperative: A Research Agenda.* Island Press, Washington, DC, USA.

Naveh, Z. 1982. Landscape ecology as an emerging branch of human ecosystem science. *In* A. MacFadyen and E. D. Ford, eds. *Advances in Ecological Research*, pp. 189–237. Academic Press, New York, New York, USA.

Naveh, Z., and A. S. Lieberman. 1984. *Landscape Ecology, Theory and Application.* Springer-Verlag, New York, New York, USA.

Naveh, Z., and A. S. Lieberman. 1990. *Landscape Ecology, Theory and Application* (2nd edition). Springer-Verlag, New York, New York, USA.

Neilson, R. P. 1995. A model for predicting continental-scale vegetation distribution and water balance. *Ecological Applications* 5:362–385.

Neilson, R. P., and R. J. Drapek. 1998. Potentially complex biosphere responses to transient global warming. *Global Change Biology* 4:505–521.

Neilson, R. P., and L. H. Wullstein. 1983. Biogeography of two southwest American oaks in relation to atmospheric dynamics. *Journal of Biogeography* 10:275–297.

Neilson, R. P., G. A. King, and G. Koerper. 1992. Toward a rule-based biome model. *Landscape Ecology* 7:27–43.

Nemani, R., L. Pierce, and S. Running. 1993. Forest ecosystem processes at the watershed scale: sensitivity to remotely sensed leaf area index estimates. *International Journal of Remote Sensing* 14:2519–2534.

Nepstad, D., C. Uhl, and E. A. Serrao. 1990. Surmounting barriers to forest regeneration in abandoned, highly degraded pastures: a case study from Paragominas, Para, Brazil. *In* A. B. Anderson, ed. *Alternatives to Deforestation: Steps Toward Sustainable Use of the Amazon Rainforest*, pp. 215–229. Columbia University Press, New York, New York, USA.

Nero, R. W., and J. J. Magnuson. 1992. Effects of changing spatial scale on acoustic observations of patchiness in the Gulf Stream. *Landscape Ecology* 6:279–291.

Nicholson, M. C., and T. N. Mather. 1996. Methods for evaluating Lyme disease risks using geographic information systems and geospatial analysis. *Journal of Medical Entomology* 33:711–720.

Nir, D. 1983. *Man, a Geomorphological Agent*. D. Reidel, Boston, Massachusetts, USA.

Nitecki, M. H., and A. Hoffman. 1987. *Neutral Models in Biology*. Oxford University Press, Oxford, UK.

Noble, I. R., and R. Dirzo. 1997. Forests as human-dominated ecosystems. *Science* 277:522–525.

Noble, I. R., and R. O. Slatyer. 1980. The use of vital attributes to predict successional changes in plant communities subject to recurrent disturbances. *Vegetatio* 43:5–21.

Noon, B. R., and K. S. McKelvey. 1996a. A common framework for conservation planning: linking individual and metapopulation models. *In* D. R. McCullough, ed. *Metapopulations and Wildlife Conservation*, pp. 139–166. Island Press, Washington, DC, USA.

Noon, B. R., and K. S. McKelvey. 1996b. Management of the spotted owl: a case history in conservation biology. *Annual Review of Ecology and Systematics* 27:135–162.

Noss, R. F. 1987. From plant communities to landscapes in conservation inventories: a look at the Nature Conservancy (USA). *Biological Conservation* 41:11–37.

Offerman, H., V. Dale, S. Pearson, R. Bierregaard, and R. O'Neill. 1995. Effects of forest fragmentation on neotropical fauna: current research and data availability. *Environmental Reviews* 3:191–211.

Okubo, A. 1974. *Diffusion-induced Instability in Model Ecosystems: Another Possible Explanation of Patchiness*. Chesapeake Bay Institute, Technical report 86.

Okubo, A. 1975. *Ecology and Diffusion*. Tsukiji Shokan, Toyko, Japan.

Okubo, A. 1980. *Diffusion and Ecological Problems: Mathematical Models*. Springer-Verlag, Berlin, Germany.

Omernick, J. M. 1977. *Nonpoint Source–stream Nutrient Level Relationships: A Nationwide Study*. EPA-600/3-77-105. U.S. Environmental Protection Agency, Corvallis, Oregon, USA.

Omernick, J. M. 1987. Ecoregions of the conterminous United States. *Annals of the Association of American Geographers* 77:118–125.

Omernik, J. M., A. R. Abernathy, and L. M. Male. 1981. Stream nutrient levels and proximity of agricultural and forest land to streams: some relationships. *Journal of Soil and Water Conservation* 36:227–231.

O'Neill, R. V. 1979a. Natural variability as a source of error in model predictions. *In* G. S. Innis and R. V. O'Neill, eds. *Systems Analysis of Ecosystems*, pp. 23–32. International Cooperative Publishing House, Fairland, Maryland, USA.

O'Neill, R. V. 1979b. Transmutative across hierarchical levels. *In* G. S. Innis and R. V. O'Neill, eds. *Systems Analysis of Ecosystems*, pp. 58–78. International Cooperative Publishing House, Fairland, Maryland, USA.

O'Neill, R. V., C. T. Hunsaker, K. B. Jones, K. H. Riitters, J. D. Wickham, P. M. Schwartz, I. A. Goodman, B. L. Jackson, and W. S. Baillargeon. 1977. Monitoring environmental quality at the landscape scale. *BioScience* 47:513–519.

O'Neill, R. V., R. H. Gardner, L. W. Barnthouse, G. W. Suter, S. G. Hildebrand, and C. W. Gehrs. 1982. Ecosystem risk analysis: a new methodology. *Environmental Toxicology and Chemistry* 1:167–177.

O'Neill, R. V., D. L. DeAngelis, J. B. Waide, and T. F. H. Allen. 1986. *A Hierarchical Concept of Ecosystems*. Princeton University Press, Princeton, New Jersey, USA.

O'Neill, R. V., J. R. Krummel, R. H. Gardner, G. Sugihara, B. Jackson, D. L. DeAngelis, B. T. Milne, M. G. Turner, B. Zygmunt, S. Christensen, V. H. Dale, and R. L Graham. 1988a. Indices of landscape pattern. *Landscape Ecology* 1:153–162.

O'Neill, R. V., B. T. Milne, M. G. Turner, and R. H. Gardner. 1988b. Resource utilization scales and landscape pattern. *Landscape Ecology* 2:63–69.

O'Neill, R. V., A. R. Johnson, and A. W. King. 1989. A hierarchical framework for the analysis of scale. *Landscape Ecology* 3:193–205.

O'Neill, R. V., R. H. Gardner, B. T. Milne, M. G. Turner, and B. Jackson. 1991a. Heterogeneity and spatial hierarchies. *In* J. Kolasa and and S. T. A. Pickett eds. *Ecological Heterogeneity*, pp. 85–96. Springer-Verlag, New York, New York, USA.

O'Neill, R. V., S. J. Turner, V. I. Cullinan, D. P. Coffin, T. Cook, W. Conley, J. Brunt, J. M. Thomas, M. R. Conley, and J. Gosz. 1991b. Multiple landscape scales: an intersite comparison. *Landscape Ecology* 5:137–144.

O'Neill, R. V., R. H. Gardner, and M. G. Turner. 1992a. A hierarchical neutral model for landscape analysis. *Landscape Ecology* 7:55–61.

O'Neill, R. V., R. H. Gardner, M. G. Turner, and W. H. Romme. 1992b. Epidemiology theory and disturbance spread on landscapes. *Landscape Ecology* 7:19–26.

O'Neill, R. V., K. B. Jones, K. H. Riitters, J. D. Wickham, and I. A. Goodman. 1994. *Landscape Monitoring and Assessment Research Plan.* U. S. EPA 620/R–94/009, Environmental Protection Agency, Las Vegas, Nevada, USA.

O'Neill, R. V., C. T. Hunsaker, S. P. Timmins, B. L. Jackson, K. B. Jones, K. H. Riitters, and J. D. Wickham. 1996. Scale problems in reporting landscape pattern at the regional scale. *Landscape Ecology* 11:169–180.

Ontario Ministry of Natural Resources. 1996. *Forest Management Planning Manual for Ontario's Crown Forests.* Queen's Printer for Ontario, Toronto, Canada.

Opdam, P. 1987. De metapopulatie, model van een populatie in een versnipperd landschap. *Landschap* 4:289–306.

Opdam, P. 1991. Metapopulation theory and habitat fragmentation: a review of holarctic breeding bird studies. *Landscape Ecology* 5:93–106.

Opdam, P., D. Van Dorp, and C. J. F. Ter Braak. 1984. The effect of isolation on the number of woodland birds in small woods in the Netherlands. *Journal of Biogeography* 11:473–476.

Opdam, P., G. Rijsdijk, and F. Hustings. 1985. Bird communities in small woods in an

agricultural landscape: effects of area and isolation. *Biological Conservation* 34:333–352.
Opdam, P., R. Foppen, R. Reijnen, and A. Schotman. 1996. The landscape ecological approach in bird conservation: integrating the metapopulation approach into spatial planning. *Ibis* 137:S139–S146.
Orians, G. H., and J. F. Wittenberger. 1991. Spatial and temporal scales in habitat selection. *American Naturalist* 137:S29–S49.
Osborne, L. L., and D. A. Kovacic. 1993. Riparian vegetated buffer strips in water quality restoration and stream management. *Freshwater Biology* 29:243–258.
Osborne, L. L., and M. J. Wiley. 1988. Empirical relationships between land use/land cover and stream water quality in an agricultural watershed. *Journal of Environmental Management* 26:9–27.
Overpeck, J. T. D. R., and R. Goldberg. 1990. The fractal shape of riparian forest patches. *Landscape Ecology* 4:249–258.
Overton, J. M. 1975. The ecosystem modeling approach in the coniferous forest biome. *In* B. C. Patten, ed. *Systems Analysis and Simulation in Ecology*, pp. 117–138. Academic Press, New York, New York, USA.
Pace, M. L., and P. M. Groffman, editors. 1998. *Successes, Limitations and Frontiers in Ecosystem Science.* Springer-Verlag, New York, New York, USA.
Paine, R. T. 1974. Intertidal community structure: experimental studies on the relationship between a dominant competitor and its principal predator. *Oecologia* 15:93–120.
Paine, R. T. 1976. Size limited predation: an observational and experimental approach with the *Mytilus–Pisaster* interaction. *Ecology* 57:858–873.
Paine, R. T., M. Tegner, and E. A. Johnson. 1998. Compounded perturbations yield ecological surprises: everything else is business as usual. *Ecosystems* 1:535–545.
Palmer, M. W. 1992. The coexistence of species in fractal landscapes. *American Naturalist* 139:375–397.
Pan, Y. D., J. M. Melillo, A. D. McGuire, D. W. Kicklighter, L. F. Pitelka, K. Hibbard, L. L. Pierce, S. W. Running, D. S. Ojima, W. J. Parton, and D. S. Schimel. 1998. Modelled responses of terrestrial ecosystems to elevated atmospheric CO_2: a comparison of simulations by the biogeochemistry models of the Vegetation/Ecosystem Modeling and Analysis Project (VEMAP). *Oecologia* 114:389–404.
Pandey, G., S. Lovejoy, and D. Schertzer. 1998. Multifractal analysis of daily river flows including extremes for basins of five to two million square kilometers, one day to 75 years. *Journal of Hydrology* 208:62–81.
Pare, D., Y. Bergeron, and C. Camire. 1993. Changes in the forest floor of Canadian southern boreal forest after disturbance. *Journal of Vegetation Science* 4:811–818.
Parker, V. T., and S. T. A. Pickett. 1998. Historical contingency and multiple scales of dynamics within plant communities. *In* E. L. Peterson and V. T. Parker, eds. *Ecological Scale, Theory and Applications*, pp. 171–193. Columbia University Press, New York, New York, USA.
Parsons, W. F. J., D. H. Knight, and S. L. Miller. 1994. Root gap dynamics in lodgepole pine forest: nitrogen transformation in gaps of different size. *Ecological Applications* 4:354–362.
Parton, W. J., D. S. Schimel, D. V. Cole, and D. S. Ojima. 1987. Analysis of factors con-

trolling soil organic matter levels in Great Plains Grasslands. *Soil Science Society of America Journal* 51:1173–1179.

Parton, W. J., A. R. Mosier, and D. S. Schimel. 1988. Rates and pathways of nitrous oxide production in a shortgrass steppe. *Biogeochemistry* 6:45–58.

Pastor, J., and Y. Cohen. 1997. Herbivores, the functional diversity of plant species, and the cycling of nutrients in boreal ecosystems. *Theoretical Population Biology*, 51:165–179.

Pastor, J., and R. J. Naiman. 1992. Selective foraging and ecosystem processes in boreal forests. *American Naturalist* 139:690–705.

Pastor, J., R. J. Naiman, B. Dewey, and P. McInnes. 1988. Moose, microbes and the boreal forest. *Bioscience* 38:770–777.

Pastor, J., B. Dewey, R. J. Naiman, P. F. McInnes, and Y. Cohen. 1993. Moose browsing and soil fertility in the boreal forests of Isle Royale National Park. *Ecology* 74:467–480.

Pastor, J., R. Moen, and Y. Cohen. 1997. Spatial heterogeneities, carrying capacity, and feedbacks in animal–landscape interactions. *Journal of Mammalogy* 78:1040–1052.

Pastor, J., Y. Cohen, and R. Moen. 1999. Generation of spatial patterns in boreal forest landscapes. *Ecosystems* 2:439–450.

Pastor, J., B. Dewey, R. Moen, D. J. Mladenoff, M. White, and Y. Cohen. 1998. Spatial patterns in the moose–forest–soil ecosystem on Isle Royale, Michigan, USA. *Ecological Applications*, 8:411–424.

Patten, B. C. 1971. A primer for ecological modeling and simulation with analog and digital computers. *In* B. C. Patten, ed. *Systems Analysis and Simulation in Ecology*, Volume 1, pp. 3–121. Academic Press, New York, New York, USA.

Patterson, W. A., and D. R. Foster. 1990. Tabernacle Pines: the rest of the story. *Journal of Forestry* 88:23–25.

Pausas, J. G., M. P. Austin, and I. R. Noble. 1997. A forest simulation model for predicting Eucalypt dynamics and habitat quality for arboreal marsupials. *Ecological Applications* 7:921–933.

Pearson, S. M. 1993. The spatial extent and relative influence of landscape-level factors on wintering bird populations. *Landscape Ecology* 8:3–18.

Pearson, S. M., and R. H. Gardner. 1997. Neutral models: useful tools for understanding landscape patterns. *In* J. A. Bisonnette, ed. *Wildlife and Landscape Ecology: Effects of Pattern and Scale*, pp. 215–230. Springer-Verlag, New York, New York, USA.

Pearson, S. M., M. G. Turner, L. L. Wallace, and W. H. Romme. 1995. Winter habitat use by large ungulates following fires in northern Yellowstone National Park. *Ecological Applications* 5:744–755.

Pearson, S. M., M. G. Turner, R. H. Gardner, and R. V. O'Neill. 1996. An organism-based perspective of habitat fragmentation. *In* R. C. Szaro, ed. *Biodiversity in Managed Landscapes: Theory and Practice*, pp. 77–95 Oxford University Press, Covelo, California, USA.

Pearson, S. M., A. B. Smith, and M. G. Turner. 1998. *Forest Fragmentation, Land Use, and Cove-forest Herbs in the French Broad River Basin*. Castanea 63:382–395.

Pearson, S. M., M. G. Turner, and J. B. Drake. 1999. Simulating land-cover change and species' habitats in the Southern Appalachian Highlands and the Olympic Peninsula. *Ecological Applications* 9:1288–1304.

Peet, R. K. 1992. Community structure and ecosystem function. *In* D. C. Glenn-Lewin, R.

K. Peet, and T. T. Veblen, eds. *Plant Succession*, pp. 103–151. Chapman and Hall, New York, New York, USA.

Peet, R. K., and N. L. Christensen. 1980. Succession: a population process. *Vegetatio* 43: 131–140.

Peitgen, H. O., H. Jurgens, and D. Saupe. 1992. *Chaos and Fractals: New Frontiers of Science.* Springer-Verlag, Berlin, Germany.

Peng, C. H., M. J. Apps, D. T. Price, I. A. Nalder, and D. H. Halliwell. 1998. Simulating carbon dynamics along the Boreal Forest Transect Case Study (BFTCS) in central Canada—1. Model testing. *Global Biogeochemical Cycles* 12:381–392.

Perera, A. H., D. L. Euler, and I. D. Thompson, eds. 2000. *Ecology of a Managed Terrestrial Landscape.* UBC Press, Vancouver, British Columbia, Canada.

Peterjohn, W. T., and D. L. Correll. 1984. Nutrient dynamics in an agricultural watershed: observations on the role of a riparian forest. *Ecology* 65:1466–1475.

Peterson, C. J., and W. P. Carson. 1996. Generalizing forest regeneration models: the dependence of propagule availability on disturbance history and stand size. *Canadian Journal of Forest Research* 26:45–52.

Peterson, C. J., and S. T. A. Pickett. 1995. Forest reorganization: a case study in an old-growth forest catastrophic blowdown. *Ecology* 76:763–774.

Peterson, D. L., and V. T. Parker. 1998. Dimensions of scale in ecology, resource management, and society. *In* D. L. Peterson and V. T. Parker, eds. *Ecological Scale: Theory and Applications*, pp. 499–522. Columbia University Press, New York, New York, USA.

Petit, D. R. 1989. Weather-dependent use of habitat patches by wintering birds. *Journal of Field Ornithology* 60:241–247.

Pickett, S. T. A. 1976. Succession: an evolutionary interpretation. *American Naturalist* 110:107–119.

Pickett, S. T. A. 1998. Natural processes. *In* M. J. Mac, P. A. Opler, P. Doran, and C. Haecker, eds. *Status and Trends of Our Nation's Biological Resources*, Volume 1, pp. 11–36. National Biological Service, Washington, DC, USA.

Pickett, S. T. A., and J. N. Thompson. 1978. Patch dynamics and the design of nature reserves. *Biological Conservation* 13:27–37.

Pickett, S. T. A., and P. S. White, editors. 1985. *The Ecology of Natural Disturbance and Patch Dynamics.* Academic Press, New York, New York, USA.

Pickett, S. T. A., and M. L. Cadenasso. 1995. Landscape ecology: spatial heterogeneity in ecological systems. *Science* 269:331–334.

Pickett, S. T. A., S. C. Collins, and J. J. Armesto. 1987a. Models, mechanisms and pathways of succession. *Botanical Review* 53:335–371.

Pickett, S. T. A., S. C. Collins, and J. J. Armesto. 1987b. A hierarchical consideration of causes and mechanisms of succession. *Vegetatio* 69:109–114.

Pickett, S. T. A., J. Kolasa, J. J. Armesto, and S. L. Collins. 1989. The ecological concept of disturbance and its expression at various hierarchical levels. *Oikos* 54:129–136.

Pickett, S. T. A., J. Kolasa, and C. G. Jones. 1994. *Ecological Understanding. The Nature of Theory and the Theory of Nature.* Academic Press, New York, New York, USA.

Pickett, S. T. A., W. R. Burch, Jr., and J. M. Grove. 1999. Interdisciplinary research: maintaining the constructive impulse in a culture of criticism. *Ecosystems* 2:302–307.

Pierce, L. L., and S. W. Running. 1995. The effects of aggregating sub-grid land surface variation on large-scale estimates of net primary production. *Landscape Ecology* 10:239–253.

Pierce, L. L., S. W. Running, and J. Walker. 1994. Regional-scale relationships of leaf area index to specific leaf area and leaf nitrogen. *Ecological Applications* 4:313–321.

Pittsburgh Community Renewal Project. 1962. *Employment, Income, and Population Submodels*. Progress Report No. 4. Department of City Planning, Pittsburgh, Pennsylvania, USA.

Platt, J. R. 1964. Strong inference. *Science* 146:347–353.

Platt, T., and K. L. Denman. 1975. Spectral analysis in ecology. *Annual Review of Ecology and Systematics* 6:189–210.

Platt, W. J., and I. M. Weis. 1985. An experimental study of competition among fugitive prairie plants. *Ecology* 66:708–720.

Plotnick, R. E., and R. H. Gardner. 1993. Lattices and landscapes. *In* R. H. Gardner, ed. *Lectures on Mathematics in the Life Sciences: Predicting Spatial Effects in Ecological Systems*, pp. 129–157. American Mathematical Society, Providence, Rhode Island, USA.

Plotnick, R. E., and K. L. Prestegaard. 1993. Fractal analysis of geologic time series. *In* N. Lam and L. DeCola, eds. *Fractals in Geography*, pp. 207–224. Prentice Hill, Upper Saddle River, New Jersey, USA.

Plotnick, R. E., R. H. Gardner, and R. V. O'Neill. 1993. Lacunarity indices as measures of landscape texture. *Landscape Ecology* 8:201–211.

Pollock, M. M. 1998. Biodiversity. *In* R. J. Naiman and R. E. Bilby, eds. *River Ecology and Management*, pp. 430–452. Springer-Verlag, New York, New York, USA.

Powell, T. M., P. J. Richerson, T. M. Dillon, B. A. Agee, B. J. Dozier, D. A. Godden, and L. O. Myrup. 1975. Spatial scales of current speed and phytoplankton biomass fluctuations in Lake Tahoe. *Science* 189:1088–1090.

Pulliam, H. R. 1988. Sources, sinks and population regulation. *American Naturalist* 132:652–661.

Pulliam, H. R., and B. J. Danielson. 1991. Sources, sinks, and habitat selection: a landscape perspective on population dynamics. *American Naturalist* 132:S50–S66.

Pulliam, H. R., and J. B. Dunning. 1994. Demographic processes: population dynamics on heterogeneous landscapes. *In* G. K. Meffe and C. R. Carroll, eds. *Principles of Conservation Biology*, pp. 179–205. Sinauer Associates, Inc., Sunderland, Massachusetts.

Pulliam, H. R., J. B. Dunning, and J. Liu. 1992. Population dynamics in complex landscapes: a case study. *Ecological Applications* 2:165–177.

Quinn, J. F., and A. E. Dunham. 1983. On hypothesis testing in ecology and evolution. *American Naturalist* 122:602–617.

Radcliffe, J. 1973. The initial geographical spread of host–vector and carrier-borne epidemics. *Journal of Applied Probability* 10:703–717.

Ranney, J. W., M. C. Bruner, and J. B. Levenson. 1981. The importance of edge in the structure and dynamics of forest islands. *In* R. L. Burgess and D. M. Sharpe, eds. *Forest Island Dynamics in Man-dominated Landscapes*, pp. 67–95. Springer-Verlag, New York, New York, USA.

Rastetter, E. B., A. W. King, B. J. Cosby, G. M. Hornberger, R. B. O'Neill, and J. E. Hobbie. 1992. Aggregating fine-scale ecological knowledge to model coarser-scale attributes of ecosystems. *Ecological Applications* 2:55–70.

Reddingius, J., and P. J. Den Boer. 1970. Simulation experiments illustrating stabilization of animal numbers by spreading of risk. *Oecologia* 5:240–284.

Reibsame, W. E., W. J. Parton, K. A. Galvin, I. C. Burke, L. Bohren, R. Young, and E. Knop. 1994. Integrated modeling of land use and cover change. *BioScience* 44:350–356.

Reich, P. B., D. F. Grigal, J. D. Aber, and S. T. Gower. 1997. Nitrogen mineralization and productivity in 50 hardwood and conifer stands on diverse soils. *Ecology* 78:335–347.

Reiners, W. A., and G. E. Lang. 1979. Vegetational patterns and processes in the balsam fir zone, White Mountains, New Hampshire. *Ecology* 60:403–417.

Remillard, M. M., G. K. Gruendling, and D. J. Bogucki. 1987. Disturbance by beaver (*Castor canadensis* Kuhl) and increased landscape heterogeneity. *In* M. G. Turner, ed. *Landscape Heterogeneity and Disturbance*, pp. 103–123. Springer-Verlag, New York, New York, USA.

Renkin, R. A., and D. G. Despain. 1992. Fuel moisture, forest type and lightning-caused fire in Yellowstone National Park. *Canadian Journal of Forest Research* 22:37–45.

Rescia, A. J., M. F. Schmitz, P. M. deAgar, C. L. dePablo, and F. D. Pineda. 1997. A fragmented landscape in northern Spain analyzed at different spatial scales: implications for management. *Journal of Vegetation Science* 8:343–352.

Richards, D., L. B. Johnson, and G. Host. 1996. Landscape-scale influences on stream habitats and biota. *Canadian Journal of Fisheries and Aquatic Science* 53 (Suppl. 1):295–311.

Richards, J. F. 1990. Land transformation. *In* B. L. Turner III et al., eds. *The Earth as Transformed by Human Action*, pp. 163–178. Cambridge University Press, Cambridge, UK.

Ricotti, M. E., and E. Zio. 1999. Neural network approach to sensitivity and uncertainty analysis. *Reliability Engineering and System Safety* 64:59–71.

Riebsame, W. E., W. B. Meyer, and B. L. Turner II. 1994. Modeling land use and cover as part of global environmental change. *Climatic Change* 28:45–64.

Riera, J. L., J. J. Magnuson, J. R. Vande Castle, and M. D. MacKenzie. 1998. Analysis of large-scale spatial heterogeneity in vegetation indices among North American landscapes. *Ecosystems* 1:268–282.

Riera, J. L., J. J. Magnuson, T. K. Kratz, and K. E. Webster. 2000. A geomorphic template for the analysis of lake district applied to the Northern Highland Lake District, Wisconsin, USA. *Freshwater Biology*, 43:301–318.

Riitters, K. H., and J. D. Wickham. 1995. *A Landscape Atlas of the Chesapeake Bay Watershed*. Environmental Research Center, Tennessee Valley Authority, Norris, Tennessee, USA.

Riitters, K. H., R. V. O'Neill, C. T. Hunsaker, J. D. Wickham, D. H. Yankee, S. P. Timmons, K. B. Jones, and B. L. Jackson. 1995. A factor analysis of landscape pattern and structure metrics. *Landscape Ecology* 10:23–40.

Riitters, K. H., R. V. O'Neill, and K. B. Jones. 1997. Assessing habitat suitability at multiple scales: a landscape-level approach. *Biological Conservation* 61:191–202.

Ripley, B. D. 1978. Spectral analysis and the analysis of pattern in plant communities. *Journal of Ecology* 66:965–981.

Ripley, B. D. 1981. *Spatial Statistics*. John Wiley & Sons, New York, New York, USA.

Risch, S. J., D. Andow, and M. A. Altieri. 1983. Agro-ecosystem diversity and pest control: data, tentative conclusions and new research directions. *Environmental Entomology* 12:625–629.

Risser, P. G., J. R. Karr, and R. T. T. Forman. 1984. *Landscape Ecology: Directions and Approaches*. Special Publication Number 2. Illinois Natural History Survey, Champaign, Illinois, USA.

Roberts, D. W. 1996. Landscape vegetation modeling with vital attributes and fuzzy systems theory. *Ecological Modeling* 90:175–184.

Roberts, M. R., and F. S. Gilliam. 1995. Patterns and mechanisms of plant diversity in forested ecosystems—implications for forest management. *Ecological Applications* 5:969–977.

Roe, F. G. 1951. *The North American Buffalo*. University of Toronto Press, Toronto, Canada.

Roese, J. H., K. L. Risenhoover, and L. J. Folse. 1991. Habitat heterogeneity and foraging efficiency: an individual-based model. *Ecological Modeling* 57:133–143.

Roff, D. A. 1974a. Spatial heterogeneity and the persistence of populations. *Oecologia* 15:240–258.

Roff, D. A. 1974b. The analysis of a population model demonstrating the importance of dispersal in a heterogeneous environment. *Oecologia* 15:259–275.

Roland, J. 1993. Large-scale forest fragmentation increases the duration of tent caterpillar outbreak. *Oecologia* 93:25–30.

Romme, W. H. 1982. Fire and landscape diversity in subalpine forests of Yellowstone National Park. *Ecological Monographs* 52:199–221.

Romme, W. H., and D. G. Despain. 1989. Historical perspective on the Yellowstone fires of 1988. *BioScience* 39:695–699.

Romme, W. H., and D. H. Knight. 1982. Landscape diversity: the concept applied to Yellowstone Park. *BioScience* 32:664–670.

Romme, W. H., E. E. Everham, L. E. Frelich, M. A. Moritz, and R. E. Sparks. 1998. Are large, infrequent disturbances qualitatively different from small, frequent disturbances? *Ecosystems* 1:524–534.

Rose, K. A., A. L. Brenkert, R. B. Cook, R. H. Gardner, and J. P. Hettelingh. 1991a. Systematic comparison of ILWAS, MAGIC, and ETD watershed acidification models: 1. Mapping among model inputs and deterministic results. *Water Resources Research* 27:2577–2598.

Rose, K. A., A. L. Brenkert, R. B. Cook, R. H. Gardner, and J. P. Hettelingh. 1991b. Systematic comparison of ILWAS, MAGIC, and ETD watershed acidification models: 2. Monte Carlo analysis under regional variability. *Water Resources Research* 27:2591–2603.

Rose, K. A., E. P. Smith, R. H. Gardner, A. L. Brenkert, and S. M. Bartell. 1991c. Parameter sensitivities, Monte Carlo filtering, and model forecasting under uncertainty. *Journal of Forecasting* 10:117–133.

Rosenberg, D. K., B. R. Noon, and E. C. Meslow. 1997. Biological corridors: form, function, and efficacy. *BioScience* 47:677–687.

Rossi, R. E., D. J. Mulla, A. G. Journel, and E. H. Franz. 1992. Geostatistical tools for modeling and interpreting ecological spatial dependence. *Ecological Monographs* 62:277–314.

Rowan, R., N. Knowlton, A. Baker, and J. Jara. 1997. Landscape ecology of algal symbionts creates variation in episodes of coral bleaching. *Nature* 388:265–269.

Rowe, J. S. 1961. The level-of-integration concept and ecology. *Ecology* 42:420–427.

Rowe, J. S., and G. W. Scotter. 1973. Fire in the boreal forest. *Quaternary Research* 3:444–464.

Runkle, J. R. 1982. Patterns of disturbance in some old-growth mesic forests of eastern North America. *Ecology* 63:1533–1546.

Runkle, J. R. 1985. Disturbance regimes in temperate forests. *In* S. T. A. Pickett and P. S. White, eds. *The Ecology of Natural Disturbance and Patch Dynamics*, pp. 17–34. Academic Press, New York, New York, USA.

Runkle, J. R. 1991. Gap dynamics of old-growth eastern forests: management implications. *Natural Areas Journal* 11:19–25.

Running, S. W. 1990. Remote sensing of terrestrial primary productivity. *In* R. J. Hobbs and H. A. Mooney, eds. *Remote sensing of biosphere functioning*, pp. 65–85. Springer-Verlag, New York, New York, USA.

Running, S. W. 1994. Testing forest-BGC ecosystem process simulations across a climatic gradient in Oregon. *Ecological Applications* 4:238–247.

Running, S. W., and J. C. Coughlan. 1988. A general model of forest ecosystem processes for regional applications. I. Hydrologic balance, canopy gas exchange and primary production processes. *Ecological Modeling* 42:125–154.

Running, S. W., and E. R. Hunt, Jr. 1993. Generalization of a forest ecosystem process model for other biomes, BIOME-BGC, and an application for global-scale models. *In* J. R. Ehlinger and C. Field, eds. *Scaling Physiological Processes: Leaf to Globe*, pp. 141–157. Academic Press, San Diego, California, USA.

Running, S. W., R. R. Nemani, and R. D. Hungerford. 1987. Extrapolation of synoptic meterological data in mountainous terrain, and its use for simulating forest evapotranspiration and photosynthesis. *Canadian Journal of Forest Research* 17:472–483.

Running, S. W., R. R. Nemani, D. L. Peterson, L. E. Band, D. F. Potts, L. L. Pierce, and M. A. Spanner. 1989. Mapping regional forest evapotranspiration and photosynthesis by coupling satellite data with ecosystem simulation. *Ecology* 70:1090–1101.

Russell, R. W., G. L. Hunt, Jr., K. O. Coyle, and R. T. Cooney. 1992. Foraging in a fractal environment: spatial patterns in a marine predator–prey system. *Landscape Ecology* 7:195–209.

Russo, J. M., and J. W. Zack. 1997. Downscaling GCM output with a mesoscale model. *Journal of Environmental Management* 49:19–29.

Ruzicka, M., and L. Miklos. 1990. Basic premises and methods in landscape ecological planning and optimization. *In* I. S. Zonneveld and R. T. T. Forman, eds. *Changing Landscapes: An Ecological Perspective*, pp. 233–260. Springer-Verlag, New York, New York, USA.

Rykiel, E. J. 1996. Testing ecological models: the meaning of validation. *Ecological Modeling* 90:229–244.

Rykiel, E. J., R. N. Coulson, P. J. H. Sharpe, T. F. H. Allen, and R. O. Flamm. 1988. Disturbance propagation by bark beetles as an episodic landscape phenomenon. *Landscape Ecology* 1:129–139.

Sala, O. E., W. J. Parton, L. A. Joyce, and W. K. Lauenroth. 1988. Primary production of the central grassland region of the United States. *Ecology* 69:40–45.

Santelmann, M., K. Freemark, D. White, J. Nassauer, M. Clark, B. Danielson, J. Eilers, R. Cruse, S. Galatowitsch, S. Polasky, K. Vache, and J. Wu. *In press*. Applying ecological principles to land-use decision making in agricultural watersheds. *In* V. H. Dale and R. Haeuber, eds. *Applying Ecological Principles to Land Management*. Springer-Verlag, New York, New York, USA.

Saunders, D. and R. J. Hobbs, editors. 1991. *Nature Conservation 2: The Role of Corridors*. Surrey Beatty & Sons, Chipping Norton, N.S.W., Australia.

Saunders, D., R. J. Hobbs, and C. R. Margules. 1991. Biological consequences of ecosystem fragmentation: a review. *Conservation Biology* 5:18–32.

Saupe, D. 1988. Algorithms for random fractals. *In* H. O. Peitgen and D. Saupe, eds. *The Science of Fractal Images*, pp. 71–133. Springer-Verlag, New York, New York, USA.

Scheffer, M., and R. J. D. Boer. 1995. Implications of spatial heterogeneity for the paradox of enrichment. *Ecology* 76:2270–2277.

Schimel, D. S., VEMAP Participants, and B. H. Braswell. 1997. Continental scale variability in ecosystem processes: models, data, and the role of disturbance. *Ecological Monographs* 67:251–271.

Schippers, P., J. Verboom, J. P. Knaapen, and R. C. van Apeldoorn. 1996. Dispersal and habitat connectivity in complex heterogeneous landscapes: an analysis with a GIS-based random walk model. *Ecography* 19:97–106.

Schmidt, K. 1993. Winter ecology of nonmigratory Alpine red deer. *Oecologia* 95:226–233.

Schneider, D. C. 1994. *Quantitative Ecology: Spatial and Temporal Scaling*. Academic Press, San Diego, California, USA.

Schneider, D. C. 1998. Applied scaling theory. *In* D. L. Peterson and V. L. Parker, eds. *Ecological Scale*, pp. 253–269. Columbia University Press, New York, New York, USA.

Schneider, D. C., and Piatt, J. F. 1986. Scale-dependent correlation of seabirds with schooling fish in a coastal ecosystem. *Marine Ecology—Progress Series* 32:237–246.

Schreiber, K. F. 1990. The history of landscape ecology in Europe. *In* I. S. Zonneveld and R. T. T. Forman, eds. *Changing Landscapes: An Ecological Perspective*, pp. 21–33. Springer-Verlag, New York, New York, USA.

Schulze, E.-D., and H. A. Mooney, editors. 1993. *Biodiversity and Ecosystem Function*. Springer-Verlag, New York, New York, USA.

Schumaker, N. H. 1996. Using landscape indices to predict habitat connectivity. *Ecology* 77:1210–1225.

Scott, J. M., J. J. Jacobi, and J. E. Estes. 1987. Species richness: a geographic approach to protecting future biological diversity. *BioScience* 37:782–788.

Scott, J. M., F. Davis, B. Csuti, R. Noss, B. Butterfield, C. Groves, H. Anderson, S. Caicco, F. D'Erchia, T. C. Edwards, Jr., J. Ulliman, and R. G. Wright. 1993. Gap analysis: a geographic approach to protection of biodiversity. *Wildlife Monographs No. 123*.

Scott, J. M., T. H. Tear, and F. W. Davis, editors. 1996. *Gap Analysis: A Landscape Approach to Biodiversity Planning*. American Society for Photogrammetry and Remote Sensing. Bethesda, Maryland, USA.

Seaber, P. R., F. P. Kapinos, and G. L. Knapp. 1984. *State Hydrologic Unit Maps*. U.S. Geological Survey, Denver, Colorado, USA.

Sedell, J. R., and J. L. Froggatt. 1984. Importance of streamside forests to large rivers: the isolation of the Willamette River, Oregon, USA, from its floodplain by snagging and streamside forest removal. *Verhandlungen des Internationalen Vereins der Limnologie* 22:1824–1834.

Segal, L. A, and S. A. Levin. 1976. Applications of nonlinear stability theory to the study of the effects of diffusion on predator–prey interactions. *Topics in Statistical Mechanics and Biophysics. Proceedings of the AIP Conference* 27:123–152.

Sellers, P. J. 1985. Canopy reflectance, photosynthesis and transpiration. *International Journal of Remote Sensing* 6:1335–1372.

Sellers, P. J. 1987. Canopy reflectance, photosynthesis and transpiration. II. The role of biophysics in the linearity of their interdependence. *Remote Sensing and Environment* 21:143–183.

Sellers, P. J. 1994. A global 1-degrees-by-1-degrees NDVI data set for climate studies. 2. The generation of global fields of terrestrial biophysical parameters from the NDVI. *International Journal of Remote Sensing* 15: 3519–3545.

Senft, R. L., M. B. Coughenour, D. W. Bailey, L. R. Rittenhouse, O. E. Sala, and D. M. Swift. 1987. Large herbivore foraging and ecological hierarchies. *BioScience* 37:789–799.

Shafer, M. L. 1985. The metapopulation and species conservation: the special case of the Northern Spotted Owl. *In* R. J. Gutierrez and A. B. Carey, eds. *Ecology and Management of the Spotted Owl in the Pacific Northwest,* pp. 86–99. United States Forest Service Technical Report PNW-185.

Shaver, G. R., K. J. Knadelhoffer, and A. E. Giblin. 1991. Biogeochemical diversity and element transport in a heterogeneous landscape, the north slope of Alaska. *In* M. G. Turner and R. H. Gardner, eds. *Quantitative Methods in Landscape Ecology*, pp. 105–125. Springer-Verlag, New York, New York, USA.

Shugart, H. H., and I. R. Noble. 1981. A computer model of succession and fire response of the high altitude Eucalyptus forest of the Brindabella Range, Australian Capital Territory. *Australian Journal of Ecology* 6:149–164.

Shugart, H. H., Jr., and D. C. West. 1981. Long-term dynamics of forest ecosystems. *American Scientist* 69:647–652.

Siemann, E., and J. H. Brown. 1999. Gaps in mammalian body size distributions reexamined. *Ecology* 80:2788–2792.

Silbernagel, J., S. R. Martin, M. R. Gale, and J. Chen. 1997. Prehistoric, historic, and present settlement patterns related to ecological hierarchy in the Eastern Upper Peninsula of Michigan, USA. *Landscape Ecology* 12:223–240.

Simberloff, D. S. 1974. Equilibrium theory of island biogeography and ecology. *Annual Review of Ecology and Systematics* 5:161–182.

Simberloff, D. S. 1978. Using island biogeographic distributions to determine if colonization is stochastic. *American Naturalist* 112:713–719.

Simberloff, D., and J. Cox. 1987. Consequences and costs of conservation corridors. *Conservation Biology* 1:63–71.

Simberloff, D., and E. O. Wilson. 1969. Experimental zoogeography of islands. The colonization of empty islands. *Ecology* 50:278–296.

Simberloff, D., and E. O. Wilson. 1970. Experimental zoogeography of islands. A two-year record of colonization. *Ecology* 50:278–316.

Simpson, J. B. 1974. Glacial migration of plants: island biogeography evidence. *Science* 185:698–700.

Sisk, T. D., N. M. Haddad, and P. R. Ehrlich. 1997. Bird assemblages in patchy woodlands: modeling the effects of edge and matrix habitats. *Ecological Applications* 7:1170–1180.

Skellam, J. G. 1951. Random dispersal in theoretical populations. *Biometrika* 38:196–218.

Skinner, C. N. 1995. Change in spatial characteristics of forest openings in the Klamath Mountains of northwestern California, USA. *Landscape Ecology* 10:219–228.

Sklar, F. H., and R. Costanza. 1990. The development of dynamic spatial models for landscape ecology: a review and prognosis. *In* M. G. Turner and R. H. Gardner, eds. *Quantitative Methods in Landscape Ecology*, pp. 239–288. Springer-Verlag, New York, New York, USA.

Slocombe, D. S. 1993. Implementing ecosystem-based management. *BioScience* 43:612–622.

Slud, P. 1976. Geographic and climatic relationships of avifaunas with special reference to comparative distribution in the Neotropics. *Smithsonian Contributions in Zoology* 212:1–149.

Small, M. F., and M. L. Hunter. 1988. Forest fragmentation and avian nest predation in forested landscapes. *Oecologia* 76:62–64.

Smith, A. T. 1980. Temporal changes in insular populations of the pika (*Ochotona princeps*). *Ecology* 61:8–13.

Smith, C. A., editor. 1976. *Regional Analysis*. Academic Press, New York, New York, USA.

Smith, M., T. Keevin, P. Mettler-McClure, and R. Barkau. 1998. Effect of the flood of 1993 on *Boltonia decurrens*, a rare floodplain plant. *Regulated Rivers Research Management* 14:191–202.

Smuts, J. 1926. *Holism and Evolution*. Macmillan, London, UK.

Sollins, P., G. Spycher, and C. Topik. 1983. Processes of soil organic matter accretion at a mudflow chronosequence, Mt. Shasta, California. *Ecology* 64:1273–1282.

Soranno, P. A., S. L. Hubler, S. R. Carpenter, and R. C. Lathrop. 1996. Phosphorus loads to surface waters: a simple model to account for spatial pattern of land use. *Ecological Applications* 6:865–878.

Soranno, P. A., K. E. Webster, J. L. Riera, T. K. Kratz, J. S. Baron, P. A. Bukaveckas, G. W. Kling, D. S. White, N. Caine, R. C. Lathrop, and P. R. Leavitt. 1999. Spatial variation among lakes within landscapes: ecological organization along lake chains. *Ecosystems* 2:395–410.

Soule, M. E., and D. Simberloff. 1986. What do genetics and ecology tell use about the design of nature reserves? *Biological Conservation* 2:75–92.

Sousa, W. P. 1984. The role of disturbance in natural communities. *Annual Review of Ecology and Systematics* 15:353–391.

Southworth, F., V. H. Dale, and R. V. O'Neill. 1991. Contrasting patterns of land use in Rondonia, Brazil: simulating the effects on carbon release. *International Social Science Journal* 43:681–698.

Spanner, M. A., L. L. Pierce, and S. W. Running. 1990. The seasonality of AVHRR data of temperate coniferous forests: relationship with leaf area index. *Remote Sensing of Environment* 33:97–112.

Sparks, R. E. 1995. Need for ecosystem management of large rivers and floodplains. *BioScience* 45:168–182.

Sparks, R. E., J. C. Nelson, and Y. Yin. 1998. Naturalization of the flood regime in regulated rivers. *BioScience* 48:706–720.

Spies, T. A., W. J. Ripple, and G. A. Bradshaw. 1994. Dynamics and pattern of a managed coniferous forest landscape in Oregon. *Ecological Applications* 4:555–568.

Spies, T. A., and M. G. Turner. 1999. Dynamic forest mosaics. *In* M. Hunter, ed. *Maintaining Biodiversity in Forest Ecosystems*, pp. 95–160. Cambridge University Press, New York, New York, USA.

Sprugel, D. G. 1976. Dynamic structure of wave-generated *Abies balsamea* forests in the northeastern United States. *Journal of Ecology* 64:889–911.

Sprugel, D. G. 1985. Natural disturbance and ecosystem energetics. *In* S. T. A. Pickett and P. S. White, eds. *The Ecology of Natural Disturbance and Patch Dynamics*, pp. 335–352. Academic Press, New York, New York, USA.

Sprugel, D. G., and F. H. Bormann. 1981. Natural disturbance and the steady state in high-altitude balsam fir forests. *Science* 211:390–393.

Stanton, M. L. 1983. Spatial patterns in the plant community and their effects upon insect search. *In* S. Ahmad, ed. *Herbivore Insects: Host-seeking Behavior and Mechanisms*, pp. 125–157. Academic Press, New York, New York, USA.

Starfield, A. M., and A. L. Bleloch. 1986. *Building Models for Conservation and Management.* Macmillan Publishing Company, New York, New York, USA.

Starfield, A. M., and F. S. Chapin III. 1996. Model of transient changes in arctic and boreal vegetation in response to climate and land use change. *Ecological Applications* 6:842–864.

Stauffer, D. 1985. *Introduction to Percolation Theory.* Taylor and Francis, London, UK.

Stauffer, D., and A. Aharony. 1992. *Introduction to Percolation Theory* (2nd edition). Taylor & Francis, London, UK.

Steele, J. S. 1974a. Stability of plankton ecosystems. *In* M. B. Usher and M. H. Williamson, eds. *Ecological Stability*, pp. 179–191. Chapman and Hall, London, UK.

Steele, J. S. 1974b. Spatial heterogeneity and population stability. *Nature* 248:83.

Steele, J. S. 1989. The ocean "landscape." *Landscape Ecology* 3:185–195.

Steger, W. A. 1964. Pittsburgh urban renewal simulation model. *Journal of the Institute of Planners* 30:51–54.

Stouffer, P. C., and R. O. Bierregaard, Jr. 1995. Use of Amazonian forest fragments by understory insectivorous birds. *Ecology* 76:2429–2445.

Strong, D. R. 1980. Null hypotheses in ecology. *Synthese* 43:271–285.

Sugihara, G., and R. M. May. 1990. Applications of fractals in ecology. *Trends in Ecology and Evolution* 5:79–86.

Sukachef, V. N. 1944. Principles of genetical classification in biogeocoenology (Russ). *Zh. Obshch. Biol. (USSR)* 6.

Sukachef, V. N. 1945. Biogeocoenology and phytocoenology. *Readings of the Academy of Sciences USSR* 4:g. (Russ.)

Suter, G. W. 1993. *Ecological Risk Assessment*. Lewis Publishers, Boca Raton, Florida, USA.

Sutherland, J. P. 1974. Multiple stable points in natural communities. *American Naturalist* 108:859–873.

Swain, M. D., and T. C. Whitmore. 1988. On the definition of ecological species groups in tropical rain forests. *Vegetatio* 75:81–86.

Swank, W. T., and D. A. Crossley, Jr., editors. 1988. *Forest Hydrology and Ecology at Coweeta*. Springer-Verlag, New York, New York, USA.

Swanson, F. J., T. K. Kratz, N. Caine, and R. G. Woodmansee. 1988. Landform effects on ecosystem patterns and processes. *BioScience* 38:92–98.

Swanson, F. J., S. L. Johnson, S. V. Gregory, and S. A. Acker. 1998. Flood disturbance in a forested mountain landscape. *BioScience* 48:681–689.

Swartzman, G. L., and S. P. Kaluzny. 1987. *Ecological Simulation Primer*. Macmillan Publishing Company, New York, New York, USA.

Swetnam, T. W., and A. M. Lynch. 1993. Multicentury, regional-scale patterns of western spruce budworm outbreaks. *Ecological Monographs* 63:399–424.

Swift, B. L. 1974. Status of riparian ecosystems in the United States. *Water Resource Bulletin* 20:223–228.

Tadic, B. 1998. Disorder and self-organization in physical systems. *Solid State Phenomena* 61(2):17.

Tansley, A. G. 1935. The use and abuse of vegetational concepts and terms. *Ecology* 16:284–307.

Taylor, A. D. 1990. Metapopulation, dispersal, and predator–prey dynamics: an overview. *Ecology* 71:429–433.

Terborgh, J. 1975. Faunal equilibria and the design of wildlife preserves. *In* F. B. Golley and E. Medina, eds. *Tropical Ecology Systems*, pp. 369–380. Springer-Verlag, New York, New York, USA.

Theobald, D. M., N. T. Hobbs, T. Bearly, J. Zack, and W. E. Riebsame. 2000. Including biological information in local land-use decision-making: designing a system for conservation planning. *Landscape Ecology* 15:35–45.

Thomas, J. W., E. D. Forsman, J. B. Lint, E. C. Meslow, B. R. Noon, and J. Verner. 1990. *A Conservation Strategy for the Northern Spotted Owl: A Report of the Interagency Scientific Committee to Address the Conservation of the Northern Spotted Owl*. Portland, Oregon. U.S. Department of Agriculture, Forest Service, U.S. Department of Interior, Bureau of Land Management, Fish and Wildlife Service, National Park Service.

Thomas, K. A., and F. W. Davis. 1996. Applications of GAP analysis data in the Mojave Desert of California. *In* J. M. Scott, T. H. Tear, and F. W. Davis, eds. *Gap Analysis: A Landscape Approach to Biodiversity Planning*, pp. 209–219. American Society for Photogrammetry and Remote Sensing, Bethesda, Maryland, USA.

Thornton, P. K., and P. G. Jones. 1998. A conceptual approach to dynamic agricultural land-use modelling. *Agricultural Systems* 57:505–521.

Tilman, D. 1994. Competition and biodiversity in spatially structured habitats. *Ecology* 75:2–16.

Tilman, D., and P. Kareiva, editors. 1997. *Spatial Ecology: The Role of Space in Population Dynamics and Interspecific Interactions*. Monographs in Population Biology. Princeton University Press, Princeton, New Jersey, USA.

Tilman, D., R. M. May, C. L. Lehman, and M. A. Nowak. 1994. Habitat destruction and the extinction debt. *Nature* 371:65–66.

Tilman, D., C. L. Lehman, and P. Kareiva. 1997. Population dynamics in spatial habitats. *In* D. Tilman and P. Kareiva, eds. *Spatial Ecology: The Role of Space in Population Dynamics and Interspecific Interactions*, pp. 3–20. Princeton University Press, Princeton, New Jersey, USA.

Tjallingii, S. P., and A. A. de Veer, editors. 1982. *Perspectives in Landscape Ecology*. Proceedings of the International Congress of the Netherlands Society for *Landscape Ecology*. PUDOC, Wageningen, The Netherlands.

Toft, C. A., and T. W. Schoener. 1983. Abundance and diversity of orb spiders on 106 Bahamian islands: biogeography at an intermediate level. *Oikos* 41:411–426.

Trimble, S. W. 1974. *Man-induced Soil Erosion on the Southern Piedmont, 1700–1970*. Soil Conservation Society of America, Ankeny, Iowa, 180 p.

Troll, C. 1939. *Luftbildplan und okologische Bodenforschung*. Zeitschrift der Gesellschaft fur Erdkund, Berlin, Germany, pp. 241–298.

Troll, C. 1968. Landschaftsokologie. *In* R. Tuxen, ed. *Pflanzensoziologie und Landschaftsokologie*, pp. 1–21. Berichte das Internalen Symposiums der Internationalen Vereinigung fur Vegetationskunde. Sotzenau/Weser, The Hague, Netherlands.

Tucker, D. J., J. R. G. Townshend, and T. E. Goff. 1985. African land cover classification using satellite data. *Science* 227:369–374.

Tucker, K., S. P. Rushton, R. A. Sanderson, E. B. Martin, and J. Blaiklock. 1997. Modeling bird distributions—a combined GIS and Bayesian rule-based approach. *Landscape Ecology* 12:77–93.

Turchin, P. 1998. *Quantitative Analysis of Animal Movement*. Sinauer, Sunderland, Massachusetts, USA.

Turner, B. L., and W. B. Meyer. 1993. Environmental change: the human factor. *In* M. J. McDonnell and S. T. A. Pickett, eds. *Humans as Components of Ecosystems*, pp. 40–50. Springer-Verlag, New York, New York, USA.

Turner, C. L., J. M. Blair, R. J. Schartz, and J. C. Neel. 1997. Soil N and plant responses to fire, topography, and supplemental N in tallgrass prairie. *Ecology* 78:1832–1843.

Turner, M. G. 1987a. Spatial simulation of landscape changes in Georgia: a comparison of 3 transition models. *Landscape Ecology* 1:29–36.

Turner, M. G., editor. 1987b. *Landscape Heterogeneity and Disturbance*. Springer-Verlag, New York. New York, USA.

Turner, M. G. 1987c. Land use changes and net primary production in the Georgia, USA, landscape: 1935–1982. *Environmental Management* 11:237–247.

Turner, M. G. 1989. Landscape ecology: the effect of pattern on process. *Annual Review of Ecology and Systematics* 20:171–197.

Turner, M. G. 1990. Spatial and temporal analysis of landscape patterns. *Landscape Ecology* 4:21–30.

Turner, M. G., and S. P. Bratton. 1987. Fire, grazing and the landscape heterogeneity of a Georgia barrier island. *In* M. G. Turner, ed. *Landscape Heterogeneity and Disturbance*, pp. 85–101. Springer-Verlag, New York, New York, USA.

Turner, M. G., and S. R. Carpenter. 1999. Tips and traps in interdisciplinary research. *Ecosystems* 2:275–276.

Turner, M. G., and V. H. Dale. 1998. Comparing large, infrequent disturbances: what have we learned? Introduction for special feature. *Ecosystems* 1:493–496.

Turner, M. G,. and R. V. O'Neill. 1995. Exploring aggregation in space and time. *In* C. G. Jones and J. H. Lawton, eds. *Linking Species and Ecosystems*, pp. 194–208. Chapman and Hall, New York, New York, USA.

Turner, M. G., and W. H. Romme. 1994. Landscape dynamics in crown fire ecosystems. *Landscape Ecology* 9:59–77.

Turner, M. G., and C. L. Ruscher. 1988. Changes in landscape patterns in Georgia, USA. *Landscape Ecology* 1:241–251.

Turner, M. G., R. Costanza, and F. H. Sklar. 1989a. Methods to compare spatial patterns for landscape modeling and analysis. *Ecological Modeling* 48:1–18.

Turner, M. G., R. V. O'Neill, R. H. Gardner, and B. T. Milne. 1989b. Effects of changing spatial scale on the analysis of landscape pattern. *Landscape Ecology* 3:153–162.

Turner, M. G., R. H. Gardner, V. H. Dale, and R. V. O'Neill. 1989c. Predicting the spread of disturbance across heterogeneous landscapes. *Oikos* 55:121–129.

Turner, S. J., R. V. O'Neill, W. Conley, M. R. Conley, and H. C. Humphries. 1991. Pattern and scale: statistics for landscape ecology. *In* M. G. Turner and R. H. Gardner, eds. *Quantitative Methods in Landscape Ecology*, pp. 17–41. Springer-Verlag, New York, New York, USA.

Turner, M. G., Y. Wu., W. H. Romme, and L. L. Wallace. 1993a. A landscape simulation model of winter foraging by large ungulates. *Ecological Modeling* 69:163–184.

Turner, M. G., W. H. Romme, R. H. Gardner, R. V. O'Neill, and T. K. Kratz. 1993b. A revised concept of landscape equilibrium: disturbance and stability on scaled landscapes. *Landscape Ecology* 8:213–227.

Turner, M. G., Y. Wu, W. H. Romme, L. L. Wallace, and A. Brenkert. 1994a. Simulating winter interactions between ungulates, vegetation and fire in northern Yellowstone Park. *Ecological Applications* 4:472–496.

Turner, M. G., W. H. Hargrove, R. H. Gardner, and W. H. Romme. 1994b. Effects of fire on landscape heterogeneity in Yellowstone National Park, Wyoming. *Journal of Vegetation Science* 5:731–742.

Turner, M. G., G. J. Arthaud, R. T. Engstrom, S. J. Hejl, J. Liu, S. Loeb, and K. McKelvey. 1995a. Usefulness of spatially explicit animal models in land management. *Ecological Applications* 5:12–16.

Turner, M. G., R. H. Gardner, and R. V. O'Neill. 1995b. Ecological dynamics at broad scales. *BioScience*: Supplement S-29 to S-35.

Turner, M. G., D. N. Wear, and R. O. Flamm. 1996. Land ownership and land-cover change in the Southern Appalachian Highlands and the Olympic Peninsula. *Ecological Applications* 6:1150–1172.

Turner, M. G., V. H. Dale, and E. E. Everham III. 1997a. Fires, hurricanes and volcanoes: comparing large-scale disturbances. *BioScience* 47:758–768.

Turner, M. G., W. H. Romme, R. H. Gardner, and W. W. Hargrove. 1997b. Effects of patch size and fire pattern on early post-fire succession on the Yellowstone Plateau. *Ecological Monographs* 67:411–433.

Turner, M. G., S. M. Pearson, W. H. Romme, and L. L. Wallace. 1997c. Landscape heterogeneity and ungulate dynamics: what spatial scales are important? *In* J. A. Bissonette, ed. *Wildlife and Landscape Ecology*, pp. 331–348. Springer-Verlag, New York, New York, USA.

Turner, M. G., S. R. Carpenter, E. J. Gustafson, R. J. Naiman, and S. M. Pearson. 1998a. Land use. *In* M. J. Mac, P. A. Opler, P. Doran, and C. Haecker, eds. *Status and Trends of Our Nation's Biological Resources*, Volume 1, pp. 37–61. National Biological Service, Washington, DC, USA.

Turner, M. G., W. L. Baker, C. Peterson, and R. K. Peet. 1998b. Factors influencing succession: lessons from large, infrequent natural disturbances. *Ecosystems* 1:511–523.

Turner, M. G., T. R. Crow, J. Liu, D. Rabe, C. F. Rabeni, P. Soranno, W. Taylor, K. Vogt, and J. A. Wiens. *In press*. *In* J. Liu. and W. Taylor, eds. *Integrating Landscape Ecology into Natural Resource Management*. Cambridge University Press, Cambridge, Massachusetts, USA.

Upton, G., and B. Fingleton. 1985a. *Spatial Data Analysis by Example, volume 1: Point Pattern and Quantitative Data*. John Wiley & Sons, Chichester, UK.

Upton, G., and B. Fingleton. 1985b. *Spatial Data Analysis by Example, volume 2: Categorical and Directional Data*. John Wiley & Sons, Chichester, UK.

Urban, D. L., R. V. O'Neill, and H. H. Shugart. 1987. Landscape ecology. *BioScience* 37:119–127.

Urban, D. L., G. B. Bonan, T. M. Smith, and H. H. Shugart. 1991. Spatial applications of gap models. *Forest Ecology and Management* 42:95–110.

Van Cleve, K. J., J. Yarie, R. Erickson, and C. T. Dryness. 1993. Nitrogen mineralization and nitrification in successional ecosystems on the Tanana River floodplain, interior Alaska. *Canadian Journal of Forest Research* 23:970–978.

van der Maarel, E. 1993. Some remarks on disturbance and its relations to diversity and stability. *Journal of Vegetation Science* 3:733–736.

Vandermeer, J. H. 1973. On the regional stabilization of locally unstable predator–prey relationships. *Journal of Theoretical Biology* 41:161–170.

Van Dorp, D., and P. F. M. Opdam. 1987. Effects of patch size, isolation, and regional abundance on forest bird communities. *Landscape Ecology* 1:59–73.

Van Dyne, G. M. 1969. *The Ecosystem Concept in Natural Resource Management*. Academic Press, New York, New York, USA.

Van Horne, B. 1983. Density as a misleading indicator of habitat quality. *Journal of Wildlife Management* 47:893–901.

Van Horne, B., and J. A. Wiens. 1991. *Forest Bird Habitat Suitability Models and the Development of General Habitat Models*. U.S. Department of the Interior, Fish and Wildlife Service, Fish and Wildlife Research 8. Washington, DC, USA.

Van Wagner, C. E. 1978. Age-class distribution and the forest fire cycle. *Canadian Journal of Forest Research* 8:220–227.

Van Wagner, C. E. 1983. Fire behavior in northern conifer forests and shrublands. *In* R. W. Wein and D. A. MacLean, eds. *The Role of Fire in Northern Circumpolar Ecosystems*, pp. 65–80. John Wiley & Sons, New York, New York, USA.

VEMAP Members. 1995. Vegetation/ecosystem modeling and analysis project (VEMAP): Comparing biogeography and biogeochemistry models in a continental-scale study of terrestrial ecosystem responses to climate change and CO_2 doubling. *Global Biogeochemical Cycles* 9:407–437.

Verboom, J. A., P. Schotman, P. Opdam, and J. A. J. Metz. 1991. European nuthatch metapopulations in a fragmented agricultural landscape. *Oikos* 61:149–156.

Verner, J., K. S. McKelvey, B. R. Noon, R. J. Gutierrez, G. I. Gould, Jr., and T. W. Beck. 1992. *The California Spotted Owl: A Technical Assessment of Its Current Status.* U.S. Forest Service General Technical Report PSW-GTR-133. Pacific Southwest Research Station, Albany, California, USA.

Villa, F., O. Rossi, and F. Sartore. 1992. Understanding the role of chronic environmental disturbance in the context of island biogeographic theory. *Environmental Management* 16:653–666.

Vitousek, P. M., and R. W. Howarth. 1991. Nitrogen limitation on land and in the sea: how can it occur? *Biogeochemistry* 13:87–115.

Vitousek, P. M., and P. A. Matson. 1985. Disturbance, nitrogen availability, and nitrogen losses in an intensively managed loblolly pine plantation. *Ecology* 66: 1360–1376.

Vitousek, P. M., and J. M. Melillo. 1979. Nitrate losses from disturbed forests: patterns and mechanisms. *Forest Science* 25:605–619.

Vitousek, P. M., H. A. Mooney, J. Lubchenco, and J. M. Melillo. 1997b. Human domination of Earth's ecosystems. *Science* 277:494–499.

Vitousek, P. M., J. R. Gosz, C. C. Grier, J. M. Melillo, W. A. Reiners, and R. L. Todd. 1979. Nitrate losses from disturbed ecosystems. *Science* 204:469–474.

Vitousek, P. M., J. D. Aber, R. W. Howarth, G. E. Likens, P. A. Matson, D. W. Schindler, W. H. Schlesinger, and D. G. Tilman. 1997a. Human alteration of the global nitrogen cycles: sources and consequences. *Ecological Applications* 7:737–750.

Von Bertalanffy, L. 1968. *General System Theory, Foundations, Development and Applications.* George Braziller, New York, New York, USA.

Von Bertalanffy, L. 1969. Change or law. *In* A. Koestler and J. R. Smythies, eds. *Beyond Reductionism: The Alpbach Symposium*, pp. 56–84. George Braziller, New York, New York, USA.

Von Humboldt, A. 1807. *Ideen zu Einer Geographie der Pflanzen Nebst Einem Gemalde der Tropenlaender.* Tubingen, Germany.

Vos, C. C., and P. Opdam. 1993. *Landscape Ecology of a Stressed Environment.* Chapman & Hall, London, UK.

Vuilleumier, F. 1970. Insular biogeography in continental regions. I. The northern Andes of South America. *American Naturalist* 104:373–388.

Vuilleumier, F. 1973. Insular biogeography in continental regions. II. Cave faunas from Tesin, southern Switzerland. *Systematic Zoology* 22:64–76.

Walker, B., W. Steffen, J. Canadell, and J. Ingram. 1999. *The Terrestrial Biosphere and Global Change: Implications for Natural and Managed Ecosystems.* Cambridge University Press, Cambridge, UK.

Wallace, L. L., M. G. Turner, W. H. Romme, R. V. O'Neill, and Y. Wu. 1995. Scale of heterogeneity of forage production and winter foraging by elk and bison. *Landscape Ecology* 10:75–83.

Wallach, D., and M. Genard. 1998. Effect of uncertainty in input and parameter values on model prediction error. *Ecological Modeling* 105:337–345.

Walley, F. L., C. Van Kessel, and D. J. Pennock. 1996. Landscape-scale variability of N mineralization in forest soils. *Soil Biology and Biochemistry* 28:383–391.

Wallin, D. O., F. J. Swanson, and B. Marks. 1994. Landscape pattern response to changes in pattern generation rules: land-use legacies in forestry. *Ecological Applications* 4:569–580.

Ward, D., and D. Saltz. 1994. Foraging at different spatial scales: dorcas gazelles foraging for lilies in the Negev desert. *Ecology* 75:45–58.

Warming, E. 1925. *Oecology of Plants; An Introduction to the Study of Plant-communities.* Oxford, London, UK.

Watras, C. J., K. A. Morrison, and J. S. Host. 1995. Concentration of mercury species in relationship to other site-specific factors in the surface waters of northern Wisconsin lakes. *Limnology and Oceanography* 40:556–565.

Watt, A. S. 1947. Pattern and process in the plant community. *Journal of Ecology* 35:1–22.

Watt, K. E. F. 1968. *Ecology and Resource Management.* McGraw-Hill, New York, New York, USA.

Watts, B. D. 1996. Landscape configuration and diversity hotspots in wintering sparrows. *Oecologia* 108:512–517.

Wear, D. N. 1999. Challenges to interdisciplinary discourse. *Ecosystems* 2:299–301.

Wear, D. N., and P. Bolstad. 1998. Land-use changes in southern Appalachian landscapes: spatial analysis and forecast evaluation. *Ecosystems* 1:575–594.

Wear, D. N., and R. O. Flamm. 1993. Public and private disturbance regimes in the Southern Appalachians. *Natural Resource Modeling* 7:379–397.

Wear, D. N., M. G. Turner, and R. O. Flamm. 1996. Ecosystem management with multiple owners: landscape dynamics in a Southern Appalachian watershed. *Ecological Applications* 6:1173–1188.

Wear, D. N., M. G. Turner, and R. J. Naiman. 1998. Institutional imprints on a developing forested landscape: implications for water quality. *Ecological Applications* 8:619–630.

Webb, N. R., and A. H. Vermaat. 1990. Changes in vegetational diversity on remnant heathland fragments. *Biological Conservation* 53:253–264.

Webster, K. E., T. K. Kratz, C. J. Bowser, J. J. Magnuson, and W. J. Rose. 1996. The influence of landscape position on lake chemical response to drought in northern Wisconsin, USA. *Limnology and Oceanography* 41:977–984.

Webster, K. E., P. A. Soranno, S. B. Baines, T. K. Kratz, C. J. Bowser, P. J. Dillon, P. Campbell, E. J. Fee, and R. E. Hecky. 2000. Structuring features of lake districts: geomorphic and landscape controls on lake chemical responses to drought. *Freshwater Biology* 43:499–515.

Wegner, J., and G. Merriam. 1979. Movements by birds and small mammals between a wood and adjoining farm habitats. *Journal of Applied Ecology* 16:349–357.

Weiss, S. B., D. D. Murphy, and R. R. White. 1988. Sun, slope, and butterflies: topographic determinants of habitat quality for *Euphydryas editha. Ecology* 69:1486–1496.

Weller, D. E., T. E. Jordan, and D. L. Correll. 1998. Heuristic models for material discharge from landscapes with riparian buffers. *Ecological Applications* 8:1156–1169.

Whitcomb, B. L., R. F. Whitcomb, and D. Bystrak. 1977. Island biogeography and "habitat islands" of eastern forests: III. Long-term turnover and effects of selective logging on the avifauna of forest fragments. *American Birds* 31:17–23.

Whitcomb, R. F. 1977. Island biogeography and habitat islands of eastern forests. *American Birds* 31:3–5.

White, D., P. G. Minotti, M. J. Barczak, J. C. Sifneos, K. E. Freemark, M. V. Santelmann, C. F. Steinitz, A. R. Kiester, and E. M. Preston. 1997. Assessing risks to biodiversity from future landscape change. *Conservation Biology* 11:349–360.

White, M. A., and D. J. Mladenoff. 1994. Old-growth forest landscape transitions from pre-European settlement to present. *Landscape Ecology* 9:191–206.

White, P. S. 1979. Pattern, process, and natural disturbance in vegetation. *Botanical Review* 45:229–299.

White, P. S., and S. T. A. Pickett. 1985. Natural disturbance and patch dynamics: an introduction. *In* S. T. A. Pickett and P. S. White, eds. *The Ecology of Natural Disturbance and Patch Dynamics*, pp. 3–13. Academic Press, New York, New York, USA.

Whitney, G. G. 1986. Relation of Michigan's presettlement pine forests to substrate and disturbance history. *Ecology* 67:1548–1559.

Whitney, G. G., and W. J. Somerlot. 1985. A case study of woodland continuity and change in the American Midwest. *Biological Conservation* 31:265–287.

Whittaker, R. H. 1952. A study of summer foliage insect communities in the Great Smoky Mountains. *Ecological Monographs* 22:1–44.

Whittaker, R. H. 1956. Vegetation of the Great Smoky Mountains. *Ecological Monographs* 26:1–80.

Whittaker, R. H., and S. A. Levin. 1977. The role of mosaic phenomena in natural communities. *Theoretical Population Biology* 12:117–139.

Wickham, J. D., and K. H. Riitters. 1995. Sensitivity of landscape metrics to pixel size. *International Journal of Remote Sensing* 16:3585–3594.

Wickham, J. D., R. V. O'Neill, K. H. Riitters, T. G. Wade, and K. B. Jones. 1997. Sensitivity of selected landscape pattern metrics to land-cover misclassification and differences in land-cover composition. *Photogrammetric Engineering and Remote Sensing* 63:397–402.

Wiens, J. A. 1976. Population responses to patchy environments. *Annual Review of Ecology and Systematics* 7:81–120.

Wiens, J. A. 1984. On understanding a nonequilibrium world: myth and reality in community patterns and processes. *In* D. R. Strong, Jr., D. Simberloff, L. G. Abele, and A. B. Thistle, eds. *Ecological Communities: Conceptual Issues and the Evidence*, pp. 439–457. Princeton University Press, Princeton, New Jersey, USA.

Wiens, J. A. 1989a. Spatial scaling in ecology. *Functional Ecology* 3:385–397.

Wiens, J. A. 1989b. *The Ecology of Bird Communities. Volume 1. Foundations and Patterns.* Cambridge University Press, Cambridge, UK.

Wiens, J. A. 1995. Landscape mosaics and ecological theory. *In* L. Hansson, L. Fahrig, and G. Merriam, eds. *Mosaic Landscapes and Ecological Processes*, pp. 1–26. Chapman & Hall, London, UK.

Wiens, J. A. 1997. Metapopulation dynamics and landscape ecology. *In* I. Hanski and M. E. Gilpin, eds. *Metapopulation Biology*, pp. 43–62. Academic Press, New York, New York, USA.

Wiens, J. A. 1999. The science and practice of landscape ecology. *In* J. M. Klopatek and R. H. Gardner, eds. *Landscape Ecological Analysis: Issues and Applications*, pp. 372–383. Springer-Verlag, New York, New York, USA.

Wiens, J. A., and B. T. Milne. 1989. Scaling of "landscapes" in landscape ecology, or landscape ecology from a beetle's perspective. *Landscape Ecology* 3:87–96.

Wiens, J. A., N. C. Stenseth, B. Van Horne, and R. A. Ims. 1993. Ecological mechanisms and landscape ecology. *Oikos* 66:369–380.

Wiens, J. A., T. O. Crist, K. A. With, and B. T. Milne. 1995. Fractal patterns of insect movement in microlandscape mosaics. *Ecology* 76:663–666.

Wiens, J. A., R. L. Schooley, and R. D. Weeks. 1997. Patchy landscapes and animal movements: do beetles percolate? *Oikos* 78:257–264.

Wienstein, D. A., and H. H. Shugart. 1983. Ecological modeling of landscape dynamics. *In* H. A. Mooney and M. Godron, eds. *Disturbance and Ecosystems*, pp. 29–45. Springer-Verlag, New York, New York, USA.

Wilcove, D. S., C. H. McLellan, and A. P. Dobson. 1986. Habitat fragmentation in the temperate zone. *In* M. E. Soule, ed. *Conservation Biology: The Science of Scarcity and Diversity*, pp. 237–256. Sinauer Assocication, Sunderland, Massachusetts, USA.

Wilkin, D. C., and R. W. Jackson. 1983. Non-point water quality contributions from land use. *Journal of Environmental Systems* 13(22):127–136.

Williams, C. B. 1964. *Patterns in the Balance of Nature.* Academic Press, New York, New York, USA.

Williams, M. 1989. *Americans and Their Forests: A Historical Geography.* Cambridge University Press, New York, New York, USA.

Williamson, M. 1981. *Island Populations.* Oxford University Press, Oxford, UK.

Willson, M. F., S. M. Gende, and B. H. Marston. 1998. Fishes and the forest. *BioScience* 48:455–462.

Wilson, J. B., M. T. Sykes, and R. K. Peet. 1995. Time and space in the community structure of a species-rich limestone grassland. *Journal of Vegetation Science* 6:729–740.

With, K. A. 1994a. Using fractal analysis to assess how species perceive landscape structure. *Landscape Ecology* 9:25–36.

With, K. A. 1994b. Onogenetic shifts in how grasshoppers interact with landscape structure: an analysis of movement patterns. *Functional Ecology* 8:477–485.

With, K. A. 1997a. The application of neutral landscape models in conservation biology. *Conservation Biology* 11:1069–1080.

With, K. A. 1997b. The theory of conservation biology. *Conservation Biology* 11:1436–1440.

With, K. A., and T. O. Crist. 1995. Critical thresholds in species' responses to landscape pattern. *Ecology* 76:2446–2459.

With, K. A., and A. W. King. 1997. The use and misuse of neutral landscape models in ecology. *Oikos* 97:219–229.

With, K. A., and A. W. King. 1999a. Dispersal success on fractal landscapes: a consequence of lacunarity thresholds. *Landscape Ecology* 14:73–82.

With, K. A., and A. W. King. 1999b. Extinction thresholds for species in fractal landscapes. *Conservation Biology* 13:314–326.

With, K. A., R. H. Gardner, and M. G. Turner. 1997. Landscape connectivity and population distributions in heterogeneous environments. *Oikos* 78:151–169.

With, K. A., S. J. Cadaret, and C. Davis. 1999. Movement responses to patch structure in experimental fractal landscapes. *Ecology* 80:1340–1353.

Wohl, D. L., J. B. Wallace, and J. L. Meyer. 1995. Benthic macroinvertebrate community structure, function and production with respect to habitat type, reach and drainage basin in the southern Appalachians (USA). *Freshwater Biology* 34:447–464.

Wolfram, S. 1984. Cellular automata as models of complexity. *Nature* 311:419–424.

Wolman, M. G. 1967. Cycle of sedimentation and erosion in urban river channels. *Geografiska Annaler*, 49A:385–395.

Wood, D. M., and R. del Moral. 1993. Colonizing plants on the pumice plains, Mount St. Helens, Washington. *American Journal of Botany* 75:1228–1237.

Wright, H. E., Jr., and M. L. Heinselman. 1973. Introduction to symposium on the ecological role of fire in natural coniferous forests of western and northern America. *Quaternary Research* 3:319–328.

Wu, J., and S. A. Levin. 1994. A spatial patch dynamic modeling approach to pattern and process in an annual grassland. *Ecological Monographs* 64:447–464.

Wu., J., and O. L. Loucks. 1995. From balance of nature to hierarchical patch dynamics: a paradigm shift in ecology. *Quarterly Review of Biology* 70:439–466.

Yamamura, N. 1976. A mathematical approach to spatial distribution and temporal succession in plant communities. *Bulletin of Mathematical Biology* 38:517–526.

Yarie, J. 1981. Forest fire cycles and life tables: a case study from interior Alaska. *Canadian Journal of Forest Research* 11:554–562.

Zackrisson, O. 1977. Influence of forest fires on the North Swedish boreal forest. *Oikos* 29:22–32.

Zak, D. R., K. S. Pregitzer, and G. E. Host. 1987. Landscape variation in nitrogen mineralization and nitrification. *Canadian Journal of Forest Research* 16:1258–1263.

Zak, D. R., G. E. Host, and K. S. Pregitzer. 1989. Regional variability in nitrogen mineralization, nitrification, and overstory biomass in northern Lower Michigan. *Canadian Journal of Forest Research* 19:1521–1526.

Zedler, P. H., and F. G. Goff. 1973. Size-association analysis of forest successional trends in Wisconsin. *Ecological Monographs* 43:79–94.

Zhang, R., J. D. Hamerlinck, S. P. Gloss, and L. Munn. 1996. Determination of nonpoint-source pollution using GIS and numerical models. *Environmental Quality* 25:411–418.

Zhou, G., and A. M. Liebhold. 1995. Forecasting the spatial dynamics of gypsy moth outbreaks using cellular transition models. *Landscape Ecology* 10:177–189.

Ziegler, B. P. 1977. Persistence and patchiness of predator–prey systems induced by discrete event population exchange mechanisms. *Journal of Theoretical Biology* 67:687–713.

Zonneveld, I. S. 1982. Land(scape) ecology, a science or a state of mind. *In* S. P. Tjallingii and A. A. de Veer, eds. *Perspectives in Landscape Ecology*, pp. 9–15. Proceedings of the International Congress of The Netherlands Society of Landscape Ecology. PUDOC, Wageningen, The Netherlands.

Zonneveld, I. S. 1990. Scope and concepts of landscape ecology as an emerging science. *In* I. S. Zonneveld and R. T. T. Forman, eds. *Changing Landscapes: An Ecological Perspective*, pp. 1–20. Springer-Verlag, New York, New York, USA.

Zonneveld, I. S. 1995. *Land Ecology*. SPB Academic Publishing, Amsterdam, The Netherlands.

Index

Abies, 82
Abies amabilis, 187
Abies balsamea, 281
Abies lasiocarpa, 78, 188
Abiotic legacies, definition, 176
Aboveground net primary production (ANPP), 253, 255
Acer pennsylvanicum, 180
Acer saccharum, 87, 121, 162
Acid deposition, 26, 316
Adaptive management, 15
Adeiges picea, 85
Adiantum pedatum, 222
Adirondack Mountains, New York, 316–317
Advanced very high resolution radiometer (AVHRR), 251
Aerial photography, 10, 48, 85, 95, 144, 318
Aerial survey, 170
Aesculus sylvatica, 162
Africa, 16, 251
Aggregation errors, 103
Agroecosystem, 5, 165, 256
Aimophilia aestivalis, 241
Alaska, USA, 5, 258, 266
Albedo, 251
Alces alces, 281
Alder, mountain, 216
Allegheny Plateau, Ohio, 256
Allerton Park, Illinois, workshop, 12
Alligator mississippiensis, 282
Allometric rules, 222
Alnus sinuate, 216
Amazonian forests, 230
American alligator, 282
American crocodile, 283
Ammodramus maritimus, 282
Anadramous fish, 278, 308
Anderson classification system, 97
Anisotropy, 111
Anolis communities, 138

Anoxia, 180, 274
APACK, 109
Aphis sambuci, 55
Appalachian Mountains, 4, 121, 298–299
Applied ecology, 321
Ardea alba, 283
Ardea herodias, 283
Army ants, 230
Artemesia, 78
Arthropod, 5, 37
Aspen, 78, 170, 172, 281
Atchafalaya Basin, Louisiana, 260
ATLSS, 282–284
Australia, 239
Australian dry-land communities, 147
Australian wheatbelt, 302–304
Autecology, 179
Autocorrelation, 126, 138, 150, 152

Bachman's Sparrow, 241–242
Badger, 72
Bahamian islands, 207
Bald Cyprus, 88
Bald Eagle, 278
Balsam fir, 281
Balsam wooly adelgid, 85
Bark beetle, 85
Bayesian model, 61
Bears, 278–279
Beaver, 85
Beech, 87
Beetle, 221, 235
Beetle, cereal leaf, 205
Bellwort, 222
Berberis nervosa, 216
Betula alleghaniensis, 121, 180
Betula lutea, 162, 281
Betula papyrifera, 82
Biodiversity, 201, 218–219, 291, 303
  definition, 201
Biogeocenology, 10
Biogeochemical cycling, 253, 257
Biogeography, 12, 13
Biomass, 250–251
Biome, 73
Biotic interactions, influence on landscape pattern, 83–86
Biotic legacies, definition, 176
Biotic residuals, 179–181
Birch, 82, 121, 281
Bird assemblages, 138
Birds, 230, 233
Bison, 85, 226–227, 239
Bison bison, 85, 226
Body mass, frequency distribution, 223–224
Border Lakes, Michigan, 122
Boreal forest, 71, 183, 191, 224, 280–281
Boundary effects, 244
Boundary shape, 231, 245
Boundary Waters Canoe Area, 185–186, 191–193
Bouteloua gracilis, 5
Bouteloua-Artemisia, 231
Brachyramphus marmoratus, 308
Brazil, 4, 304–305
Breeding Bird Survey (BBS), 238
Britain, 42
British Columbia, Canada, 178, 310
Buchloe dactyloides, 5
Budworm outbreaks, 172
Butterfly
  cabbage, 205
  checkerspot, 217

Cadiz Township, Wisconsin, 94
Calcium, 263–264
California, USA, 173, 220, 310, 312, 318
Canada, 232, 258
Canis lupus lycaon, 214
Canopy gaps, 19
Cape Sable Seaside Sparrow, 282, 283
Carbon, 251, 255–257, 264, 304
Carbon dioxide, 75–76, 79, 257
Carbon monoxide, 257
Cascade Mountains, Oregon, 318
Castor Canadensis, 85
Cellular automata models, 65
Central place theory, 16
Century model, 260
Cervus elaphus, 226, 231
Cesium contamination, 151–152
Chesapeake Bay, 104, 269, 317–318
Chile, 6
China, 6
Chipmunks, 233

Chlorophyll, 264
Choristoneura fumiferana, 85
Choristoneura occidentalis, 172
Climate
  definition, 72
  influence on landscape pattern, 72–74, 91
Climate change, 72–80, 188
  responses of biota to, 75–80
Clusters, 139–140, 148
  formation, 141
  number, 144
  size, 142
Coevolution, 19
Colorado Front Range, USA, 172
Colorado Natural Diversity Information Source (NDIS), 296–297
Colorado, USA, 5, 6, 292–294, 256, 260
Community formation, 137
Community structure, 137
Competition, 18, 72, 83–84, 91, 207
Complex adaptive systems, 329
Complex systems, 19
Configuration, definition of, 3
Connectivity (*also see* Habitat connectivity), 138, 146, 153–154, 167, 244
  and percolation, 19
  definition, 3
Conservation, 26, 234–235
  design, 205, 246
  planning, 207
Conservation biology, 20, 203, 221, 246, 293, 331
Constancy, 189–190
Contagion, 147, 227
Coral bleaching, 20
Correlogram, 127–129, 131
Corridor, 3, 12, 229, 233, 235–237, 239
  definition of, 3
Cotton rat, 223
Cover type, definition, 3
Coweeta, North Carolina, 250
Critical threshold, 29, 31, 141–143, 146, 153, 167, 173–174, 227
  definition of, 29, 31
Crocodylus acutus, 283
Cropland abandonment, 256
Crown Forest Sustainability Act (CFSA), 312
Curdled maps, 154
Curdled systems, 148

Deciduous forest, 71, 232
Decomposition, 261
Deer, 231
Deer mouse, 223
Deforestation, 87, 291
  in the USA, 89
Deme, definition, 210
Denitrification, 259, 274
Denver, Colorado, 296
Desert grasses, 213
Desmognathus ochrophaeus, 216
Differential equations, 51
Diffusion coefficient, 204, 246
Diffusion, physical, 204
Diffusive instability, 84
Dimensional analysis, 58
Dispersal, 18, 181
  as cause of spatial pattern, 72–73
  in response to climate change, 77
  interaction with spatial heterogeneity, 17
Dissolved organic carbon (DOC), 264, 278
Disturbance, 49, 65, 72, 82, 91, 155, 157–199, 250, 286, 291, 296, 328
  and biogeochemical cycling, 259
  and nitrogen leaching, 258
  chronic, 206
  clear-cuts, 166
  definition of, 90, 159
  disease, 170
  downbursts, 164–165
  dynamics, 153
  effects, 146
  epidemics, 165, 170, 172
  fire (*also see* Fire), 160, 172, 175, 178, 183
  floods, 176, 178, 180
  forest harvest, 183
  frequency, 179, 181
  frequency, definition, 161
  human, 159, 165
  hurricanes, 163–165, 176–178

Disturbance (*continued*)
ice storms, 164
influence of topographic position, 82
insects, 164
intensity, 160, 170, 181
intensity, definition, 161
lightning strikes, 163–165, 174
logging, 172
mosaic, 174, 178
natural, 176, 179, 231
natural disturbance regime, 159, 163
patch, 180
pathogens, 163, 166
pattern, 162
pests, 170
regime, 29, 77, 146, 159–160, 161–162, 164, 184, 186–187, 196, 198
regime, definition, 161
residuals, definition, 161
return interval, definition, 161
roads, 165
rotation period, definition, 161
severity, 160
severity, definition, 161
sheep grazing, 172
size, 160, 179–181
size, definition, 161
spatial effects, 181
tornadoes, 164, 176–178, 180
volcano, 160
windstorms, 163–164, 173, 175, 180, 183, 197
Diversity, 137
Douglas-fir forest, 168

Eastern hemlock, 162
Ecological neighborhoods, 39, 223, 225, 226, 244
Economics, 7
Ecosystem
definition of, 249
concept, 10
processes, 329
science, 322
Ecosystem ecology, focus of, 249
Ecosystem management, 8, 289
Ecosystem models, 253, 259–261
Ecotone, 83–84, 245
definition of, 71
Edge, 138–139, 144
definition of, 3
effects, 152, 236, 240, 296
Elder aphid, 55
Elephants, 85
Elk, 226–227, 231, 239
Elm, 77
Endpoint, 315
Energy flow, 250
England, 218
Environmental gradients, 72
Epidemiology, 169–170
Equilibrium, 189, 328
Equisetum telmateia, 216
Eudocimus albus, 283
Europe, 6, 10–12, 14,17, 86, 87, 117, 205, 233
Eutrophication, 268
Evapotranspiration, 250, 253, 259, 261
Everglades, Florida, 282, 284
Exposure-habitat modification, 315
Extent, 3, 43
definition of, 28–29
Extinction debt, 209
Extinction rate, 206–207
Extinction threshold, 209, 212
Extrapolate, 19, 26, 29, 31, 44, 49
and ecosystem processes, 259
definition of, 29

Fagus grandifolia, 87, 162, 180
Felis concolor coryi, 283
Fencerows, 232–233
Fern, licorice, 217
Fern, maidenhair, 221
Fertilizer, 268
Field mapped data, 96
Fire (*also see* Disturbance), 77, 82, 90, 162, 183, 259–260, 291, 312
and carbon emission, 257
crown fire, 165, 173–174, 193, 197
fire-dominated landscapes, 191
frequency, 184, 188, 257
regime, 77, 184, 188
spread, 173
suppression, 172, 184

FIRE-BGC, 261
Fish, 20
Floodplains, clearing of, 272
Floods (*see also* Disturbance), 90
Florida Bay, 282
Florida panther, 283
Florida, USA, 282
Focal species, 302
Forcing function, 58
Forecasting, 302
Forest tent caterpillar, 170
Forest
  boreal, 258
  clear-cutting, 250, 259, 269
  disturbance, 177
  fragmentation, 170
  gap, 142, 162, 182
  interior habitat, 216
  management, 121, 307
  mosaic, 191
  old-growth, 121, 164, 307–312
  secondary, 256
FOREST-BGC, 253–254, 260
Forest tent caterpillar, 173
Forestry, 7
Founder's effect, 72
Four-neighbor rule, 234
Fractal dimension, 5, 18–19, 27, 42, 44, 116, 127, 150, 185
  Brownian motion, 150
  fractal analysis, 228
  fractal landscapes, 149–151
  fractal maps, 150–152
  geometry, 18, 19, 136, 148
  use in scaling up, 42
Fraction of absorbed photosynthetically active radiation (FPAR), 251
Fragmentation, definition of, 3
Fragmented landscapes, 121
FRAGSTATS, 109
Front Range, Colorado, 6, 292–294
Fugitive species, 72–73

Galapagos archipelago, 32
Gambel oak, 5
Gap analysis, 219, 318
General circulation models (GCMs), 42, 48, 253
General epidemic theory, 170
General systems theory, 56
Geographic information system (GIS), 9, 10, 15, 34, 48
  raster format, 97
  sources of error, 97, 100
  suggested texts, 21
  vector format, 97
Geography, 7, 15, 20
Georgia, USA, 39, 144, 256, 270
Geostatistics, 125
German, 10
Glacial-interglacial cycles, 74
Glacier National Park, USA, 165
Global change, 26, 257
Global warming, 251, 316
Gordon Conference, 136, 143
Gradient analysis, 13
Grain (resolution), 3, 43, 97, 102, 104, 106
  definition of, 28–29, 30
Grape, Cascade Oregon, 216
GRASS, 109
Grasshoppers, 223
Grasslands, 231
Grays Lake, Idaho, 78
Great Blue Heron, 283
Great Britain, 55
Great Egret, 283
Great Lakes Region, 122
Great Plains, North America, 253, 255
Great Smoky Mountains, 13–14, 101, 194–195
Greece, 86–87
Green-winged orchid, 55
Greenhouse gases, 75–76
Ground squirrels, 72
Gulf of Mexico, 282

H. J. Andrews, Oregon, 250
Habitat connectivity, 152, 229, 232, 234, 245–246, 296
Habitat destruction, 208–209
Habitat fragmentation, 13, 26, 65, 88, 201, 206, 220, 235, 291, 298, 329
  definition, 201
  habitat loss, 318
Habitat quality, 207

Habitat utilization, 153
Halesia carolina, 162
Haliaeetus leucocephalus, 278
Hardwood forests, 194
Hawaii, USA, 164
Hawaiian archipelago, 32
Hazard, 315
Hemlock, 77, 121, 183
Heterogeneity, definition of, 3
Hierarchical neutral map, 149
Hierarchical structure, 27, 147–148
Hierarchy, 160, 261
  definition of, 29, 34
Hierarchy theory, 6, 27, 34, 38, 43
Holling's hypothesis (body size distribution), 138, 223
Holocene, 78, 86
Holon, 35–36, 160–161
  definition of, 29, 34
Home range, 236
Homogenous spatial distribution, 83
Horsetail, 216
Hubbard Brook, New Hampshire, 250
Huckleberry, mountain, 216
Humans
  effects on landscapes, 15
  impacts on landscapes, 65
  influence on landscape patterns, 6, 86–90
  population growth, 327, 330
  prehistoric, 86
  relationship with landscape, 12
Hurricanes (*also see* Disturbance), 82
Hylocichla mustelina, 216

Idaho, USA, 6, 78
Illinois, USA, 12, 88, 202, 268
Indian Ocean, 75
Infinite cluster, 141
Initial conditions, 58
Insect movement, 19
Interdisciplinary studies, 306
  importance of, 20
International Biological Program (IBP), 48
Island biogeography, 13, 137, 203, 205–207, 220, 221, 243
Isle Royale, Michigan, 281

JABOWA/FORET models, 48, 65
Junco, Dark-eyed, 239
Juniperus, 78

Kansas, USA, 253
Keystone species, 84, 91

Labrador, Canada, 82
Lacunarity analysis, 127
Lake chains, 264
Lake district, 262
Lake Mendota, Wisconsin, 270–271
Lake Okeechobee, Florida, 282
Lake Superior, 281
Land classification, 10, 98
Land cover, definition of, 86
  classification, 98
  land-cover type, 142, 150
Land evaluation, 10
Land management decisions, 7, 159, 289
Land ownership, effect on landscape pattern, 121
Land planning, 10, 20, 218
Land use, 285, 318, 328
  and ecosystem process, 255
  and net primary production, 256
  and water flow, 282
  and water quality, 268–278
  definition of, 86
  ecological implications of`, 291–292
  influence on lakes, 262
  presettlement, 272
  scenarios, 318
Land-use change, 16, 86, 219, 290–307, 329
  in the USA, 88–90
  land-use conversion, 146
  models, 296–301, 306
  worldwide, 291
Landform
  definition of, 72
  effects on landscape pattern, 72, 80–83, 91
Landsat, 48, 121
  Multispectral Scanner (MSS), 30
  Thematic Mapper (TM), 28, 251
Landscape
  agricultural, 71
  cultural, 88

definition of, 7
intuitive sense, 1
organism's perspective, 1
Landscape architecture, 7, 10, 20
Landscape composition, 71
metrics of, 108
Landscape context, 236, 245
Landscape design, 15
Landscape diversity, 104, 192
Landscape ecology
as a viewpoint, 10
central themes, 331
coining of term, 10
conceptual basis of, 12
definition of, 2–4, 21–22
difference from other areas of ecology, 4
influence of scale theory on, 9
influence of systems ecology on, 56
influence of technology on, 9
roots of, 10
society for, 11
terminology, 3, 12
Landscape equilibrium, 188–196
Landscape fragmentation, 142, 151
Landscape heterogeneity, 165–166
Landscape management, 15, 289
Landscape metrics (*see* Metrics), 147, 313–314
Landscape models (*also see* Models), 64–66
Landscape mosaic, 88, 175, 178
Landscape pattern, 94, 174, 186, 188, 146, 147, 155
causes of, 18, 91
measurement (*see* Metrics)
mechanisms to explain, 18
state space for, 123
Landscape planning, 15
Landscape position, 165
and ecosystem process, 256
definition, 162
influence on lakes, 261–265
Landscape restoration, 302–303
Landscape structure, 155
Landscape texture, 19
Landslides, 83
Larix occidentalis, 80
Latitude, 73–74
LC_{max}, 144
Leaf area index (LAI), 251, 253, 254
Lepidoptera, 172
Leslie matrix models, 48
Level of organization, 27, 29, 34
definition, 27, 29
Life-history, 179
Lightning (*also see* Disturbance), 82, 85
Limnologists, 261
Little Ice Age, 75, 77
Little Tennessee River Basin, 216, 218
Local extinction, 220
Lodgepole pine, 6, 78, 102, 104–105, 180, 192
Lotka-Volterra models, 48
Louisiana, USA, 260

MacArthur Award Address, 25
MacPherson Township, Ontario, 171
Madison, Wisconsin, 272
Magnesium, 263
Malacosoma disstra, 170
Map
boundaries, 153
dimensions, 153
extent, 105, 140
map overlay techniques, 15
Mapping, 10
Marbled Murrelet, 308
Markov chain, 61
Marsupials, 239
Matrix, 3, 12, 220
definition of, 3
Mechanism, definition of, 52
Mediterranean, 4, 6, 12, 86, 87
Mercury, 270
Mesic forests, 221
Mesic landforms, 184
Metapopulation, 65, 203, 208–209, 220, 311, 331
definition, 210
dynamics, 146
frogs, 220
models, 13, 219, 221, 243
snails, 220
Metric
average patch size, 114, 122
connectivity, 111, 113, 123
contagion, 111–112, 122, 124

Metric (*continued*)
 diversity, 110
 dominance, 110, 122
 edge based, 314
 fractal, 115–16, 118–119,122, 124,
 fragmentation, 113
 interpatch distance, 123
 p_i, 108
 patch area, 113
 patch perimeter, 113
 patch shape, 118
 patch-based, 112
 proximity, 114
 relative richness, 109
Metrics, 328–329
 significant changes in, 107
 summary, 132
Michigan, USA, 164, 259, 268
Microlandscape, 49, 228
 example, 5
Mid-Atlantic region, USA, 317–318
Midpoint displacement method, 150
Midwest, USA, 75, 83, 89, 121, 215
Migration, 17
Migratory pathways, 231
Milankovitch cycles, 74, 76
Mining, 231
Minneapolis-St. Paul, Minnesota, 269
Minnesota, USA, 77, 185, 214, 269
Mississippi River, 177, 273
Mississippi, USA, 88, 118
Modeling, need for training in, 323
Models
 abstract, 48
 analytical vs. simulation, 50
 and ecosystem science, 259–261
 and GIS output, 61, 65
 approaches to defining model system, 59
 as component of landscape ecology, 20
 at different scales, 261
 BACHMAP, 241
 Bayesian, 61
 calibration, 58, 62
 cell-based, 242
 cellular automata, 65
 Century model, 260
 classification of, 50–56
 corroboration, 58
 definition of, 47, 67
 deterministic vs. stochastic, 50
 disturbance, 52
 documentation, 61
 dynamic vs. static, 51
 ECOLECON, 241
 ecosystem simulation, 253
 empirical, 52–53
 error propagation, 66
 extrapolation, 183
 Fire-BGC, 184
 for future landscape scenarios, 91
 for land use, 296–301, 306
 forest gap, 183
 individual-based, 241–242, 282–284
 JABOWA/FORET, 65
 LANDIS, 183–184
 landscape, 64–66
 linked to GIS, 253, 260
 Markov chain, 61, 193
 mechanistic, 52–53
 metapopulation, 207, 209, 243
 meteorological, 163
 nonpoint pollution, 270
 parameter estimation, 62
 physical, 47–48
 process-based, 52–53
 purposes of, 48
 regional scale, 15–16, 260
 representation of time, 51
 riparian buffer, 275
 risk assessment, 317
 role in landscape ecology, 49
 simulation, 281
 spatial, 54–56, 281–285
 spatial, definition of, 54
 spatially explicit, 20, 203, 209, 245, 260–261
 spatially explicit population, 311
 specifying scales of, 59
 spotted owl, 241
 STELLA, 61
 steps in building, 56–64, 67–68
 stochastic, 296, 317
 timeline of development, 48
 topographic exposure, 163
 validation, 58, 62
 verification, 58, 61

Mojave Desert, California, 318
Monroe County, Pennsylvania, USA, 219
Montana, USA, 253
Montane forest zone, 216–217
Moose, 85, 281
Moran's I, 129
Mount St. Helens, Washington, USA, 159, 164, 176
Mouse, western harvest, 223
Mt. Ranier, USA, 187
Multiple regression, use in detecting scale-dependence, 39
Multiple stable states, 18, 83
Multiscale analysis, 39
Muskrats, 205
Mycteria americana, 283
Mytilus californianus, 84

Natchez, Mississippi, 118
Native Americans, 87
  land-use practices, 88
Natural experiment, 10
Nature reserves, designing, 13
Nearest-neighbor rule, 143–144
Neighborhood rule, 141, 144, 146
Neighborhood size, 225
Neotropical migrants, 239
Net primary production (NPP), 251, 253, 256, 260, 284–285
Netherlands, 11–12, 15
Neutral landscape model (NLM), 5, 108, 136–138, 143–147, 153–155, 227, 234–235,
Neutral model, 5, 137, 148, 150, 153, 166
New England, USA, 82, 89
New Hampshire, USA, 250
New Mexico, USA, 39, 231
New York, USA, 176, 316
Nitrate, 258–259
Nitrification, 258–259
Nitrogen, 60, 249, 251, 256–257, 267, 274, 275
Nitrogen mineralization, 258, 261, 285
Nitrogen saturation, 316
Nonlinear dynamics, 329
Nonpoint source pollution, 267, 270, 316
Normalized difference vegetation index (NDVI), 251, 253
North America, 10–12, 15,17, 56, 75–77, 85, 87, 88, 205, 250–251, 264
North Carolina, USA, 162, 216, 218, 250, 263
North Temperate Lakes LTER site, Wisconsin, 263
Northern Cardinal, 237, 239
Northern Spotted Owl, 308–312
Null hypothesis, 135–137
Nutrient cycling, 250
Nutrient dynamics, models of, 260–261

Oak, 76–77, 87, 183
Oak Ridge National Laboratory, 136
Odocoileus hemionus, 231
Odocoileus virginianus, 283
Ohio, USA, 176, 256, 258
Olea europaea, 87
Olives, 87
Olympic Peninsula, Washington, 121, 309
Ontario Ministry of Natural Resources (OMNR), 312
Ontario, Canada, 170–171, 176, 178, 308, 312
Orchis morio, 55
Orchis spectabilis, 216–217
Oregon, USA, 120, 250, 273, 318
Oxygen isotope ratio, 75
Ozone, 316–317

Pacific Northwest, USA, 80, 89, 188, 219, 241, 308–312
Paired-quadrat technique, 127
Paleoecology, 27
  definition of, 72
Parameter, definition of, 58
Parids, 237–238
Patch, 12, 71, 139, 229
  age, 185
  composition, 160
  defining for analysis, 106
  definition, 3, 210
  density, 104, 125
  duration, relative, 225
  dynamics, 158
  edge, 166

Patch (*continued*)
heterogeneity, 179, 245
identification, 106
isolation, 220
isolation, relative, 225
model, 210
mosaic, stable, 193
number, 139, 148, 166
persistence, 160
shape, 138, 148, 231
size, 104, 125, 138, 140, 148, 160, 166, 179–181, 220, 223, 229, 225, 245, 296, 309
size, relative, 225
source, 13
structure, 15, 154, 228
structure, definition of, 3
Pattern analysis, 97
Pattern, multiple scales of, 38
Paulownia tomentosa, 216
p_c, 141–142, 148
Pennsylvania, USA, 176–177, 180, 318
Percolating cluster, 141
Percolation theory, 18–19, 136, 141, 143, 146
Percolation threshold, 141
Perimeter-area ratio, 124, 229
Pesticide, 268
Phase difference, 72
Phenacomys intermedius, 217
Phosphorus, 249, 267, 274–275, 279
Photosynthesis, 53, 253–254, 261
Phytosociology, 12–13
Picea, 82
Picea engelmannii, 78
Picea glauca, 281
Pink noise, 41
Pinus albicaulis, 79
Pinus contorta, 6, 78, 180
Pinus flexilis, 78
Pinus ponderosa, 80
Pinus strobus, 87
Pinus-Juniperus, 231
Pinyon-juniper, 231
Pioides borealis, 242
Pipilo erythrophthalmus, 237
Pisaster ochraceous, 84
Plankton, 84
Pleistocene, 78
Pocono Mountains, Pennsylvania, 318
Pollen analysis, 78
Polypodium glycyhiza, 217
Population
colonization, 207
extinction, 205, 207
models, spatially explicit, 240–243
population growth models, 50–52
population growth, intrinsic rate, 205
population viability analysis, 311
sink, 213–214
source, 213–214
stability of, 18
theory, 17
Populus tremuloides, 78, 170, 281
Predation, 83, 91, 278
Predator-prey interactions, 17, 204
Predator-prey models, 84
Princess tree, 216
Pseudo-replication, 32, 101
Pseudotsuga menziessii, 78
Pyrrhuloxia cardinalis, 237

Quasi-steady-state landscape, 191
Quaternary, 77
Quercus, 76
Quercus gambelii, 5, 80
Quercus rubra, 87

r.le, 109
Raccoon, 226
Random landscape, 129
Random maps, 19, 136, 152
Range of natural variability, 313
Reaction-diffusion models, 84
Red noise, 41
Reference environment, 315
Refuges, from predators, 17
Regional planning, 7
Regional risk assessment, 314–319
Regional science, 16
Relative constancy, 191
Relative patch area, 124
Remote imagery, 10, 28, 95, 253, 318

Landsat, 28, 30, 48, 121, 251
SPOT, 28
Remote sensing, 25, 95, 251
suggested texts, 21
Resolution, definition of, 29
Resource management, 20, 159, 218
Resource utilization, 18
Respiration, 53, 261
Rhododendron catawbiense, 216
Rhododendron, Catawba, 216
Riparian buffer, conceptual model, 276
Riparian buffers, 272–276, 311–312
Riparian forest, 8
Riparian vegetation, 85, 274
Riparian zones, 302
Risk assessment, goals, 314–315
River-floodplain, 178
Road networks, 120
Rocky Mountains, USA, 78–80, 172, 293
Rodents, 220
Roman Empire, 15
Rondonia, Brazil, 4, 304–305
Rostrhamus sociabilis, 283
RULE, 109
Russia, 10

Sagavanirktok River, Alaska, 266
Saginaw River, Michigan, 268
Salamander, mountain dusky, 216
Salt River Basin, Illinois, 268
Sangre de Cristo Mountains, 172
Santiago, Chile, 6
Satellite imagery (*also see* Remote imagery), 15, 253, 318
Scale, 25, 313, 328
absolute, definition of, 29, 31
and abiotic variation, 261
and fractal geometry, 19
cartographic, 30
cartographic, definition of, 29, 43
concepts of, 8
concepts related to land use, 293–295
defining for pattern analysis, 102–106
definition, 3, 27, 29, 43
effects on ecosystem processes, 260
extent, 28–29
grain, 28–29
identifying proper, 6, 38–39
importance in landscape ecology, 26
landscape pattern and water quality, 276–278
large and small vs. fine and broad, 30–31
multiscale analysis, 39
problems, 32–34, 43
relative, definition of, 29, 31
scale coverage problem, 33
scale linkage problem, 33
scale standardization problem, 34
shift in relative importance of variables with, 36
spatial, 5
terminology, 27–31
theory of, 8
Scale dependance, 199, 228, 313, 328
definition of, 30
scale-dependent threshold, 232
Scaling, 261, 329
Scaling up, 40–43
direct extrapolation, 41
extrapolation by expected value, 41
lumping, 40
use of fractals in, 42
use of models in, 41
Scirpus americanus, 78
Sea of Cortez, 32
Sedimentation model, 152
Self-organized criticality, 18, 19
Self-similarity, 19, 42, 116
Self-similarity, applications of concepts, 19
Semivariogram, 128, 130–131
Sensitivity analysis, 58, 63
Shifting mosaic steady-state, 190–192
Showy orchis, 216–217
Shrew, southeastern, 216
Sierra Nevada Mountains, 310, 312
Silica, 264
Silver bell, 162
Simple ratio vegetation index (SR), 251
Simulation, definition of, 51
Sink, defined in modeling, 58
Small mammals, 223, 232

Smithsonian Institution, 230
Snail Kite, 283
Socioeconomic theory, 16
Soil carbon, 255
Soil catena, 80
Soil organic carbon, 260
Soil organic matter, 37, 253, 255–256
Soil patterns, 117
Soil, water-holding capacity, 253
Solomon Islands, 32
Sorex longirostris, 216
Source, defined in modeling, 58
Source-sink, 152, 213, 246
  definition, 210
Southeast USA, 89, 123–124
Southern Appalachian Mountains, USA, 164, 217, 221
Sparrow, field, 239
Sparrow, White-throated, 237, 239
Spatial analysis software, 109
Spatial configuration, 4, 71
Spatial contagion, 151
Spatial heterogeneity, 202–203, 221, 240
Spatial models, 184, 193
  definition, 54
Spatial pattern, 203, 239
  general causes of, 72–73
Spatial statistics, 25, 39, 125
  uses of, 124–126
Spatially explicit model, 210, 212
Species-area curve, 31–32
Species distributions, 152
Species diversity, 146, 230
Species richness, 31–32, 206, 230
Spectral analysis, 41, 127
Spiders, 207
Spruce budworm, 85, 173
Spruce-fir forest, 82
Squirrel, red, 217
St. Augustin River Valley, Labrador, 82
Stability, 189, 196
Starling, eastern, 216
Starling, European, 218
State variable, definition of, 58
Stationary process, 191
Statistics, multivariate, 124
STELLA, 61
Stochastic constancy, 191
Storms, 90
Streams, 20
Strix occidentalis caurina, 308
Sturnus vulgaris, 216
Succession, 90–91, 179, 180–182, 256, 260
Successional trajectories, 181
Sugar maple, 87, 121, 162
Summit County, Colorado, 296
Sweden, 191
Sylvania Wilderness, Michigan, 122
Synchrony, 265
Systems analysis, 16
Systems ecology, 56

Tamiasciurus hudsonicus, 217
Targhee National Forest, 6
Taxodium distichum, 88
Temperate forest, 183
Tennessee River, 104
Tennessee, USA, 39, 60, 152, 250, 162
Tent caterpiller outbreaks, 171
Terms, in risk assessment, 315
Theoretical ecology, 17
Threshold, 154, 329
  definition, 234
  threshold of connectivity, 235
  in habitat abundance, 311
Thryothorus ludovicianus, 237
Tillia heterophylla, 162
Timber wolves, 214
Time-space state space, 27–28
Tionesta Scenic Area, 177
Titmice, 237
Topographic map, 150
Topographic position, 164
  and nitrogen dynamics, 258
Topography, 146, 149
  effect on landscape patterns, 117
  effects on vegetation, 80
  influences on climate, 73, 81
Towhee, Rufous-Sided, 237, 239
Treefall gaps, 179
Tropical forest, 181
Tropical slash-burn agriculture, 304

Truncation effect, 140–141, 153
Truncation error, 105
Tsuga canadensis, 121, 162
Tsuga heterophylla, 188
Tsuga mertensiana, 187

Umbrella species, 302
Uncertainty analysis, 63
Ungulate, 39, 50, 226–227
United States, 12, 13, 16, 36, 83, 89–90, 101, 122, 124, 253, 273, 274, 289, 298, 312, 318
Upper Great Lakes Region, USA, 164
Upper Peninsula, Michigan, USA, 214
Urban development, 16
Urban dynamics, 16
Urbanization, 256, 269, 282
Ursus americanus, 278
Ursus arctos, 278
US Bureau of Land Management, 8, 308
US Environmental Protection Agency, 308, 318
US Forest Service, 8, 308
US General Land Office Survey, 96
US Geological Survey, 97, 318
US National Marine Fisheries Service, 308
US National Park Service, 8, 308
Uvularia grandiflora, 222

Vaccinium membranaceum, 216
Variable
  definition, 58
  driving, 58, 59
  external, 58
  output, 58
Vegetation patterns, determinants of, 13
Volcanic eruption (*see* Disturbance), 90
Vole, heather, 217
Vulture, 221

Wadis, 213
Walker Branch Watershed, Tennessee, 60, 250
Washington, USA, 39
Water quality, 250, 265–278, 292, 318
Watershed dynamics, 250
Western hemlock zone, 216–217
Western spruce budworm, 172
Wetlands, 8
Wetlands, losses of, 272–273
White basswood, 162
White Ibis, 283
White pine, 77, 87
White-footed mice, 233
White-tailed deer, 233, 283
Whitebark pine, 102, 104–105
Wildlife ecology, 7
Wildlife habitat models, 314
Willamette National Forest, 318
Willamette River, Oregon, 273
Wintering birds, 215
Wisconsin, USA, 94, 101, 124, 214, 263–264, 270, 272, 278
Wolf pack, 215
Wood Stork, 283
Wood Thrush, 216
Woodpecker, pileated, 217
Woodpecker, red-cockaded, 242
Wren, Carolina, 237–238
Wyoming, USA, 79

Xeric landforms, 184

Yellow birch, 162, 183
Yellow buckeye, 162
Yellowstone fires, 159, 176, 178, 180
Yellowstone National Park, 4, 12, 50, 79, 94, 105, 110, 164, 175, 192, 226, 239, 241

Zonotrichia albicollis, 237

About the Authors

Monica G. Turner is a Professor in the Department of Zoology, University of Wisconsin-Madison. She received her B.S. in Biology from Fordham University, Bronx, New York, in 1980, and her Ph.D. in Ecology from the University of Georgia under the direction of Frank B. Golley in 1985. Following a two-year postdoc at Georgia with Eugene P. Odum, she was a research scientist at Oak Ridge National Laboratory for seven years before joining the University of Wisconsin faculty in 1994. Her research interests focus on the causes and effects of natural disturbances, particularly fire in Yellowstone National Park; movement and habitat use patterns of large ungulates; the causes of land-use change and its ecological legacies; the intersection between ecosystem and landscape ecology; and interactions between terrestrial and aquatic systems. Her work is characterized by the use of field studies as well as modeling and theory. Turner co-organized the first U.S. landscape ecology symposium in 1986 and served the U.S. chapter of the International Association of Landscape Ecology (US-IALE) as Program Chair from 1986–1989, Councilor-at-Large from 1990–1992, Local Meeting Host in 1993, and President from 1994–1996. She is also active in the Ecological Society of America and has served on advisory committees for numerous organizations, including the National Science Foundation and National Research

Council. She is co-editor-in-chief of the journal *Ecosystems* and serves on the editorial board of *Ecological Applications, BioScience, and Landscape Ecology*.

ROBERT H. GARDNER is a Professor at the Appalachian Laboratory of the University of Maryland Center for Environmental Science. He received a B.A. in Biology from Taylor University, an M.A. in Biology from William and Mary, and a Ph.D. in Zoology from North Carolina State University. His research interests and publications are varied, including topics related to the statistical analysis of model simulations, risk analysis, and issues of scale. Of special interest have been theoretical and empirical studies of landscape disturbances. Gardner has served on the editorial board of several journals, including editor-in-chief of *Landscape Ecology* from 1996–1999; has been an active participant in the Global Change and Terrestrial Ecosystems research of the International Geosphere-Biosphere Program (IGBP); and has served on a variety of committees for the Ecological Society of America and US-IALE.

ROBERT V. O'NEILL has recently retired as Corporate Fellow in the Environmental Sciences Division at Oak Ridge National Laboratory. He remains an adjunct Professor at the University of Tennessee. He received his Ph.D. from the University of Illinois in 1967. His interest in landscape ecology dates back to the late 1970s and has resulted in over 75 publications. He was a participant in the workshop at Allerton Park, Illinois, that catalyzed the development of landscape ecology in the United States and served as Councilor-at-Large for US-IALE from 1989–1994. He is a Fellow of the American Association for the Advancement of Science and has served on the editorial boards of *Ecology*, Columbia University Press, *Ecosystems*, and *Conservation Ecology*. He has received a number of awards, including the Technical Achievement Award, Martin Marietta Corporation 1987; Distinguished Statistical Ecologist, International Association for Statistical Ecology, 1994; Robert H. MacArthur Award, Ecological Society of America, 1998; Bronze Medal, EPA-ORD 1999; Twentieth Century Distinguished Service Award, 1999, Ninth Lukacs Symposium, Bowling Green State University. He is currently President of the U.S. Chapter of the International Society for Ecological Economics.

Each of the authors has been honored to receive the Distinguished Landscape Ecologist award from US-IALE: Gardner in 1994, O'Neill in 1995, and Turner in 1998. We believe that this recognition derived largely from having been part of such a rewarding and productive long-term collaboration.